Developments in Mathematics

VOLUME 37

Series Editors:
Krishnaswami Alladi, *University of Florida, Gainesville, FL, USA*
Hershel M. Farkas, *Hebrew University of Jerusalem, Jerusalem, Israel*

More information about this series at http://www.springer.com/series/5834

Geoffrey Mason • Ivan Penkov • Joseph A. Wolf
Editors

Developments and Retrospectives in Lie Theory

Geometric and Analytic Methods

 Springer

Editors
Geoffrey Mason
Department of Mathematics
University of California, Santa Cruz
Santa Cruz, CA, USA

Ivan Penkov
Jacobs University
Bremen, Germany

Joseph A. Wolf
Department of Mathematics
University of California, Berkeley
Berkeley, CA, USA

ISSN 1389-2177   ISSN 2197-795X (electronic)
ISBN 978-3-319-34875-9   ISBN 978-3-319-09934-7 (eBook)
DOI 10.1007/978-3-319-09934-7
Springer Cham Heidelberg New York Dordrecht London

Printed on acid-free paper

Springer is part of Springer Science+Business Media (www.springer.com)

Lie Groups, Lie Algebras and their Representations

Third Announcement -- October 1991

This is to announce a new program *Lie Groups, Lie Algebras and Their Representations* starting in the academic year 1991-92. We plan to meet at various University of California campuses for one weekend every other month during the year. The purpose of the program is to communicate results and ideas rather than to deliver polished presentations. The tentative format of the meetings is

Saturday:	one late morning talk and two afternoon talks; these scheduled in advance.
Sunday:	two morning talks and one early afternoon talk; some may be scheduled in advance and the rest will be scheduled Saturday.

The meetings are scheduled as follows:

Location	Dates	Local Organizer	e-mail address
Berkeley	October 19, 20	Joseph Wolf	jawolf@cartan.berkeley.edu
Riverside	December 7, 8	Ivan Penkov	penkov@ucrmath.ucr.edu
Davis	February 8, 9 ??	Alice Fialowski	fialowski@bolyai.ucdavis.edu
UCLA	April 4, 5	Robert Blattner	blattner@math.ucla.edu
UC Santa Cruz	May 30, 31	Daniel Goldstein	danny@cats.ucsc.edu

The program for the first meeting, Saturday October 19 and Sunday October 20, is

time	speaker	title
Sat 11:00	Joseph Wolf	Lie structures on direct limit groups and their completions (joint work with L. Natarajan and E. Rodríguez-Carrington)
Sat 02:00	Dmitri Fuchs	On the cohomology of Lie algebras of vector fields in the line and of Hamiltonian vector fields in the plane
Sat 04:00	Nicolai Reshetikhin	q-holonomic systems and quantum affine Lie algebras, I
Sun 09:00	Ivan Penkov	Infinite dimensional irreducible modules over finite dimensional Lie superalgebras
Sun 10:30	Yan Soibelman	Algebras of functions on compact quantum groups
Sun 01:00	Nicolai Reshetikhin	q-holonomic systems and quantum affine Lie algebras, II

These talks will take place in 9 Evans Hall on the Berkeley campus. The one hour scheduled time of any talk will be extended by the informal discussion that takes place during that talk.

There are no registration fees. No support is currently available. We hope to minimize participants' expenses by car-pooling and various informal arrangements for staying over Saturday night.

People interested in participating, or speaking and participating, at a meeting, should contact the local organizer for that meeting. People interested in helping with the local organization should contact both the local organizer and Joseph Wolf. If you wish to be on the mailing list for further announcements, and for meeting programs/schedules, write to Joseph Wolf.

Announcement of the First Conference, October 1991

Preface

The West Coast of the United States has a longstanding tradition in Lie theory, although before 1991 there had been no systematic cooperation between the various strongholds. In 1991, partially inspired by the arrival of some well-known Lie theorists from eastern Europe, the situation changed. A new structure emerged: a seminar that would meet at various University of California campuses three or four times a year. The purpose of the seminar was to foster contacts between researchers and graduate students at the various campuses by facilitating the sharing of ideas prior to formal publication. This idea quickly gained momentum, and became a great success. It was enthusiastically supported by graduate students. A crucial feature of the entire endeavor was the feeling of genuine interest for the work of colleagues and the strong desire to collaborate.

The first meeting of the new seminar Lie Groups, Lie Algebras and their Representations was held in Berkeley on October 19 and 20, 1991. On the second day of the seminar, excitement was made even more memorable by the historic Berkeley–Oakland fire, which we observed from Evans Hall. The original announcement for that meeting is reprinted here on page v. The phrase "The purpose of the program is to communicate results and ideas rather than to deliver polished presentations" quickly became, and still is, the guiding principle of the seminar. We never restricted ourselves to Lie Theory *per se*, and speakers from geometry, algebra, complex analysis, and other adjacent areas were often invited.

NSF travel grants were crucial to the success of the seminar series. These grants funded travel for speakers, graduate students and postdoctoral researchers, and we thank the National Science Foundation for its continued support.

Over the years our idea became widely popular. On occasion the seminar took place in Salt Lake City, in Stillwater, Oklahoma, and in Eugene, Oregon. In addition, colleagues from other regional centers of Lie theory and related areas picked up on our idea and created their own meeting series. This is how the "Midwest Lie Theory Seminars", the "Midwest Group Theory Seminar", the "Southeastern Lie Theory Workshops", and other regional series emerged.

The California Lie Theory Seminar has now been alive and well for 23 years. Joe Wolf has always played a central role in the seminar, along with Geoff Mason and Ivan Penkov. When Ivan left for Germany in 2004, Susan Montgomery and Milen Yakimov joined the team of organizers.

Over the course of these 23 years, at about 20 talks per year, some 450 talks have been hosted by the seminar. It seemed unrealistic to give an overview of all of the topics covered over all these years, and similarly unrealistic to try to publish a comprehensive set of volumes. Rather, we settled on two retrospective volumes containing work representative of the seminar as a whole. We started with a list of participants who spoke more than once in the seminar, and invited them to submit work relevant to their seminar talks. For obvious reasons we did not hear from everyone. Nevertheless, there was a strong response, and the reader of these Volumes will find 26 research papers, all of which received strong referee reports, in the greater area of Lie Theory. We decided to split the papers into two volumes: "Algebraic Methods" and "Geometric/Analytic Methods". We thank Springer, the publisher of these volumes, and especially Ann Kostant and Elizabeth Loew, for their cooperation and assistance in this project.

This is the Geometric/Analytic Methods volume.

Santa Cruz, CA, USA   Geoffrey Mason
Bremen, Germany   Ivan Penkov
Berkeley, CA, USA   Joseph A. Wolf

Contents

Group Gradings on Lie Algebras
and Applications to Geometry: II

Yuri Bahturin, Michel Goze, and Elisabeth Remm

Abstract This paper is devoted to some applications of the theory of group gradings on Lie algebras to two topics of differential geometry, such as generalized symmetric manifolds and affine structures on nilmanifolds.

Key words Symmetric manifolds • Nilmanifolds • Affine connections • Filiform Lie algebras • Graded algebras

Mathematics Subject Classification (2010): 17B20, 17B30, 17B70, 17B40, 53C35, 53C05, 57S25.

1 Introduction

This paper is a sequel to [4]. Here we discuss two topics in differential geometry, closely related to the theory of gradings on Lie algebras which we developed in that earlier paper. These topics are as follows: the generalization of the theory of symmetric manifolds to the case where the local symmetries form the group $\mathbb{Z}_2 \times \mathbb{Z}_2$ and Milnor's problem about the existence of affine structure on nilmanifolds. This latter topic quickly leads to the necessity of describing abelian group gradings on filiform nilpotent Lie algebras. As we showed in [4], in the case of filiform Lie algebras of nonzero rank, all abelian group gradings of a filiform Lie algebra are isomorphic to the coarsenings of a standard grading produced by the action of the maximal torus. In this paper, we classify these gradings up to equivalence (see Sect. 2) and also produce some results on filiform Lie algebras of zero rank, also called characteristically nilpotent.

Y. Bahturin (✉)
Department of Mathematics and Statistics, Memorial University of Newfoundland,
St. Johns, NL, Canada A1C5S7
e-mail: bahturin@mun.ca

M. Goze • E. Remm
LMIA, Université de Haute Alsace, 4 rue des Frères Lumière, 68093 Mulhouse, France
e-mail: michel.goze@uha.fr; elisabeth.remm@uha.fr

© Springer International Publishing Switzerland 2014
G. Mason et al. (eds.), *Developments and Retrospectives in Lie Theory*,
Developments in Mathematics 37, DOI 10.1007/978-3-319-09934-7_1

1

The results on the first topic have been reported to the Lie Theory Workshop at UC Berkeley in 2006. Since there was some later development in this area, we describe these results in some detail in Sect. 3.

In this geometric part the field of coefficients $\mathbb{K}$ has characteristic zero, and starting from Sect. 3.3, $\mathbb{K}$ is $\mathbb{R}$, or $\mathbb{C}$, and for a real Lie algebra $\mathfrak{g}$ its complexification $\mathfrak{g} \otimes \mathbb{C}$ will be denoted $\mathfrak{g}_{\mathbb{C}}$.

2 Gradings on Filiform Algebras of Nonzero Rank, Up to Equivalence

Recall that two gradings of an algebra are equivalent if there is an automorphism of this algebra permuting the components of the grading. In what follows we use a theorem in [4] according to which any grading of a filiform Lie algebra of nonzero rank is isomorphic, hence equivalent, to a grading where the elements of an adapted (case L_n and A_n^p) or quasi adapted (case Q_n and B_n^p) basis are homogeneous. Let $\mathcal{G}$ be the grading group and, for each $i = 1, 2, \ldots, n$, d_i denote the degree of the ith element of this basis. So if $\{X_1, X_2, \ldots, X_n\}$ is the adapted basis of $\mathfrak{g}$, then $d_i = \deg X_i$, $i = 1, 2, \ldots, n$, in the case of L_n and A_n^p and $d_i = \deg Y_i$, $i = 1, 2, \ldots, n$, in the case of Q_n and B_n^p, where $Y_1 = X_1 + X_2$ and $Y_i = X_i$, for $i = 2, \ldots, n$. Since $\mathfrak{g}$ is generated by X_1, X_2 (respectively, Y_1, Y_2) it follows that knowing $a = d_1$ and $b = d_2$ automatically gives values for the remaining d_i, $i = 3, \ldots, n$.

2.1 Gradings on L_n and A_n^p

Let $\{X_1, X_2, \ldots, X_n\}$ be an adapted basis of a filiform Lie algebra $\mathfrak{g}$ of type L_n or A_n^p. Since $[X_1, X_i] = X_{i+1}$ for $i = 2, \ldots, n-1$, we know that $d_i = a^{i-2}b$ for $i = 3, \ldots, n$. However, if $\mathfrak{g}$ is of the type A_n^p, we already have $b = a^{p+1}$. So in this case, $d_i = a^{p+i-1}$. In the case of L_n, the universal group of any grading is the factor-group of the free abelian group with free basis a, b by the relations satisfied by a, b, while in the case of A_n^p this is a factor-group of the free abelian group of rank 1 or rank 2 but we have to consider, in the latter case, that $b = a^{p+1}$.

Let us first consider the case of L_n. Any grading is a coarsening of the standard grading, which we denote by Γ_{st}. If all d_i are pairwise different then there is no coarsening and we have the standard grading.

Case 1. If $d_1 = d_l$, for $2 \leq l \leq n$, then the grading is a coarsening of the grading

$$\Gamma_0^l : \mathfrak{g} = \langle X_2 \rangle \oplus \cdots \oplus \langle X_1, X_l \rangle \oplus \cdots \oplus \langle X_n \rangle.$$

Since this is indeed a grading of $\mathfrak{g}$, our claim follows. The universal group of this grading is the factor-group of the free abelian group generated by a, b by a single relation $a = a^{l-2}b$, that is, the group $\mathbb{Z}$.

Case 2. If $d_i = d_j$, for $2 \leq i < j \leq n$, then the grading is a coarsening of a grading

$$\Gamma_k^0 : \mathfrak{g} = \langle X_1 \rangle \oplus [X_2]_k \oplus \cdots \oplus [X_{k+1}]_k.$$

Here $[X_i]_k$ is the span of the set of all X_j, $2 \leq j \leq n$ such that k divides $i - j$. This easily follows if we choose k to be the least positive with $d_2 = d_{2+k}$.

The universal group of this grading is the factor-group of the free abelian group generated by a, b by a single relation $b = a^k b$, that is, the group $\mathbb{Z}_k \times \mathbb{Z}$.

Any further coarsening of Γ_0^l is clearly also the coarsening of a Γ_k^0, so we can restrict ourselves to consider only the coarsenings of the latter grading.

Case 3. Any proper coarsening of Γ_k^0, which does not decrease k, is equivalent to one of the following.

$$\Gamma_k^l : \mathfrak{g} = [X_2]_k \oplus \cdots \oplus (\langle X_1 \rangle \oplus [X_l]_k) \oplus \cdots \oplus [X_{k+1}]_k.$$

Indeed, any further coarsening of $[X_l]_k$ will decrease k so we have to assume that $d_1 = d_l$, for $2 \leq l \leq n$, proving our claim. The universal group of this grading is the factor group of the group $\mathbb{Z}_k \times \mathbb{Z}$ by additional relation $a = a^{l-2}b$, that is, the group $\mathbb{Z}_k$.

Clearly, any further coarsening will lead to the decreasing of k, and so any grading is equivalent to one of the previous gradings.

Notice that these gradings are pairwise inequivalent. First, we have to look at the number of homogeneous components. Then it becomes clear that we only need to distinguish between the gradings with different values of the superscript parameter $l = 2, \ldots, n$. In this case, if an automorphism φ maps Γ_0^l to Γ_0^m, or Γ_k^l to Γ_k^m, where $m > l$, then $\varphi(\langle X_1, X_l \rangle) = \langle X_1, X_m \rangle$. But then $\varphi(X_{l+1}) = \varphi([X_1, X_l]) = [\varphi(X_1), \varphi(X_l)] = \alpha_{m+1}X_{m+1}$, $\varphi(X_{l+2}) = \alpha_{m+2}X_{m+2}$, etc. Finally, $\varphi(X_{n-m+l+1}) = 0$, which is impossible because $2 \leq n - m + l + 1 \leq n$.

As a result, we have the following.

Theorem 2.1. *Let $\mathfrak{g}$ be a filiform Lie algebra of the type L_n. If $\mathfrak{g}$ is $\mathcal{G}$-graded, then there exists a graded homogeneous adapted basis $\{X_1, X_2, \ldots, X_n\}$. If d_i denotes the degree of X_i, then any $\mathcal{G}$-grading is equivalent to one of the following pairwise inequivalent gradings:*

(1) Γ_{st}, $U(\Gamma_{\mathrm{st}}) = \mathbb{Z}^2$, $d_1 = (1, 0)$, $d_i = (i - 2, 1)$, $i = 2, \ldots, n$.
(2) Γ_0^l, $U(\Gamma_0^l) = \mathbb{Z}$, $2 \leq l \leq n$, $d_1 = 1$, $d_i = i - l + 1$, $i = 2, \ldots, n$.
(3) Γ_k^0, $U(\Gamma_k^0) = \mathbb{Z}_k \times \mathbb{Z}$, $1 \leq k \leq n-2$, $d_1 = (\bar{1}, 0)$, $d_i = (\overline{i - 2}, 1)$, $i = 2, \ldots, n$.
(4) Γ_k^l, $U(\Gamma_k^l) = \mathbb{Z}_k$, $1 \leq k \leq n - 2$, $2 \leq l \leq k + 1$, $d_1 = \bar{1}$, $d_i = \overline{i - l + 1}$, $i = 2, \ldots, n$.

The total number of pairwise inequivalent gradings of L_n is equal to
$$1 + (n-1) + (n-2) + (1 + 2 + \cdots + (n-2)) = \frac{(n-1)(n+2)}{2}.$$

In the case of A_n^p, we need to consider the coarsenings of the standard grading, which we denote here by Γ_{st}, where $U(\Gamma_{\text{st}}) \cong \mathbb{Z}$, with free generator a, $d_1 = a$, $d_i = a^{i+p-1}$, where $i = 2, \ldots, n$. Clearly, in this case, any proper coarsening leads to relations $a^i = a^j$, hence $a^{i-j} = e$, for different $1 \le i, j \le n$. If m is the greatest common divisor of all such $i - j$, then the universal group is $\mathbb{Z}_m$. Any value of m between 1 and $n + p - 2$ is possible. Indeed, if $p \le m \le n + p - 2$, then $d_1 = d_{m-p+2}$. If $1 \le m \le n - 2$, then $d_2 = d_{m+2}$. But $1 \le p \le n - 4 < n - 2$, and so any m between 1 and $n + p - 2$ is available. Let us denote the grading corresponding to m by $\Gamma(m)$. If $1 \le m \le p - 1$, $\Gamma(m)$ is similar to Γ_m^0 of L_n, then

$$\Gamma(m) : \mathfrak{g} = \langle X_1 \rangle \oplus [X_2]_m \oplus \cdots \oplus [X_{m+1}]_m.$$

If $n - 1 \le m \le n + p - 1$ and $l = m - p + 2$, then $\Gamma(m)$ is similar to Γ_0^l of L_n:

$$\Gamma(m) : \mathfrak{g} = \langle X_2 \rangle \oplus \cdots \oplus \langle X_1, X_l \rangle \oplus \cdots \oplus \langle X_n \rangle.$$

If $p \le d \le n - 2$, then we have the gradings similar to Γ_m^l of L_n:

$$\Gamma(m) : \mathfrak{g} = [X_2]_m \oplus \cdots \oplus (\langle X_1 \rangle \oplus [X_l]_m) \oplus \cdots \oplus [X_{m+1}]_m.$$

Here l is a number between 2 and n such that $l + p - 1 \equiv 1 \bmod m$. The pairwise inequivalence of all the above gradings follows, as in the case of L_n.

Theorem 2.2. *Let $\mathfrak{g}$ be a filiform Lie algebra of the type A_n^p. If $\mathfrak{g}$ is $\mathcal{G}$-graded, then there exists a graded homogeneous adapted basis $\{X_1, X_2, \ldots, X_n\}$. If d_i denotes the degree of X_i, then any $\mathcal{G}$-grading is equivalent to one of the following non equivalent gradings:*

(1) Γ_{st}, $U(\Gamma_{\text{st}}) = \mathbb{Z}$, $d_1 = 1$, $d_i = p + i - 1$, $i = 2, \ldots, n$.
(2) $\Gamma(m)$, $U(\Gamma(m)) = \mathbb{Z}_m$, $1 \le m \le n + p - 2$, $d_1 = \bar{1}$, $d_i = \overline{p + i - 1}$, $i = 2, \ldots, n$.

The total number of pairwise inequivalent gradings of A_n^p is equal to $n + p - 2$.

2.2 Gradings on Q_n and B_n^p

Let $\{Y_1, Y_2, \ldots, Y_n\}$ be a quasi-adapted basis of a filiform Lie algebra $\mathfrak{g}$ of type Q_n or B_n^p. Since $[Y_2, Y_i] = Y_{i+1}$ for $i = 2, \ldots, n - 2$, we know that $d_i = a^{i-2}b$ for $i = 3, \ldots, n - 1$. Also, $[Y_i, Y_{n-i+1}] = (-1)^{i+1}Y_n$ which assigns to d_n the value of $d_n = a^{n-3}b^2$. If $\mathfrak{g}$ is of the type B_n^p, we already have $b = a^{p+1}$. So in this case, $d_i = a^{p+i-1}$, for $2 \le i \le n-1$, and $d_n = a^{n+2p-1}$. In the case of Q_n, the universal group of any grading is the factor group of the free abelian group with free basis a, b by the relations satisfied by a, b while in the case of B_n^p, this is a factor-group of the free abelian group of rank 1 or rank 2 but we have to consider, in the latter case, that $b = a^{p+1}$.

Let us first consider the case of Q_n. We remember that $n = 2m$, for some $m \geq 2$. Now any grading is a coarsening of the standard grading, which we denote by $\overline{\Gamma}_{\text{st}}$. If all d_i are pairwise different, then there is no coarsening and we have the standard grading.

Case 1. If $d_1 = d_n$, then the grading is a coarsening of the grading

$$\overline{\Gamma}(1, n) : \mathfrak{g} = \langle Y_2 \rangle \oplus \cdots \oplus \langle Y_{n-1} \rangle \oplus \langle Y_1, Y_n \rangle.$$

Since this is indeed a grading of $\mathfrak{g}$, our claim follows. The universal group of this grading is the factor group of the free abelian group generated by a, b by a single relation $a = a^{n-3}b^2$, that is, the group $\mathbb{Z} \times \mathbb{Z}_2$.

Case 2. If $d_1 = d_l$, for $2 \leq l \leq n - 1$, then $a = a^{l-2}b$. In this case also $d_n = a^{n-3}b^2 = a^{n-l}b = d_{n-l+2}$. Hence, for all q satisfying $l + q = n + 2$, except $l = 2$, or $q = 2$, the grading is a coarsening of one of the following gradings. If $l \neq q$, then we have

$$\overline{\Gamma}_0^l : \mathfrak{g} = \langle Y_2 \rangle \oplus \cdots \oplus \langle Y_1, Y_l \rangle \oplus \cdots \oplus \langle Y_m, Y_n \rangle \oplus \langle Y_{n-1} \rangle.$$

If $l = q = m + 1$, then we have

$$\overline{\Gamma}_0^{m+1} : \mathfrak{g} = \langle Y_2 \rangle \oplus \cdots \oplus \langle Y_1, Y_{m+1}, Y_n \rangle \oplus \cdots \oplus \langle Y_{n-1} \rangle.$$

Notice that this grading is a coarsening of $\overline{\Gamma}(1, n)$.

In the exceptional cases, $l = 2$ or $q = 2$, we have the following.

If $l = 2$, then we have

$$\overline{\Gamma}_0^2 : \mathfrak{g} = \langle Y_1, Y_2 \rangle \oplus \cdots \oplus \langle Y_1, Y_l \rangle \oplus \cdots \oplus \langle Y_n \rangle.$$

If $q = 2$, then we have

$$\overline{\Gamma}_0^n : \mathfrak{g} = \langle Y_2, Y_n \rangle \oplus \langle Y_3 \rangle \oplus \cdots \oplus \langle Y_{n-1} \rangle \oplus \langle Y_1 \rangle.$$

Since all these are indeed gradings of $\mathfrak{g}$, our claim follows. The universal group of this grading is the factor-group of the free abelian group generated by a, b by a single relation $a = a^{l-2}b$, that is, the group $\mathbb{Z}$.

Case 3. If $d_i = d_j$, for $2 \leq i < j \leq n - 1$, then the grading is a coarsening of a grading

$$\overline{\Gamma}_k^0 : \mathfrak{g} = \langle Y_1 \rangle \oplus [Y_2]_k \oplus \cdots \oplus [Y_{k+1}]_k \oplus \langle Y_n \rangle,$$

where k is such that $1 \leq k \leq n - 3$. Here $[Y_i]_k$ is the span of the set of all Y_j, $2 \leq j \leq n$, such that k divides $i - j$. This easily follows if we choose k to be the least positive with $d_2 = d_{2+k}$.

The universal group of this grading is the factor-group of the free abelian group generated by a, b by a single relation $b = a^k b$, that is, the group $\mathbb{Z}_k \times \mathbb{Z}$.

Any coarsening of $\overline{\Gamma}(1, n)$ (Case 1) is either $\overline{\Gamma}_0^{m+1}$ or is a coarsening of some $\overline{\Gamma}_k^0$. Any coarsening of a grading $\overline{\Gamma}_0^l$ (Case 2) is also a coarsening of some $\overline{\Gamma}_k^0$, so we can restrict ourselves to consider only the coarsenings of the latter gradings. This shows that any grading is either standard, or equivalent to one of the gradings in Cases 1–3 or is a proper coarsening of a grading $\overline{\Gamma}_k^0$. Let us choose $\overline{\Gamma}_k^0$ with minimal possible k.

Case 4. Any proper coarsening of $\overline{\Gamma}_k^0$, which does not change k, is equivalent to one of the following:

$$\overline{\Gamma}(1, n)_k : \mathfrak{g} = [Y_2]_k \oplus \cdots \oplus [Y_{k+2}]_k \oplus \langle Y_1, Y_n \rangle,$$

where $1 \leq k \leq n - 3$, or

$$\overline{\Gamma}_k^l : \mathfrak{g} = [Y_2]_k \oplus \cdots \oplus (\langle Y_1 \rangle \oplus [Y_l]_k) \oplus \cdots \oplus ([Y_q]_k \oplus \langle Y_n \rangle) \oplus \cdots \oplus [Y_{k+1}]_k,$$

where $1 \leq k \leq n - 3, 2 \leq l, q \leq k + 1, l + q \equiv n + 2 \bmod k$.

The universal group of $\overline{\Gamma}(1, n)_k$ is $\mathbb{Z}_k \times \mathbb{Z}_2$, whereas, in the case of $\overline{\Gamma}_k^l$, the universal group is $\mathbb{Z}_k$.

Indeed, any further coarsening of $[Y_l]_k$ will decrease k, so we have to assume that $d_1 = d_l$, for $2 \leq l \leq n$, proving our claim. Clearly, any further coarsening will lead to further decreasing of k, and so any grading is equivalent to one of the previous gradings.

Notice that these gradings are pairwise inequivalent. First, the gradings with different universal groups are not equivalent. As a result, we only need to distinguish between the gradings in the sets $\overline{\Gamma}_0^l$, $2 \leq l \leq n$ (the universal group $\mathbb{Z}$) and $\overline{\Gamma}_k^l$, $2 \leq l \leq k + 1$ (the universal group $\mathbb{Z}_k$). This is done exactly in the same way, except for the cases where Y_1 is in a component which is not two-dimensional. However, such a grading can only be mapped to a grading with the same property because Y_1 is the only element among Y_i with $(\mathrm{ad} Y_i)^{n-2} \neq 0$.

As a result, we have the following.

Theorem 2.3. *Let $\mathfrak{g}$ be a filiform Lie algebra of the type Q_n. If $\mathfrak{g}$ is $\mathcal{G}$-graded, then there exists a graded homogeneous quasi-adapted basis $\{Y_1, Y_2, \ldots, Y_n\}$. If d_i denotes the degree of Y_i, then any $\mathcal{G}$-grading is equivalent to one of the following pairwise equivalent gradings:*

(1) $\overline{\Gamma}_{\mathrm{st}}$, $U(\overline{\Gamma}_{\mathrm{st}}) = \mathbb{Z}^2$, $d_1 = (1, 0)$, $d_i = (i - 2, 1), 2 \leq i \leq n - 1, d_n = (n - 3, 2)$.
(2) $\overline{\Gamma}(1, n)$, $U(\overline{\Gamma}(1, n)) = \mathbb{Z} \times \mathbb{Z}_2$, $d_1 = (1, \overline{0}) = d_n, d_i = (i - 2, \overline{1}), 2 \leq i \leq n - 1$.

(3) $\overline{\Gamma}_0^l$, $U(\overline{\Gamma}_0^l) = \mathbb{Z}$, $2 \leq l \leq n$, $d_1 = 1$, $d_i = i - l + 1$, $2 \leq i \leq n - 1$, $d_n = n - 2l + 3$.

(4) $\overline{\Gamma}_k^0$, $U(\overline{\Gamma}_k^0) = \mathbb{Z}_k \times \mathbb{Z}$, $1 \leq k \leq n - 3$, $d_1 = (\overline{1}, 0)$, $d_i = (\overline{i-2}, 1)$, $d_n = (\overline{n-3}, 2)$.

(5) $\overline{\Gamma}(1, n)_k$, $U(\overline{\Gamma}(1, n)_k) = \mathbb{Z}_k \times \mathbb{Z}_2$, $1 \leq k \leq n - 3$, $d_1 = (\overline{1}, \overline{0}) = d_n$, $d_i = (\overline{i-2}, \overline{1})$, $2 \leq i \leq n - 1$.

(6) $\overline{\Gamma}_k^l$, $U(\overline{\Gamma}_k^l) = \mathbb{Z}_k$, $1 \leq k \leq n - 3$, $2 \leq l \leq k + 1$, $d_1 = \overline{1}$, $d_i = \overline{i - l + 1}$, $d_n = n - 2l + 3$.

The total number of pairwise inequivalent gradings of Q_n is equal to $1 + 1 +$

$$(n - 1) + (n - 3) + (n - 3) + (1 + 2 + \cdots + (n - 3)) = \frac{(n - 1)(n + 2)}{2} - 1.$$

In the case of B_n^p, we need to consider the coarsenings of the standard grading, which we denote here by Γ_{st}, where $U(\Gamma_{st}) \cong \mathbb{Z}$, with free generator a, $d_1 = a$, $d_i = a^{i+p-1}$, where $i = 2, \ldots, n - 1$, $d_n = a^{n+2p-1}$. Clearly, in this case, any proper coarsening leads to relations $a^i = a^j$, hence $a^{i-j} = e$, for different $1 \leq i, j \leq n$. If m is the greatest common divisor of all such $i - j$, then the universal group is $\mathbb{Z}_m$. Any value of m between 1 and $n + p - 3$ is possible, similar to the case of A_n^p. One more possible isolated value for m appears if we choose $d_1 = d_n$. Then $m = n + 2p - 3$. Let us denote the grading corresponding to m by $\overline{\Gamma}(m)$.

If $1 \leq m \leq p - 1$, $\overline{\Gamma}(m)$ is similar to $\overline{\Gamma}_m^0$ of Q_n, then

$$\overline{\Gamma}(m) : \mathfrak{g} = \langle Y_1 \rangle \oplus [Y_2]_m \oplus \cdots \oplus [Y_{m+1}]_m \oplus \langle Y_n \rangle.$$

If $p \leq m \leq n - 3$, then we have gradings similar to $\overline{\Gamma}_m^l$ of Q_n:

$$\overline{\Gamma}(m) : \mathfrak{g} = [Y_2]_m \oplus \cdots \oplus (\langle Y_1 \rangle \oplus [Y_l]_m) \oplus \cdots \oplus ([Y_q]_m \oplus \langle Y_n \rangle) \oplus \cdots \oplus [Y_{m+1}]_m.$$

If $n - 2 \leq m \leq n + p - 3$, $\overline{\Gamma}(m)$ is similar to $\overline{\Gamma}_0^l$ of Q_n, then

$$\overline{\Gamma}(m) : \mathfrak{g} = \langle Y_2 \rangle \oplus \cdots \oplus \langle Y_1, Y_l \rangle \oplus \cdots \oplus \langle Y_m, Y_n \rangle \oplus \langle Y_{n-1} \rangle.$$

Here $l = m - p + 2$.

If $m = n + 2p - 3$, then $\overline{\Gamma}(m)$ is similar to $\overline{\Gamma}(1, n)$:

$$\overline{\Gamma}(n + 2p - 3) : \mathfrak{g} = \langle Y_2 \rangle \oplus \cdots \oplus \langle Y_{n-1} \rangle \oplus \langle Y_1, Y_n \rangle.$$

The pairwise inequivalence of all the above gradings follows, as in the case of Q_n.

Theorem 2.4. *Let $\mathfrak{g}$ be a filiform Lie algebra of the type B_n^p. If $\mathfrak{g}$ is $\mathcal{G}$-graded, then there exists a graded homogeneous adapted basis $\{Y_1, Y_2, \ldots, Y_n\}$. If d_i denotes the degree of Y_i, then any $\mathcal{G}$-grading is equivalent to one of the following non equivalent gradings:*

(1) $\overline{\Gamma}_{st}$, $U(\overline{\Gamma}_{st}) = \mathbb{Z}$, $d_1 = 1$, $d_i = p + i - 1$, $i = 2, \ldots, n - 1$, $d_n = n + 2p - 2$.
(2) $\overline{\Gamma}(m)$, $U(\overline{\Gamma}(m)) = \mathbb{Z}_m$, $1 \le m \le n + p - 3$ or $m = n + 2p - 3$, $d_1 = \overline{1}$, $d_i = p + i - 1$, $i = 2, \ldots, n$, $d_n = \overline{n + 2p - 2}$.

The total number of pairwise inequivalent gradings of B_n^p is equal to $n + p - 3$.

2.3 Characteristically Nilpotent Lie Algebras

Definition 2.1. A finite-dimensional $\mathbb{K}$-Lie algebra $\mathfrak{g}$ is called characteristically nilpotent if any derivation of $\mathfrak{g}$ is nilpotent. It is called characteristically unipotent if the group $\mathrm{Aut}(\mathfrak{g})$ of the automorphisms of $\mathfrak{g}$ is unipotent.

A characteristically nilpotent Lie algebra has rank 0 and the Lie algebra of derivations Der $(\mathfrak{g})$ is nilpotent (but not necessarily characteristically nilpotent). In this case also Aut $(\mathfrak{g})$ is nilpotent. However, this group does not have to be unipotent. It is unipotent if $\mathfrak{g}$ is characteristically unipotent. Of course, if $\mathfrak{g}$ is characteristically unipotent, it is characteristically nilpotent.

Examples. (1) The simplest example [1], denoted by $\mathfrak{n}_{7,4}$ in terminology of [17], is
  seven-dimensional and given by

$$\begin{cases} [X_1, X_i] = X_{i+1}, \ 2 \le i \le 6, \\ [X_2, X_3] = -X_6, \\ [X_2, X_4] = -[X_5, X_2] = -X_7, \\ [X_3, X_4] = X_7. \end{cases}$$

This Lie algebra is filiform. Any automorphism of $\mathfrak{n}_{7,4}$ is unipotent. It is defined on the generators X_1, X_2 by

$$\begin{cases} \sigma(X_1) = X_1 + a_2 X_2 + a_3 X_3 + a_4 X_4 + a_5 X_5 + a_6 X_6 + a_7 X_7, \\ \sigma(X_2) = X_2 + b_3 X_3 + \frac{b_3^2 - a_2}{2} X_4 + b_5 X_5 + b_6 X_6 + b_7 X_7. \end{cases}$$

It follows that Aut $(\mathfrak{n}_{7,4})$ is a ten-dimensional unipotent Lie group and $\mathfrak{n}_{7,4}$ is also characteristically unipotent. Let us note that any $\sigma \in$ Aut $(\mathfrak{n}_{7,4})$ of finite order is equal to the identity.

(2) The first example of characteristically nilpotent Lie algebra was given by
  Dixmier and Lister [9]. It can be written as

$$\begin{cases} [X_1, X_2] = X_5, \ [X_1, X_3] = X_6, \ [X_1, X_4] = X_7, \ [X_1, X_5] = -X_8, \\ [X_2, X_3] = X_8, \ [X_2, X_4] = X_6, \ [X_2, X_6] = -X_7, \ [X_3, X_4] = -X_5, \\ [X_3, X_5] = -X_7, \ [X_4, X_6] = -X_8. \end{cases}$$

It is an eight-dimensional nilpotent Lie algebra of nilindex 3, and it is not filiform. Let us note that nil-index 3 is the lowest possible nil-index for a characteristically nilpotent Lie algebra. Its automorphisms group Aut $(\mathfrak{g})$ is not unipotent. For example, the linear map given by

$$\begin{cases} \sigma(X_1) = X_5, \ \sigma(X_5) = X_1, \\ \sigma(X_2) = X_7, \ \sigma(X_7) = X_2, \\ \sigma(X_4) = X_8, \ \sigma(X_8) = X_4, \\ \sigma(X_3) = -X_3, \ \sigma(X_6) = -X_6 \end{cases}$$

is a non-unipotent automorphism of $\mathfrak{g}$ of order 2. In the case where char $\mathbb{K} \neq 2$, σ defines a nontrivial $\mathbb{Z}_2$-grading Γ of $\mathfrak{g}$:

$$\Gamma: \quad \mathfrak{g} = \langle X_1 + X_5, X_2 + X_7, X_4 + X_8 \rangle \oplus \langle X_1 - X_5, X_2 - X_7, X_4 - X_8, X_3, X_6 \rangle.$$

2.4 Structure of Characteristically Nilpotent Lie Algebras

It is known from [13] that any filiform $(n + 1)$-dimensional Lie algebra over an algebraic field of characteristic 0 is defined by its Lie bracket μ with $\mu = \mu_0 + \psi$ where μ_0 is the Lie multiplication of L_{n+1} and ψ a 2-cocycle of $Z^2(L_{n+1}, L_{n+1})$ satisfying $\psi \circ \psi = 0$, that is, ψ is also a $(n + 1)$-dimensional Lie multiplication. Let us consider the natural $\mathbb{Z}$-grading of L_{n+1}:

$$L_{n+1} = \bigoplus_{i \in \mathbb{Z}} L_{n+1,i},$$

where $L_{n+1,1}$ is generated by e_0, e_1 and $L_{n+1,i}$ by e_i for $i = 2, \ldots, n$, and other subspaces are zero. This grading induces a $\mathbb{Z}$-grading in the spaces of cochains of the Chevalley–Eilenberg complex of L_{n+1}:

$$C_p^k(L_{n+1}, L_{n+1}) = \{\phi \in C_k(L_{n+1}, L_{n+1}), \ \phi(L_{n+1,i_1}, \ldots, L_{n+1,i_k})$$
$$\subset L_{n+1,i_1 + \ldots + i_k + p}\}.$$

Since $d(C_p^k(L_{n+1}, L_{n+1})) \subset C_p^{k+1}(L_{n+1}, L_{n+1})$, we deduce a grading in the spaces of cocycles and coboundaries. Let

$$H_p^k(L_{n+1}, L_{n+1}) = Z_p^k(L_{n+1}, L_{n+1}) / B_p^k(L_{n+1}, L_{n+1})$$

be the corresponding grading in the Chevalley–Eilenberg cohomological spaces of L_{n+1}. We put

$$F_0 H^k(L_{n+1}, L_{n+1}) = \bigoplus_{p \in \mathbb{Z}} H_p^k(L_{n+1}, L_{n+1}),$$

$$F_1 H^k(L_{n+1}, L_{n+1}) = \bigoplus_{p \geq 1} H_p^k(L_{n+1}, L_{n+1}).$$

We have

Proposition 2.1. *Let* $\psi_{k,s}$, $1 \leq k \leq n - 1$, $2k \leq s \leq n$ *be the 2-cocycle in* $C^2(L_{n+1}, L_{n+1})$ *defined by*

- $\psi_{k,s}(e_k, e_{k+1}) = e_s$,
- $\psi_{k,s}(e_i, e_{i+1}) = 0$ *if* $i \neq k$,
- $\psi_{k,s}(e_i, e_j) = 0$ *if* $i > k$,
- $\psi_{k,s}(e_i, e_j) = (-1)^{k-i} C_{j-k-1}^{k-i} e_{i+j+s-2k-1}$, $1 \leq i \leq k < j - 1 \leq n - 1$, $0 \leq i + j - 2k - 1 \leq n - s$

Then the family of $\psi_{k,s}$ *with* $1 \leq [n/2] - 1$, $4 \leq s \leq n$, *forms a basis of* $F_1 H^2(L_{n+1}, L_{n+1})$.

Note that the cocycles $\psi_{k,s}$ also satisfy $\psi_{k,s}(e_k, e_j) = e_{j+s-k-1}$ when $k < j$.

Proposition 2.2. *If* $\mu = \mu_0 + \psi$ *is the Lie multiplication of a filiform* $(n + 1)$-*dimensional Lie algebra, then* $[\psi] \in F_1 H^2(L_{n+1}, L_{n+1})$ *if* n *is even or* $\psi \in F_1 H^2(L_{n+1}, L_{n+1}) + [\psi_{(n-1)/2,n}]$ *if* n *is odd, where* $[\psi]$ *denote the class in* $H^2(L_{n+1}, L_{n+1})$ *of the 2-cocycle* ψ.

Examples. (1) Any seven-dimensional filiform Lie algebra can be written as $\mu = \mu_0 + \psi$ with

$$\psi = a_{1,4}\psi_{1,4} + a_{1,5}\psi_{1,5} + a_{1,6}\psi_{1,6} + a_{2,6}\psi_{2,6}.$$

(2) Any eight-dimensional filiform Lie algebra can be written as $\mu = \mu_0 + \psi$ with

$$\psi = a_{1,4}\psi_{1,4} + a_{1,5}\psi_{1,5} + a_{1,6}\psi_{1,6} + a_{1,7}\psi_{1,7} + a_{2,6}\psi_{2,6} + a_{2,7}\psi_{2,7} + a_{3,7}\psi_{3,7}.$$

(3) Any nine-dimensional filiform Lie algebra can be written as $\mu = \mu_0 + \psi$ with

$$\psi = a_{1,4}\psi_{1,4} + a_{1,5}\psi_{1,5} + a_{1,6}\psi_{1,6} + a_{1,7}\psi_{1,7} + a_{1,8}\psi_{1,8} + a_{2,6}\psi_{2,6} + a_{2,7}\psi_{2,7} + a_{2,8}\psi_{2,8} + a_{3,8}\psi_{3,8}.$$

(4) Any ten-dimensional filiform Lie algebra can be written as $\mu = \mu_0 + \psi$ with

$$\psi = a_{1,4}\psi_{1,4} + a_{1,5}\psi_{1,5} + a_{1,6}\psi_{1,6} + a_{1,7}\psi_{1,7} + a_{1,8}\psi_{1,8} + a_{1,9}\psi_{1,9} + a_{2,6}\psi_{2,6} + a_{2,7}\psi_{2,7} + a_{2,8}\psi_{2,8} + a_{2,9}\psi_{2,9} + a_{3,8}\psi_{3,8} + a_{3,9}\psi_{3,9} + a_{4,9}\psi_{4,9}.$$

To recognize characteristically nilpotent Lie algebras among the filiform Lie algebras, we can use the notion of a *sill* algebra. Recall that a filiform Lie algebra $\mathfrak{g}$ such that gr $(\mathfrak{g})$ is isomorphic to L_{n+1} can be written in an adapted basis as

$$[e_0, e_i] = e_{i+1}, \; i = 1 \cdots, n - 1, \; [e_i, e_j] = \sum_{r=1}^{n-i-j} a_{ij}^r e_{i+j+r}, \; 1 \leq i < j \leq n - 2.$$

A filiform Lie algebra $\mathfrak{g}$ such that gr $(\mathfrak{g})$ is isomorphic to Q_{n+1} can be written in an quasi-adapted basis:

$$[Z_0, Z_i] = Z_{i+1}, \; i = 1 \cdots, n-2, \; [Z_i, Z_{n-i}] = (-1)^i Z_n,$$

$$[Z_i, Z_j] = \sum_{r=1}^{n-i-j} b_{ij}^r Z_{i+j+r}$$

for $1 \leq i < j \leq n-2$.

Definition 2.2 ([14]). Let $\mathfrak{g}$ be a $(n+1)$-dimensional filiform Lie algebra such that gr $(\mathfrak{g})$ is isomorphic to L_{n+1}. The sill algebra of $\mathfrak{g}$ is defined by

$$[e_0, e_i] = e_{i+1}, \; i = 1, \ldots, n-1, \; [e_i, e_j] = a_{ij}^r e_{i+j+r}$$

where $r \neq 0$ is the smallest index such that $a_{ij}^r \neq 0$ for some (i, j).

If gr $(\mathfrak{g})$ is isomorphic to Q_{n+1}, then the sill algebra is defined by

(1) If $b_{ij}^{n-i-j} = 0$, then

$$[Z_0, Z_i]=Z_{i+1}, \; i=1,\ldots,n-2, \; [Z_i, Z_{n-i}]=(-1)^i Z_n, \; [Z_i, Z_j]=b_{ij}^r Z_{i+j+r}$$

for $1 \leq i < j \leq n-2$ where $r \neq 0$ is the smallest index such that $b_{ij}^r \neq 0$ for some (i, j).

(2) If $b_{ij}^{n-i-j} \neq 0$ for some (i, j), then

$$[Z_0, Z_i]=Z_{i+1}, \; i = 1,\ldots,n-2, \; [Z_i, Z_{n-i}]=(-1)^i Z_n, \; [Z_i, Z_j]=b_{ij}^{n-i-j} Z_n$$

for $1 \leq i < j \leq n-2$.

Proposition 2.3 ([26]). *A filiform Lie algebra is characteristically nilpotent if and only if it is not isomorphic to its sill algebra.*

Examples. (1) Any seven-dimensional filiform characteristically nilpotent Lie algebra can be written $\mu = \mu_0 + \psi$ with

    (a) $\psi = \psi_{1,5} + \psi_{1,6}$,
    (b) $\psi = \psi_{1,4} + \psi_{1,6}$,
    (c) $\psi = \psi_{1,5} + \psi_{2,6}$.

(2) Any eight-dimensional filiform characteristically nilpotent Lie algebra can be written as $\mu = \mu_0 + \psi$ with

    (a) $\psi = \psi_{1,4} + a_{1,5}\psi_{1,5} - a_{2,6}\psi_{2,6} + \psi_{3,7}$.
    (b) $\psi = \psi_{1,5} + a_{1,6}\psi_{1,6} + \psi_{3,7}$.
    (c) $\psi = a_{1,4}\psi_{1,4} + \psi_{2,6} + \psi_{2,7}$.
    (d) $\psi = \psi_{1,5} + \psi_{2,6}$.
    (e) $\psi = \psi_{1,4} + a_{1,6}\psi_{1,6} + \psi_{2,7}$.

(f) $\psi = a_{1,5}\psi_{1,5} + \psi_{1,6} + \psi_{2,7}$.
(g) $\psi = a_{1,4}\psi_{1,4} + \psi_{1,6} + \psi_{1,7}$.
(h) $\psi = \psi_{1,4} + \psi_{1,6}$.
(i) $\psi = \psi_{1,4} + \psi_{1,7}$.
(j) $\psi = \psi_{1,5} + \psi_{1,6}$.

2.5 $\mathbb{Z}_2$-Gradings of Filiform Characteristically Nilpotent Lie Algebras

Since characteristically nilpotent Lie algebras have zero rank, they do not admit $\mathbb{Z}$-gradings. The following shows that there exists a class of such Lie algebras admitting $\mathbb{Z}_2$-gradings.

Proposition 2.4. *Let $\mu_0 + \psi$ the multiplication of a $(n + 1)$-dimensional characteristically nilpotent Lie algebra $\mathfrak{g}$ such that gr $(\mathfrak{g})$ is isomorphic to L_{n+1}. If*

$$\psi = \sum a_{k,2s}\psi_{k,2s}$$

or

$$\psi = \sum a_{k,2s+1}\psi_{k,2s+1},$$

then this Lie algebra admits a $\mathbb{Z}_2$-grading.

Proof. Since $\psi_{k,2s}(e_i, e_j) = a e_{i+j+2s-2k-1}$, the Lie algebra $\mu_0 + \psi$ is characteristically nilpotent as soon as we have in the sum ψ two terms $\psi_{k,2s}$ and $\psi_{k',2s'}$ such that $s - k \neq s' - k'$. Let us consider two $\mathbb{Z}_2$-gradings of the vector space $\mathfrak{g}$ as follows.

- $\mathfrak{g} = \langle e_1, e_3, \ldots, e_{2p\pm1} \rangle \bigoplus \langle e_0, e_2, e_4, \ldots, e_{2p} \rangle$
- $\mathfrak{g} = \langle e_2, e_4, \ldots, e_{2p} \rangle \bigoplus \langle e_0, e_1, e_3, \ldots, e_{2p\pm1} \rangle$;

(the $\pm$ sign means that we consider the cases n odd and n even at the same time).

We consider the first vectorial decomposition. From our description of gradings on L_n we can see that this is a grading of μ_0. It is sufficient then to check that this is a grading of $\psi_{k,2s}$. A cocycle $\psi_{k,s}$ is homogeneous, that is it satisfies $\psi_{k,s}(\mathfrak{g}_i, \mathfrak{g}_j) \subset \mathfrak{g}_{i+j(\text{mod}2)}$ if s is even. In fact

$$\psi_{k,s}(e_{2i+1,2j+1}) = \lambda e_{2i+2j-2k+1+s}$$

where λ is a nonzero constant and $2i + 2j - 2k + 1 + s$ is odd if and only if s is even. Likewise

$$\psi_{k,s}(e_{2i,2j}) = \lambda e_{2i+2j-2k-1+s}$$

with $\lambda \neq 0$ and $2i + 2j - 2k - 1 + s$ is odd if and only if s is even. Since moreover

$$\psi_{k,s}(e_{2i}, e_{2j+1}) = e_{2i+2j-2k+s},$$

$2i + 2j - 2k + s$ is even as soon as s is even. Thus the existence of this grading implies that s is even.

Similarly, the vectorial decomposition of the second type, is a $\mathbb{Z}_2$-grading of $\mu_0 + \psi_{k,s}$ if s is odd. □

Proposition 2.5. *Let $\mu_0 + \psi$ the multiplication of a $(n + 1)$-dimensional characteristically nilpotent Lie algebra $\mathfrak{g}$ such that $\mathrm{gr}\,(\mathfrak{g})$ is isomorphic to Q_{n+1}. If*

$$\psi = \sum a_{k,2s} \psi_{k,2s} + \psi_{\frac{n-1}{2},n}$$

or

$$\psi = \sum a_{k,2s+1} \psi_{k,2s+1} + \psi_{\frac{n-1}{2},n},$$

then this Lie algebra admits a $\mathbb{Z}_2$-grading.

Proof. Since any $\mathbb{Z}_2$-grading on $\mathfrak{g}$ induces the same grading on $\mathrm{gr}\,(\mathfrak{g}) = Q_{n+1}$, we consider the vectorial decompositions of $\mathfrak{g}$:

- $\mathfrak{g} = \langle e_2, e_4, \ldots, e_{n-1} \rangle \oplus \langle e_0, e_1, e_3, \ldots, e_n \rangle$,
- $\mathfrak{g} = \langle e_0 + e_1, e_n \rangle \oplus \langle e_1, e_2, \ldots, e_{n-1} \rangle$,
- $\mathfrak{g} = \langle e_1, e_3, \ldots, e_{n-2} \rangle \oplus \langle e_0 + e_1, e_2, e_4, \ldots, e_{n-1}, e_n \rangle$.

If the multiplication of $\mathfrak{g}$ is given by $\mu_0 + \psi$ with $\psi = \sum a_{k,2s+1} \psi_{k,2s+1} + \psi_{\frac{n-1}{2},n}$, then the first vectorial decomposition is also a $\mathbb{Z}_2$-grading. If the multiplication of $\mathfrak{g}$ is given by $\mu_0 + \Psi$ with $\psi = \sum a_{k,2s} \psi_{k,2s} + \psi_{\frac{n-1}{2},n}$, then the third vectorial decomposition is also a $\mathbb{Z}_2$-grading. If the second vectorial decomposition is a grading, then $\mathfrak{g}$ is not characteristically nilpotent. □

Corollary 2.1. *There exists an infinite family of graded characteristically nilpotent filiform Lie algebras.*

Proof. We consider the nine-dimensional filiform Lie algebras given by

$$\mu = \mu_0 + \psi_{1,4} + \alpha \psi_{2,6} + \psi_{2,8} + \frac{3\alpha^2}{\alpha + 2} \psi_{3,8}$$

with $\alpha \neq 0$ and -2. These Lie algebras admits a $\mathbb{Z}_2$-grading and for two different values of α we have non isomorphic Lie algebras [12]. Moreover, since $\alpha \neq 0$, these Lie algebras are characteristically nilpotent. □

2.6 $\mathbb{Z}_k$-Gradings, $k > 2$, of Filiform Characteristically Nilpotent Lie Algebras

We assume that $k > 2$. A $\mathbb{Z}_k$-grading of L_n is equivalent to one of the following

$$\Gamma_k^l : L_n = \sum_{i=2}^{l-1} [X_i]_k \oplus (\langle X_1 \rangle \oplus [X_l]_k) \oplus \sum_{j=l+1}^{k+1} [X_j]_k,$$

where l is a parameter satisfying $2 \leq l \leq k + 1$. The homogeneous component of this grading corresponding to the identity of $\mathbb{Z}_k$ is $[X_{l-1}]_k$. Two cocycles ψ_{h_1,s_1} and ψ_{h_2,s_2} send an homogeneous component (in particular $[X_{l-1}]_k$) in another homogeneous component if and only if

$$s_1 - 2h_1 = s_2 - 2h_2 (\bmod k).$$

We deduce

Proposition 2.6. *Any filiform characteristically nilpotent Lie algebra* $\mathfrak{g}$ *such that* gr $(\mathfrak{g})$ *is isomorphic to* L_{n+1} *whose Lie multiplication is of the form*

$$\mu = \mu_0 + \sum_{i \in I} a_{h_i,s_i} \psi_{h_i,s_i}$$

with

$$s_i - 2h_i = s_j - 2h_j (\bmod k), \ k < n - 2$$

for any $i, j \in I$ *admits a* $\mathbb{Z}_k$-*grading.*

Proposition 2.7. *For any* k, *there exists a* $\mathbb{Z}_k$-*graded characteristically nilpotent filiform Lie algebra.*

Proof. If fact, we consider the filiform Lie algebra of dimension $n = k + 5$ given by

$$\mu = \mu_0 + \psi_{1,4} + \psi_{1,4+k},$$

that is

$$\begin{cases} \mu(e_0, e_i) = e_{i+1}, \ i = 2, \dots, k + 3, \\ \mu(e_1, e_2) = e_4 + e_{k+4}, \\ \mu(e_1, e_i) = e_{i+2}, \ i = 3, \dots, k + 2. \end{cases}$$

The Jacobi conditions are satisfied. The sill algebra is given by $\mu_0 + \psi_{1,4}$ and is not isomorphic to μ as soon as $k \neq 0$. Then this Lie algebra is characteristically nilpotent and from the previous proposition, it is $\mathbb{Z}_k$-graded.   □

3 $\mathcal{G}$-symmetric spaces

3.1 Definition

Let G be a Lie group and H a closed subgroup of G.

Definition 3.1. Let $\mathcal{G}$ be a finite abelian group. A homogeneous space $M = G/H$ is called $\mathcal{G}$-*symmetric* if

(1) The Lie group G is connected,
(2) The group G is effective on G/H (i.e., the Lie algebra $\mathfrak{h}$ of H does not contain a nonzero proper ideal of the Lie algebra $\mathfrak{g}$ of G),
(3) There is an injective homomorphism

$$\rho : \mathcal{G} \to \mathrm{Aut}\,(G)$$

such that if $G^{\mathcal{G}}$ is the closed subgroup of all elements of G fixed by $\rho(\mathcal{G})$ and $(G^{\mathcal{G}})_e$ the identity component of $G^{\mathcal{G}}$, then

$$(G^{\mathcal{G}})_e \subset H \subset G^{\mathcal{G}}.$$

Examples: Symmetric and k-symmetric Spaces. If $\mathcal{G} = \mathbb{Z}_2$, then the notion of $\mathbb{Z}_2$-symmetric spaces corresponds to the classical notion of symmetric spaces [19]. If $\mathcal{G} = \mathbb{Z}_p$ with p a prime number, we find again the p-manifolds in the sense of Ledger–Obata [21].

We denote by ρ_γ the automorphism $\rho(\gamma)$ for any $\gamma \in \mathcal{G}$. If H is connected, we have

$$\begin{cases} \rho_{\gamma_1} \circ \rho_{\gamma_2} = \rho_{\gamma_1 \gamma_2}, \\[2mm] \rho_\varepsilon = Id, \\[2mm] \rho_\gamma(g) = g, \ \forall \gamma \in \Gamma \Longleftrightarrow g \in H. \end{cases} \quad ,$$

where ε is the identity element of $\mathcal{G}$. Each automorphism ρ_γ of G, $\gamma \in \mathcal{G}$, induces an automorphism of $\mathfrak{g}$, denoted by τ_γ and given by $\tau_\gamma = (T\rho_\gamma)_e$ where $(Tf)_x$ is the tangent map of f at the point x.

Lemma 3.1. *The map* $\tau : \mathcal{G} \longrightarrow \mathrm{Aut}\,(\mathfrak{g})$ *given by*

$$\tau(\gamma) = (T\rho_\gamma)_e$$

is an injective homomorphism of groups.

Proof. Let γ_1, γ_2 be in $\mathcal{G}$. Then $\rho_{\gamma_1} \circ \rho_{\gamma_2} = \rho_{\gamma_1 \gamma_2}$. It follows that $(T\rho_{\gamma_1})_e \circ (T\rho_{\gamma_2})_e = (T\rho_{\gamma_1}\rho_{\gamma_2})_e = (T\rho(\gamma_1 \gamma_2))_e$, that is, $\tau(\gamma_1 \gamma_2) = \tau(\gamma_1)\tau(\gamma_2)$. Now let us assume that $\tau(\gamma) = Id_{\mathfrak{g}}$. Then $(T\rho_\gamma)_e = Id = (T\rho_\varepsilon)_e$. But ρ_γ is uniquely determined by the corresponding tangent automorphism of $\mathfrak{g}$. Then $\rho_\gamma = \rho_\varepsilon$ and $\gamma = \varepsilon$. $\square$

Then we have the infinitesimal version of a $\mathcal{G}$-symmetric space. A $\mathcal{G}$-symmetric Lie algebra is a triple $(\mathfrak{g}, \mathfrak{h}, \tau)$ consisting of a Lie algebra $\mathfrak{g}$, a Lie subalgebra $\mathfrak{h}$ and an injective homomorphism of group $\tau : \mathcal{G} \longrightarrow \mathrm{Aut}\,(\mathfrak{g})$ such that $\mathfrak{h}$ consists of all elements X of $\mathfrak{g}$ satisfying $\tau(\gamma)(X) = X$ for all $\gamma \in \mathcal{G}$. Thus, if G/H is a $\mathcal{G}$-symmetric space, the triple $(\mathfrak{g}, \mathfrak{h}, \tau)$ is a $\mathcal{G}$-symmetric Lie algebra. Conversely, if $(\mathfrak{g}, \mathfrak{h}, \tau)$ is a $\mathcal{G}$-symmetric Lie algebra and if G is a connected, simply connected Lie group whose Lie algebra is $\mathfrak{g}$ and H is a connected subgroup associated with $\mathfrak{h}$, then G/H is a $\mathcal{G}$-symmetric space.

3.2 $\mathcal{G}$-Grading of a $\mathcal{G}$-Symmetric Lie Algebra

Let $(\mathfrak{g}, \mathfrak{h}, \tau)$ be a $\mathcal{G}$-symmetric Lie algebra. Since $\tau : \mathcal{G} \to \mathrm{Aut}\,(\mathfrak{g})$ is an injective homomorphism, the image $\mathcal{G}_1$ of $\mathcal{G}$ is a finite abelian subgroup of $\mathrm{Aut}(\mathfrak{g})$ isomorphic to $\mathcal{G}$. For each $\gamma \in \mathcal{G}$, let $(\mathfrak{g}_{\mathbb{C}})_\gamma$ the subspace of $\mathfrak{g}_{\mathbb{C}}$ given by

$$(\mathfrak{g}_{\mathbb{C}})_\gamma = \{X \mid \gamma(X) = \tau(\gamma)(X)\},$$

The vector space decomposition $\mathfrak{g}_{\mathbb{C}} = \bigoplus_{\gamma \in \mathcal{G}} (\mathfrak{g}_{\mathbb{C}})_\gamma$ is just a standard weight decomposition under the action of an abelian semisimple group of linear transformations over an algebraically closed field.

Proposition 3.1. *If $(\mathfrak{g}, \mathfrak{h}, \tau)$ is a $\mathcal{G}$-symmetric Lie algebra, then the complex Lie algebra $\mathfrak{g}_{\mathbb{C}}$ admits a $\mathcal{G}$-grading. If $\tau(\gamma)^2 = \mathrm{Id}$ for any $\gamma \in \mathcal{G}$, then we have a $\mathcal{G}$-grading on $\mathfrak{g}$ itself.*

Explicitly, since $\mathcal{G}_1$ is a finite abelian subgroup of $\mathrm{Aut}\,(\mathfrak{g})$, one can write $\mathcal{G}_1$ as $\mathcal{G}_1 = K_1 \times \ldots \times K_p$ where K_i is a cyclic group of order r_i. Let κ_i be a generator of K_i. The automorphisms κ_i satisfy

$$\begin{cases} \kappa_i^{r_i} = \mathrm{Id}, \\ \kappa_i \circ \kappa_j = \kappa_j \circ \kappa_i, \end{cases}$$

for all $i, j = 1, \ldots, p$. These automorphisms are simultaneously diagonalizable. If ξ_i is a primitive r_i^{th} root of 1, then the eigenspaces

$$\mathfrak{g}_{s_1, \ldots, s_p} = \{X \in \mathfrak{g} \text{ such that } \kappa_i(X) = \xi_i^{s_i} X, \ i = 1, \ldots, p\}$$

give the following grading of $\mathfrak{g}$ by $\mathbb{Z}_{r_1} \times \ldots \times \mathbb{Z}_{r_p}$:

$$\mathfrak{g} = \bigoplus_{(s_1, \ldots, s_p) \in \mathbb{Z}_{r_1} \times \ldots \times \mathbb{Z}_{r_p}} \mathfrak{g}_{s_1, \ldots, s_p}.$$

Actually, the actions of a group G by automorphisms of an algebra and the gradings by G of the same algebra are closely connected in a much more general setting. This we explained in detail in [4]. Here we quickly recall the main idea.

Assume that the Lie algebra $\mathfrak{g}$ is G-graded where G is an finite abelian group. Let $\hat{G}$ be the dual group of G, that is, the group of characters of G. If we assume that a Lie algebra $\mathfrak{g}$ is G-graded, then we obtain a natural action of $\hat{G}$ by linear transformations on $\mathfrak{g}_{\mathbb{C}}$. If $\chi \in \hat{G}$ and $X \in \mathfrak{g}_\gamma$ then

$$\chi(X) = \chi(\gamma)X.$$

Since for $X \in \mathfrak{g}_{\gamma_1}$ and $Y \in \mathfrak{g}_{\gamma_2}$ we have $[X, Y] \in \mathfrak{g}_{\gamma_1\gamma_2}$, it follows that

(1) $\chi([X, Y]) = \chi(\gamma_1\gamma_2)[X, Y] = [\chi(\gamma_1)X, \chi(\gamma_2)Y] = [\chi(X), \chi(Y)],$

that is, $\hat{G}$ acts by Lie automorphisms on $\mathfrak{g}$. In this case there is a canonical homomorphism

(2) $\alpha : \hat{G} \to \mathrm{Aut}(\mathfrak{g}_{\mathbb{C}})$ given by $\alpha(\chi)(X) = \chi(X).$

If for any γ in the support of the grading, we have $\gamma^2 = 1$, then the action is defined even on $\mathfrak{g}$ itself and the above homomorphism maps $\hat{G}$ onto a subgroup of $\mathrm{Aut}(\mathfrak{g})$. Conversely, if $\hat{G}$ acts on $\mathfrak{g}_{\mathbb{C}}$ by automorphisms, then setting $(\mathfrak{g}_{\mathbb{C}})_\gamma = \{X \mid \chi(X) = \chi(\gamma)X\}$ provides $\mathfrak{g}_{\mathbb{C}}$ with a G-grading.

Proposition 3.2. *Let G be a finite abelian group and $\hat{G}$, the group of complex characters of G.*

(a) *A complex Lie algebra $\mathfrak{g}$ is G-graded if and only if the dual group $\hat{G}$ maps homomorphically onto a finite abelian subgroup of $\mathrm{Aut}(\mathfrak{g})$, by the canonical homomorphism α.*
(b) *A real Lie algebra $\mathfrak{g}$ is G-graded, with $\gamma^2 = 1$ for each γ in the support of this grading, if and only if there is a homomorphism $\alpha : \hat{G} \to \mathrm{Aut}(\mathfrak{g})$ such that $\alpha(\chi)^2 = \mathrm{id}_\mathfrak{g}$ for any $\chi \in \hat{G}$.*
(c) *In both cases above, the support generates G if and only if the canonical mapping α has trivial kernel, that is, $\hat{G}$ is isomorphic to a (finite abelian) subgroup of $\mathrm{Aut}(\mathfrak{g})$.*

From Lemma 3.1 we derive the following. Let $M = G/H$ be a G-symmetric space. Then $\mathfrak{g}$ is graded by the dual group of G. Since G is abelian, the groups G and $\hat{G}$ are isomorphic (a noncanonical isomorphism). To simplify notations, we will identify G and its dual, this permits one to speak of G-symmetric spaces and G-graded Lie algebras (in place of $\hat{G}$-graded Lie algebras).

Proposition 3.3. *If $M = G/H$ is a G-symmetric space, then the complex Lie algebra $\mathfrak{g}_{\mathbb{C}} = \mathfrak{g} \otimes \mathbb{C}$, where $\mathfrak{g}$ is the Lie algebra of G, is G-graded and if $G = \mathbb{Z}_2^k$, then the real Lie algebra $\mathfrak{g}$ of G is G-graded. The subgroup of G generated by the support of the grading is G itself.*

Proof. Indeed, by Lemma 3.1, $\alpha : \mathcal{G} \to$ Aut $(\mathfrak{g})$ is an injective homomorphism, so all our claims follow by Proposition 3.2. $\qquad\qquad\qquad\qquad\qquad\qquad\qquad\qquad\quad\square$

To study $\mathcal{G}$-symmetric spaces, we need to start with the study of $\mathcal{G}$-graded Lie algebras. But in a general case, if G is a connected Lie group corresponding to $\mathfrak{g}$, the $\mathcal{G}$-grading of $\mathfrak{g}$ or $\mathfrak{g}_{\mathbb{C}}$ does not necessarily give a $\mathcal{G}$-symmetric space G/H. Some examples are given in [7], even in the symmetric case. Still, if G is simply connected, Aut (G) is a Lie group isomorphic to Aut $(\mathfrak{g})$ and the $\mathcal{G}$-grading of $\mathfrak{g}$ determines a structure of $\mathcal{G}$-symmetric space from G.

Proposition 3.4. *Let* $\mathcal{G} = \mathbb{Z}_2^k$, *with the identity element* ε *and* $\mathfrak{g}$ *a real* $\mathcal{G}$-graded *Lie algebra such that the subgroup generated by the support of the grading equals* $\mathcal{G}$ *and the identity component* $\mathfrak{h} = \mathfrak{g}_\varepsilon$ *of the grading does not contain a nonzero ideal of* $\mathfrak{g}$. *If* G *is a connected simply connected Lie group with Lie algebra* $\mathfrak{g}$ *and* H *a Lie subgroup associated with* $\mathfrak{h}$, *then the homogeneous space* $M = G/H$ *is a* $\mathcal{G}$-symmetric space.

3.3 $\mathcal{G}$-Symmetries on a $\mathcal{G}$-Symmetric Space

Given a $\mathcal{G}$-symmetric space $(G/H, \mathcal{G})$ it is easy to construct, for each point x of the homogeneous space $M = G/H$, a subgroup of the group $\text{Diff}(M)$ of diffeomorphisms of M, isomorphic to $\mathcal{G}$, which has x as an isolated fixed point. We denote by $\overline{g}$ the class of $g \in G$ in M. If e is the identity of G, $\gamma \in \mathcal{G}$, we set

$$s_{(\gamma,\bar{e})}(\overline{g}) = \overline{\rho_\gamma(g)}.$$

If $\overline{g}$ satisfies $s_{(\gamma,\bar{e})}(\overline{g}) = \overline{g}$, then $\overline{\rho_\gamma(g)} = \overline{g}$, that is $\rho_\gamma(g) = gh_\gamma$ for $h_\gamma \in H$. Thus $h_\gamma = g^{-1}\rho_\gamma(g)$. But $\mathcal{G} \cong \hat{\mathcal{G}}$ is a finite abelian group. If p_γ is the order of γ, then $\rho_\gamma^{p_\gamma} = Id$. Then

$$h_\gamma^2 = g^{-1}\rho_\gamma(g)\rho_\gamma(g^{-1})\rho_{\gamma^2}(g) = g^{-1}\rho_{\gamma^2}(g).$$

Applying induction, and considering $(h_\gamma)^m \in H$ for any m, we have

$$(h_\gamma)^m = g^{-1}\rho_{\gamma^m}(g).$$

For $m = p_\gamma$ we obtain

$$(h_\gamma)^{p_\gamma} = e.$$

If g is near the identity element of G, then h_γ is also close to the identity and $h_\gamma^{p_\gamma} = e$ implies $h_\gamma = e$. Then $\rho_\gamma(g) = g$. This is true for all $\gamma \in \mathcal{G}$ and thus $g \in H$. It follows that $\overline{g} = \overline{e}$ and that the only fixed point of $s_{(\gamma,\bar{e})}$ is $\overline{e}$. In conclusion, the family $\{s_{(\gamma,\bar{e})}\}_{\gamma \in \mathcal{G}}$ of diffeomorphisms of M satisfy

$$\begin{cases} s_{(\gamma_1,\bar{e})} \circ s_{(\gamma_2,\bar{e})} = s_{(\gamma_1\gamma_2,\bar{e})} \\ s_{(\gamma,\bar{e})}(\bar{g}) = \bar{g}, \forall \, \gamma \in \mathcal{G} \Rightarrow \bar{g} = \bar{e}. \end{cases}$$

Thus,

$$\mathcal{G}_{\bar{e}} = \{s_{(\gamma,\bar{e})}, \, \gamma \in \mathcal{G}\}$$

is a finite abelian subgroup of $\mathrm{Diff}(M)$ isomorphic to $\mathcal{G}$, for which $\bar{e}$ is an isolated fixed point.

At another point $\overline{g_0}$ of M we put

$$s_{(\gamma,\overline{g_0})}(\bar{g}) = g_0(s_{(\gamma,\bar{e})})(g_0^{-1}\bar{g}).$$

As above, we can see that

$$\begin{cases} s_{(\gamma_1,\overline{g_0})} \circ s_{(\gamma_2,\overline{g_0})} = s_{(\gamma_1\gamma_2,\overline{g_0})} \\ s_{(\gamma,\overline{g_0})}(\bar{g}) = \bar{g}, \forall \gamma \in \mathcal{G} \Rightarrow \bar{g} = \overline{g_0}. \end{cases}$$

and

$$\mathcal{G}_{\overline{g_0}} = \{s_{(\gamma,\overline{g_0})}, \, \gamma \in \mathcal{G}\}$$

is a finite abelian subgroup of $\mathrm{Diff}(M)$ isomorphic to $\mathcal{G}$, for which $\overline{g_0}$ is an isolated fixed point.

Thus for each $\bar{g} \in M$ we have a finite abelian subgroup $\mathcal{G}_{\bar{g}}$ of $\mathrm{Diff}(M)$ isomorphic to $\mathcal{G}$, for which $\bar{g}$ is an isolated fixed point.

Definition 3.2. Let $(G/H, \mathcal{G})$ be a $\mathcal{G}$-symmetric space. For any point $x \in M = G/H$ the subgroup $\mathcal{G}_x \subset \mathrm{Diff}(M)$ is called *the group of symmetries* of M at x.

Since for every $x \in M$ and $\gamma \in \mathcal{G}$, the map $s_{(\gamma,x)}$ is a diffeomorphism of M such that $s_{(\gamma,x)}(x) = x$, the tangent linear map $(Ts_{(\gamma,x)})_x$ is in $\mathrm{GL}(T_xM)$. For every $x \in M$, we obtain a linear representation

$$S_x : \mathcal{G} \longrightarrow \mathrm{GL}(T_xM)$$

defined by

$$S_x(\gamma) = (Ts_{(\gamma,x)})_x.$$

Thus for every $\gamma \in \mathcal{G}$ the map

$$S(\gamma) : M \longrightarrow T(M)$$

defined by $S(\gamma)(x) = S_x(\gamma)$ is a $(1, 1)$-tensor on M which satisfies:

(1) the map $S(\gamma)$ is of class C^∞,
(2) for every $x \in M$,

$$\{X_x \in T_x(M) \mid S_x(\gamma)(X_x) = X_x, \forall \gamma \in \mathcal{G}\} = \{0\}.$$

In fact, this last remark is a consequence of the property : $s_{(\gamma,x)}(y) = y$ for every γ implies $y = x$.

3.4   The Reductivity of $\mathcal{G}$-Symmetric Spaces

We assume here that the group $\mathcal{G}$ is isomorphic to $\mathbb{Z}_2^k$. If $(G/H, \mathcal{G})$ be a $\mathcal{G}$-symmetric space, then the real Lie algebra $\mathfrak{g}$ of G is $\mathcal{G}$-graded. If ε is the identity element of $\mathcal{G}$, then the component $\mathfrak{h} = \mathfrak{g}_\varepsilon$ is a Lie subalgebra of $\mathfrak{g}$ corresponding to the Lie subgroup H. Let us consider the subspace $\mathfrak{m}$ of $\mathfrak{g}$:

$$\mathfrak{m} = \oplus_{\gamma \neq \varepsilon} \mathfrak{g}_\gamma.$$

Then $\mathfrak{g} = \mathfrak{h} \oplus \mathfrak{m}$ and

$$[\mathfrak{h}, \mathfrak{m}] \subset \mathfrak{m}$$

so that $\mathfrak{m}$ is an ad $\mathfrak{h}$-invariant subspace. If H is connected, then $[\mathfrak{h}, \mathfrak{m}] \subset \mathfrak{m}$ is equivalent to $(\mathrm{ad}\,H)(\mathfrak{m}) \subset \mathfrak{m}$, that is, $\mathfrak{m}$ is an ad H-invariant subspace. This property is true without any conditions on H.

Lemma 3.2. *Any $\mathcal{G}$-symmetric space $(G/H, \mathcal{G})$ is reductive.*

Proof. Let us consider the associated local $\mathcal{G}$-symmetric space $(\mathfrak{g}/\mathfrak{h}, \mathcal{G})$. We need to find a decomposition $\mathfrak{g} = \mathfrak{h} \oplus \mathfrak{m}$, such that $\mathfrak{m}$ is invariant under the adjoint action of the isotropy subgroup H or, which is the same, under the action of the isotropy subalgebra $\mathfrak{h}$. Now since $\mathfrak{h} = \mathfrak{g}_\varepsilon$ and $\mathfrak{m} = \oplus_{\gamma \neq \varepsilon}\mathfrak{g}_\gamma$ we have that $[\mathfrak{h}, \mathfrak{m}] \subset \mathfrak{m}$.   □

We now deduce from [19, Chapter X], that $M = G/H$ admits two G-invariant canonical connections denoted by ∇ and $\overline{\nabla}$. The *first canonical connection* ∇ satisfies

$$\begin{cases} R(X, Y) = -\mathrm{ad}([X, Y]_\mathfrak{h}), \ T(X, Y)_{\overline{e}} = -[X, Y]_\mathfrak{m}, \ \forall X, Y \in \mathfrak{m}, \\ \nabla T = 0, \\ \nabla R = 0, \end{cases}$$

where T and R are the torsion and the curvature tensors of ∇. The tensor T is trivial if and only if $[X, Y]_\mathfrak{m} = 0$ for all $X, Y \in \mathfrak{m}$. This means that $[X, Y] \in \mathfrak{h}$, that is $[\mathfrak{m}, \mathfrak{m}] \subset \mathfrak{h}$. If the grading of $\mathfrak{g}$ is given by $\mathcal{G} = \mathbb{Z}_2^k$ with $k > 1$, then $[\mathfrak{m}, \mathfrak{m}]$

is not a subset of $\mathfrak{h}$, and then the torsion T need not vanish. In this case another connection $\overline{\nabla}$ will be defined if one sets $\overline{\nabla}_X Y = \nabla_X Y - T(X, Y)$. This is an affine invariant torsion free connection on G/H which has the same geodesics as ∇. This connection is called the *second canonical connection* or the *torsion-free canonical connection*.

Remark. Actually, there is another way of writing the canonical affine connection of a $\mathcal{G}$-symmetric space, without any reference to Lie algebras. This is done by an intrinsic construction of $\mathcal{G}$-symmetric spaces proposed by Lutz in [22].

4 Classification of Simple $\mathbb{Z}_2^2$-Symmetric Spaces

4.1 $\mathbb{Z}_2^2$-Gradings of Classical Simple Lie Algebras

We have seen that the classification of $\mathcal{G}$-symmetric spaces $(G/H, \mathcal{G})$, when G is connected and simply connected, corresponds to the classification of Lie algebras graded by $\hat{\mathcal{G}}$ which is isomorphic to $\mathcal{G}$. Below we establish the classification of local $\mathbb{Z}_2^2$-symmetric spaces $(\mathfrak{g}, \mathcal{G})$ in the case where the corresponding Lie algebra $\mathfrak{g}$ is simple, complex and classical.

In this section $\mathcal{G} = \{\varepsilon, a, b, c\}$ is the group $\mathbb{Z}_2^2$ with identity ε and $a^2 = b^2 = c^2 = \varepsilon$, $ab = c$. We will consider $\mathbb{Z}_2^2$-gradings on a complex simple Lie algebra $\mathfrak{g}$ of type A_l, $l \geq 1$, B_l, $l \geq 2$, C_l, $l \geq 3$ and D_l, $l \geq 4$. We will describe these gradings using the results of [4]. Let us note that these gradings have been described in [3] where all the cases have been covered except for the case $\mathfrak{g} = so(8)$ which was later handled in [20] and [10] (see also [11]). Clearly, for our purposes, we need to classify these gradings up to a weak isomorphism.

To classify these gradings, we use techniques surveyed in [4]. In the case $\mathcal{G} = \mathbb{Z}_2 \times \mathbb{Z}_2$, the situation is considerably simpler than in the general case.

Recall that we always have to start with a grading on a matrix algebra $M_n(\mathbb{K})$, either Γ graded by $\mathcal{G}$ or $\overline{\Gamma}$ by $\overline{G} = \mathcal{G}/\langle a \rangle$ where a is an element of order 2 in $\mathcal{G}$. In both cases our matrix algebra is of the form of $M_m(D)$, D a graded division algebra. In the case of $\overline{G}$, we only have $D = \mathbb{K}$. In the case of Γ there are two options for D: either $D = \mathbb{K}$ or $D = M_2(\mathbb{K})$. In the latter case, if $\mathcal{G} = \{\varepsilon, a, b, c\}$, then a graded basis of D is formed by the Pauli matrices:

$$\left\{ X_\varepsilon = I, X_a = \begin{pmatrix} -1 & 0 \\ 0 & 1 \end{pmatrix}, X_b = \begin{pmatrix} 0 & 1 \\ 1 & 0 \end{pmatrix}, X_c = \begin{pmatrix} 0 & -1 \\ 1 & 0 \end{pmatrix} \right\}.$$

The graded involution of D is given by the transpose.

The Type I grading of $\mathfrak{sl}_n(\mathbb{K})$ and the gradings of the orthogonal and symplectic algebras appear as the restriction of Γ to either $\mathfrak{sl}_n(\mathbb{K})$ or to the space of skew-symmetric elements under a graded involution.

The Type II gradings appear as a refinement of $\overline{\Gamma}$ by a graded involution (in the general case one has to consider the eigenspaces of more general graded antiautomorphisms).

The graded involutions of the associative algebra $M_n(D)$ of matrices of order n over a graded division algebra D are given by $X \mapsto \Phi^{-1} X^t \Phi$, where Φ is a matrix of a sesquilinear form with coefficients in D. The description of graded involutions (that is, possible forms of Φ) was first given in [5] and can be found in [4].

In the following list we give the pairs $(\mathfrak{g}, \mathfrak{g}_\varepsilon = \mathfrak{h})$ corresponding to a $\mathbb{Z}_2^2$-grading (and not a $\mathbb{Z}_2$-grading) of simple classical complex Lie algebras.

$\mathfrak{g}$	$\mathfrak{g}_\varepsilon = \mathfrak{h}$
$sl(k_1 + k_2 + k_3)$	$sl(k_1) + sl(k_2) + sl(k_3) \oplus \mathbb{C}^2$
$sl(k_1 + k_2 + k_3 + k_4)$	$sl(k_1) + sl(k_2) + sl(k_3) + sl(k_4) \oplus \mathbb{C}^3$
$sl(2n)$	$sl(n)$
$sl(k_1 + k_2)$	$so(k_1) \oplus so(k_2)$
$sl(2(k_1 + k_2))$	$sp(2k_1) \oplus sp(2k_2)$
$sl(2n)$	$gl(n)$
$so(k_1 + k_2 + k_3),$	$so(k_1) + so(k_2) + so(k_3)$
$so(k_1 + k_2 + k_3 + k_4)$	$so(k_1) + so(k_2) + so(k_3) + so(k_4)$
$sp(2(k_1 + k_2 + k_3))$	$sp(2k_1) + sp(2k_2) + sp2(k_3)$
$sp(2(k_1 + k_2 + k_3 + k_4))$	$sp(2k_1) + sp(2k_2) + sp2(k_3) + sp(2k_4)$
$so(2k_1 + 2k_2)$	$gl(k_1) \oplus gl(k_2)$
$sp(2k_1 + 2k_2)$	$gl(k_1) \oplus gl(k_2)$
$so(2m),\ m \neq 4$	$so(m)$
$so(4m),\ m \neq 2$	$sp(2m)$
$sp(4m)$	$sp(2m)$
$sp(2m)$	$so(m)$

Consequences. The following homogeneous spaces

$SU(k_1 + k_2 + k_3)/SU(k_1) \times SU(k_2) \times SU(k_3) \times \mathbb{T}^2,$

$SU(k_1 + k_2 + k_3 + k_4)/SU(k_1) \times SU(k_2) \times SU(k_3) \times SU(k_4) \times \mathbb{T},$

$SU(2n)/SU(n),$

$SU(k_1 + k_2)/SO(k_1) \times SO(k_2),$

$SU(2k_1 + 2k_2)/Sp(2k_1) \times Sp(2k_2),$

$SO(k_1 + k_2 + k_3)/SO(k_1) \times SO(k_2) \times SO(k_3),$

$SO(k_1 + k_2 + k_3 + k_4)/SO(k_1) \times SO(k_2) \times SO(k_3) \times SO(k_4),$

$Sp(2k_1 + 2k_2 + 2k_3)/Sp(2k_1) \times Sp(2k_2) \times Sp(2k_3),$

$Sp(2k_1 + 2k_2 + 2k_3 + 2k_4)/Sp(2k_1) \times Sp(2k_2) \times Sp(2k_3) \times Sp(2k_4),$

$SO(2m)/SO(m),\ SO(4m)/Sp(2m),\ Sp(4m)/Sp(2m),\ Sp(2m)/SO(m).$

are compact nonsymmetric spaces but $\mathbb{Z}_2^2$-symmetric spaces. In particular, we obtain the classical flag manifolds which are not symmetric when they do not correspond to a Grassmanian manifold. A Riemannian study of these spaces is proposed in [16, 23].

4.2 Classification in the Exceptional Cases

The study of $\mathbb{Z}_2^2$-gradings on exceptional simple Lie algebras has been done by Kollross [20]. His approach is different: he starts with exploring the possibilities to obtain involute automorphisms on a reductive complex Lie algebra from a given root space decomposition $\mathfrak{g} = \mathfrak{g}_0 + \sum_{\alpha \in \Phi} \mathfrak{g}_\alpha$, where $\mathfrak{g}_0$ is a Cartan subalgebra. He distinguishes 3 types, as follows.

- If $\mathfrak{h}$ is a symmetric ($\mathbb{Z}_2$-graded) subalgebra of maximal rank containing $\mathfrak{g}_0$, then $\mathfrak{h} = \mathfrak{g}_0 + \sum_{\alpha \in S} \mathfrak{g}_\alpha$, $S \subset \Phi$, and the automorphism σ is given by $\sigma(X) = X$ if $X \in \mathfrak{h}$, $\sigma(X) = -X$ if $X \in \sum_{\alpha \in \Phi - S} \mathfrak{g}_\alpha$.
- The automorphism σ is an outer automorphism induced from an automorphism of the Dynkin diagram.
- The automorphism σ acts as $-\mathrm{Id}$ on a Cartan subalgebra of $\mathfrak{g}$ and send each $\mathfrak{g}_\alpha$ to $\mathfrak{g}_{-\alpha}$, where $\alpha \in \Phi$.

Using these automorphisms allows one to construct $\mathbb{Z}_2^2$-symmetric gradings on $\mathfrak{g}$. For example, if σ_1 and σ_2 are of the first type and associated with two elements g_1, g_2 belonging to a maximal torus T of G, we consider the root space decomposition of the complexified $\mathfrak{g}_{\mathbb{C}}$ of $\mathfrak{g}$ defined by the set $\{\sigma_1, \sigma_2\}$ with $\mathfrak{g}_0$ the Lie algebra of a closed subgroup of the torus T. The maximal subgroups of maximal rank in a simple compact Lie group are described by the extended Dynkin diagrams.

Examining all possibilities and proving that all cases actually occur, one obtains the following classification.

$\mathfrak{g}$	$\mathfrak{g}_\varepsilon = \mathfrak{h}$	$\mathfrak{g}$	$\mathfrak{g}_\varepsilon = \mathfrak{h}$
E_6	$so(6) \oplus \mathbb{C}$	E_7	$so(8)$
	$sp(2) \oplus sp(2)$		$su(4) \oplus su(4) \oplus \mathbb{C}$
	$sp(3) \oplus sp(1)$		$sp(4)$
	$su(3) \oplus su(3) \oplus \mathbb{C}^2$		$su(6) \oplus sp(1) \oplus \mathbb{C}$
	$su(4) \oplus sp(1) \oplus sp(1) \oplus \mathbb{C}$		$so(8) \oplus so(4) \oplus sp(1)$
	$su(5) \oplus \mathbb{C}^2$		$u(6) \oplus \mathbb{C}$
	$so(8) \oplus \mathbb{C}^2$		$so(10) \oplus \mathbb{C}^2$
	$so(9)$		F_4

$\mathfrak{g}$	$\mathfrak{g}_\varepsilon = \mathfrak{h}$
E_8	$so(8) \oplus so(8)$
	$su(8) \oplus \mathbb{C}$
	$so(12) \oplus sp(1) \oplus sp(1)$
	$E_6 \oplus \mathbb{C}^2$

$\mathfrak{g}$	$\mathfrak{g}_\varepsilon = \mathfrak{h}$
F_4	$u(3) \oplus \mathbb{C}$
	$sp(2) \oplus sp(1) \oplus sp(1)$
	$so(8)$
G_2	$\mathbb{C}^2$

4.3   Riemannian Compact $\mathbb{Z}_2^2$-Symmetric Spaces

Let $M = G/H$ be a $\mathbb{Z}_2^k$-symmetric space with G and H connected. The homogeneous space $M = G/H$ is reductive. Then there exists a one-to-one correspondence between the G-invariant pseudo-Riemannian metrics g on M and the nondegenerated symmetric bilinear form B on $\mathfrak{m}$ satisfying

$$B([Z, X], Y) + B(X, [Z, Y]) = 0$$

for all $X, Y \in \mathfrak{m}$ and $Z \in \mathfrak{g}_\varepsilon$.

Definition 4.1 ([16]). A $\mathbb{Z}_2^k$-symmetric space $M = G/H$ with $\mathrm{Ad}_G(H)$-compact is called Riemannian $\mathbb{Z}_2^k$-symmetric if M is provided with a G-invariant Riemannian metric g whose associated bilinear form B satisfies

(1) $B(\mathfrak{g}_\gamma, \mathfrak{g}_{\gamma'}) = 0$ if $\gamma \neq \gamma' \neq \varepsilon \neq \gamma$,
(2) The restriction of B to $\mathfrak{m} = \oplus_{\gamma \neq \varepsilon}\mathfrak{g}_\gamma$ is positive definite.

In this case the linear automorphisms which belong to $\hat{\Gamma}$ are linear isometries. When $k \geq 2$, the geometry of $\mathbb{Z}_2^k$-Riemannian symmetric spaces is not similar to the Riemannian symmetric case. For example, if we consider the flag manifold $SO(5)/SO(2) \times SO(2) \times SO(1)$, this homogeneous reductive space is not symmetric but $\mathbb{Z}_2^2$-symmetric. It is proved in [23] that there exists a Riemannian $\mathbb{Z}_2^2$-symmetric tensor which is not naturally reductive and satisfying the first Ledger condition. This is impossible in the symmetric case.

5   Affine Structures on Lie Algebras

5.1   General Definitions

Let M be a differential manifold. We denote by $\mathfrak{X}(M)$ the set of vector fields on M, that is, the set of differentiable section

$$X : M \to T(M)$$

of the tangent bundle of M. Recall that it is a module on the Lie algebra $\mathcal{D}(M)$ of differentiable functions on M and an $\mathbb{R}$-Lie algebra.

Definition 5.1. An affine connection on M is a $\mathbb{R}$-bilinear map

$$\nabla : \mathfrak{X}(M) \times \mathfrak{X}(M) \to \mathfrak{X}(M)$$

satisfying for any $X, Y \in \mathfrak{X}(M)$ and $f \in \mathcal{D}(M)$

- $\nabla(fX, Y) = f\nabla(X, Y)$,
- $\nabla(X, fY) = f\nabla(X, Y) + X(f)Y$.

This means that ∇ is not a $\mathcal{D}(M)$-bilinear map; it is only linear on the first argument. A vector field X such that $\nabla(X, Y) = 0$ for any $Y \in \mathfrak{X}(M)$ is called parallel for ∇.

The torsion of the affine connection ∇ is the bilinear map

$$T_\nabla : \mathfrak{X}(M) \times \mathfrak{X}(M) \to \mathfrak{X}(M)$$

given by $T_\nabla(X, Y) = \nabla(X, Y) - \nabla(Y, X) - [X, Y]$.

When $T_\nabla = 0$, that is, if ∇ is torsion free, then T_∇ defines a structure of a Lie-admissible algebra on $\mathfrak{X}(M)$ adapted to the Lie bracket. Recall that a Lie-admissible algebra is an algebra $(\mathcal{A}, \cdot)$ whose product satisfies $x \cdot y - y \cdot x$ is a Lie bracket.

A classical example of torsion free affine connection is given by the Levi-Civita connection.

The curvature of the affine connection ∇ is the trilinear map

$$R_\nabla : \mathfrak{X}(M) \times \mathfrak{X}(M) \times \mathfrak{X}(M) \to \mathfrak{X}(M)$$

given by

$$R_\nabla(X, Y, Z) = \nabla(X, \nabla(Y, Z)) - \nabla(Y, \nabla(X, Z)) - \nabla([X, Y], Z).$$

The connection is flat if $R_\nabla = 0$. Assume now that ∇ is such that $T_\nabla = 0$. Then $R_\nabla = 0$ is equivalent to

$$\nabla(X, \nabla(Y, Z)) - \nabla(\nabla(X, Y), Z) = \nabla(Y, \nabla(X, Z)) - \nabla(\nabla(Y, X), Z).$$

A Lie-admissible multiplication satisfying this identity is called *left-symmetric* (an algebra satisfying this identity is sometimes called a pre-Lie algebra). Thus a flat affine connection (that is with both the torsion and the curvature vanish identically) on M provides $\mathfrak{X}(M)$ with a left-symmetric algebra structure associated with the Lie bracket of vector fields.

If the differential manifold M admits a flat torsion free affine connection, then M admits an atlas $A = (U_i, \varphi_i)$ such that the changes of charts

$$\varphi_i \circ \varphi_j^{-1} : \varphi_j \left(U_i \cap U_j \right) \to \varphi_i \left(U_i \cap U_j \right)$$

are affine transformations of $\mathbb{R}^n$, where n is the dimension of M. We will denote by $\mathrm{Aff}_n(\mathbb{R})$ the group of affine transformations of $\mathbb{R}^n$ (considered as an affine space). The elements of this group are represented by the matrices

$$\begin{pmatrix} a & \xi \\ 0 & 1 \end{pmatrix}$$

where $a \in \mathrm{GL}_n(\mathbb{R})$ and ξ is a column vector of $\mathbb{R}^n$.

Let (M, ∇) be a differentiable manifold with an affine connection. A diffeomorphism f of M is an affine transformation if

$$f(\nabla(X, Y)) = \nabla(fX, fY)$$

for any vector field X, Y on M. This notion can be interpreted in terms of geodesics. A curve γ_t on M with $t \in]a, b[$ is a geodesic if the vector field $X = \overset{\bullet}{\gamma}_t$ is parallel along γ_t, that is

$$\nabla(X, X) = 0$$

for all t. Thus an affine transformation maps every geodesic into a geodesic.

A vector field X on M is an infinitesimal affine transformation of M if for any point $x \in M$, a local 1-parameter group of local transformations γ_t generated by X in a neighborhood of x is an affine mapping. This is equivalent to the equation:

$$[X, \nabla(Y, Z)] - \nabla(Y, [X, Z]) = \nabla([X, Y], Z)$$

for any vector fields Y, Z on M. The set of infinitesimal affine transformations of M is a Lie subalgebra $\mathfrak{a}(M)$ of the Lie algebra $\mathfrak{X}(M)$. If the connection ∇ is flat, then $\dim \mathfrak{a}(M) = n^2 + n$. Moreover, in this case we have

$$\nabla(\nabla(Y, Z), X) = \nabla(Y, \nabla(Z, X))$$

for any vector fields Y, Z on M.

To end this short presentation of the affine connection, we have to recall the notion of complete affine connection. An affine connection is called *complete* if every geodesic can be extended to a geodesic γ_t defined for any $t \in \mathbb{R}$. In this case, every infinitesimal transformation is a complete vector field. But for a complete affine connection it is not generally true that every pair of points can be joined by a geodesic. Moreover, an affine connection on a compact manifold is not necessarily complete. The first example was described by Auslander and Markus, giving a noncomplete affine connection on the one-dimensional torus. Other examples on the two-and three-dimensional tori are given in [24].

5.2 Flat Affine Connections

An affine connection ∇ on M is flat when the torsion and the curvature tensor vanish identically. Let ∇ be an affine connection on the differentiable manifold M. If R denotes its curvature tensor, we define ∇R

$$\nabla R(X_1, X_2, X_3, X_4) = \nabla(X_1, R(X_2, X_3, X_4)) - R(\nabla(X_1, X_2), X_3, X_4)$$
$$-R(X_1, \nabla(X_2, X_3), X_4) - R(X_1, X_2, \nabla(X_3, X_4)).$$

A manifold M with an affine connection such that the torsion T satisfies $T = 0$ and $\nabla R = 0$ is called affine locally symmetric. In this case, all the symmetries s_x which are local diffeomorphisms of M sending $\exp(X)$ to $\exp(-X)$ for all $X \in T_x M$ are affine transformations. This happens, in particular, when ∇ is complete.

Assume now that M is complete. Then the group of affine transformations is a Lie group. If we consider its identity component G, then M can be viewed as an homogeneous reductive space G/H and the affine connection on M coincides with the natural torsion free connection on G/H. An interesting example is a reductive space G/H that is symmetric. The Lie algebra $\mathfrak{g}$ of G admits a $\mathbb{Z}_2$-grading $\mathfrak{g} = \mathfrak{h} \oplus \mathfrak{m}$ with $\mathfrak{h}$ the Lie algebra of H and $\mathfrak{m}$ isomorphic to the tangent space (in $\bar{e}$ where e is the identity of G) of M. But in this case the curvature tensor R satisfies

$$R(X, Y, Z) = -[[X, Y], Z], \ X, Y, Z \in \mathfrak{m}$$

and the curvature tensor is trivial if and only if the Lie triple bracket $[[X, Y], Z]$ is always equal to 0. Thus it will be also interesting to look to the case G/H reductive not symmetric, for example $\mathbb{Z}_2^2$-symmetric. Recall also that if $T = 0$ and $R = 0$, the Lie algebra of the affine holonomy group is necessarily of dimension 0.

5.3 Invariant Affine Connection on Lie Groups

If M is a Lie group G, we can consider on G an affine connection ∇ satisfying

$$dL_g(\nabla(X, Y)) = \nabla(dL_g X, dL_g Y),$$

where $L_g : G \rightarrow G$ is the left shift by $g \in G$. This means that L_g is an affine diffeomorphism, for any $g \in G$. We then call ∇ left invariant. Equivalently, if X and Y are left invariant vector fields on G, then $\nabla(X, Y)$ is also left invariant. As a result, ∇ defines a bilinear map, also denoted by ∇, on the Lie algebra $\mathfrak{g}$ of G:

$$\nabla : \mathfrak{g} \times \mathfrak{g} \rightarrow \mathfrak{g}$$
$$(X, Y) \mapsto \nabla(X, Y)$$

Definition 5.2. The bilinear map ∇ is called an *affine structure* on $\mathfrak{g}$ if the underlying affine connection on G is flat and torsion free.

Thus ∇ is an affine structure on $\mathfrak{g}$ if we have

$$\begin{cases} \nabla(X, Y) - \nabla(Y, X) = [X, Y], \\ \nabla(X, \nabla(Y, Z)) - \nabla(Y, \nabla(X, Z)) = \nabla([X, Y], Z), \end{cases}$$

for any $X, Y, Z \in \mathfrak{g}$ where $[X, Y]$ is the Lie bracket of $\mathfrak{g}$.

It is proved in [18] that if the Lie algebra $\mathfrak{g}$ admits an affine structure, then

$$[\mathfrak{g}, \mathfrak{g}] \subsetneq \mathfrak{g}.$$

As an immediate consequence, we deduce no semisimple Lie algebras can be endowed with an affine structure. But there exist Lie algebras with affine structures which contain semi-simple Lie subalgebras. For example, let us consider the Lie algebra $\mathfrak{gl}(n, \mathbb{R})$. We consider the product

$$\nabla(M_1, M_2) = M_1 M_2$$

for $M_1, M_2 \in \mathfrak{gl}(n, \mathbb{R})$. We have

$$[M_1, M_2] = \nabla(M_1, M_2) - \nabla(M_2, M_1)$$

and the torsion is null. Since the product $M_1 M_2$ is associative, the curvature is also null. Thus we obtain an affine structure on $gl_n(\mathbb{R})$. Let us remark that this construction can be generalized to any associative algebra A viewed as a Lie algebra under the bracket $[X, Y] = XY - YX$; its natural affine structure is given by $\nabla(X, Y) = XY$.

Proposition 5.1 ([18]). *An affine structure on a Lie algebra is complete if the linear map*

$$R_Y : \mathfrak{g} \to \mathfrak{g}$$
$$X \mapsto \nabla(X, Y)$$

is nilpotent.

We also have a characterization of an affine structure on a Lie algebra in term of affine representations. Let $\mathrm{Aff}_n(\mathbb{R})$ the group of affine transformations of $\mathbb{R}^n$. An n-dimensional Lie group admits a left-invariant flat affine connection if and only if we have an homomorphism of Lie group

$$\varphi : G \to \mathrm{Aff}_n(\mathbb{R})$$

given by

$$\varphi(x) = (M_x, v_x)$$

with $M_x \in \text{GL}_n(\mathbb{R})$ and $v_x \in \mathbb{R}^n$, such that there exists a vector $v \in \mathbb{R}^n$ such that its orbit $\mathcal{O}(v) = \{\phi(x)(v) \mid x \in G\}$ is open and its stabilizer is discrete. Such representations are called *étale affine*. The connection ∇ is complete if and only if the action of G on $\mathbb{R}^n$ is transitive, that is, $\mathcal{O}(v) = \mathbb{R}^n$. We can state this approach in terms of representations of Lie algebra. Let $\text{aff}_n(\mathbb{R})$ be the Lie algebra of $\text{Aff}_n(\mathbb{R})$. It can be written as

$$\text{aff}_n(\mathbb{R}) = \left\{ \begin{pmatrix} M & v \\ 0 & 0 \end{pmatrix}, \ M \in gl_n(\mathbb{R}), v \in \mathbb{R}^n \right\}.$$

Let $\mathfrak{g}$ be a Lie algebra. An affine representation of $\mathfrak{g}$ is a Lie algebra homomorphism

$$\Phi : \mathfrak{g} \to \text{aff}_n(\mathbb{R}).$$

In addition, Φ is called an étale affine representation with base point v if there exists $v \in \mathbb{R}^n$ such that the mapping

$$ev_v : \mathfrak{g} \to \mathbb{R}^n$$

defined by

$$ev_v(X) = \Phi(X)v$$

for any $X \in \mathfrak{g}$ is an isomorphism. For example, consider $\Phi = (\rho, q) : \mathfrak{g} \to \text{aff}_n(\mathbb{R})$ with $\Phi(x) = \rho(x) + q(x)$. It is an affine representation if and only if $\rho : \mathfrak{g} \to gl_n(\mathbb{R})$ is a representation of $\mathfrak{g}$ and $q : \mathfrak{g} \to \mathbb{R}^n$ is a linear map satisfying $q[x, y] = \rho(x)q(y) - \rho(y)q(x)$, for all $x, y \in \mathfrak{g}$. If $\mathfrak{g}$ admits an affine structure, then the map $\rho : X \to gl(\mathfrak{g})$ given by

$$L(X) : Y \in \mathfrak{g} \to L(X)(Y) = \nabla(X, Y)$$

is a linear representation of $\mathfrak{g}$, because the tensor curvature is zero. We consider $q(X) = X$, it satisfies the previous condition because the torsion tensor is zero. Then we have an affine étale representation of $\mathfrak{g}$ in $\text{aff}_n(\mathbb{R})$. If the Lie group G of $\mathfrak{g}$ is connected and simply connected, this representation of $\mathfrak{g}$ determines an embedding of G in $\text{Aff}_n(\mathbb{R})$ and we can identify G with a subgroup of $\text{Aff}_n(\mathbb{R})$.

Proposition 5.2. *A Lie algebra $\mathfrak{g}$ admits an affine structure if and only if $\mathfrak{g}$ has an étale affine representation.*

Proof. We have seen that if $\mathfrak{g}$ admits an affine structure, then we have an affine étale representation of $\mathfrak{g}$. Conversely, let $\Phi = (\rho, q) : \mathfrak{g} \to \text{aff}_n(\mathbb{R})$ be an étale affine representation of $\mathfrak{g}$; then

$$\nabla(X, Y) = ev_v^{-1}[\rho(X)ev_v(Y)]$$

for all $X, Y \in \mathfrak{g}$ defines an affine structure on $\mathfrak{g}$. $\square$

Corollary 5.1. *If an n-dimensional Lie algebra admits an affine structure, then $\mathfrak{g}$ admits an $(n + 1)$-dimensional faithful linear representation.*

5.4   The Milnor Conjecture

Milnor asked the following: "Does every solvable Lie group admits a complete left-invariant affine structure?" This question is supported by a result due to Auslander [2]:

Theorem 5.1. *A Lie group with a complete left-invariant affine structure is solvable.*

In terms of affine structures on Lie algebras, the problem of Milnor naturally reduces to the following conjecture, called the *Milnor conjecture*:

Every nilpotent Lie algebra admits an affine structure.

In 1993, Benoist [6] presented the first counterexample to this conjecture, describing an eleven-dimensional filiform Lie algebra without affine structure. A short time afterwards, Burde and Grunewald published another counterexample, in dimension 10, probably the smallest possible dimension where we can find such counterexamples.

5.5   Affine Structures and Gradings

Proposition 5.3. *Let $\mathfrak{g}$ be a $\mathbb{Z}$-graded complex Lie algebra whose support is in $\mathbb{N}$. Then $\mathfrak{g}$ admits an affine structure.*

If $\mathfrak{g} = \bigoplus_{n \in \mathbb{Z}} \mathfrak{g}_n$, we consider the bilinear map $\nabla(X_i, X_j) = \frac{n_j}{n_i + n_j}[X_i, X_j]$ where X_i, X_j are homogeneous vectors belonging to $\mathfrak{g}_{n_i}$ and $\mathfrak{g}_{n_j}$. Then

$$\nabla(X_i, X_j) - \nabla(X_j, X_i) = \frac{n_j}{n_i + n_j}[X_i, X_j] - \frac{n_i}{n_i + n_j}[X_j, X_i] = [X_j, X_i],$$

and

$$\nabla(X_i, \nabla(X_j, X_k)) - \nabla(X_j, \nabla(X_i, X_k)) = \frac{n_k}{n_i + n_j + n_k}([X_i, [X_j, X_k]]$$
$$+ [X_j, [X_k, X_i]]),$$

and from the Jacobi identity,

$$\nabla(X_i, \nabla(X_j, X_k)) - \nabla(X_j, \nabla(X_i, X_k)) = \nabla([X_i, X_j], X_k)$$

for any $X_k \in \mathfrak{g}_{n_k}$. The map ∇ defines an affine structure on $\mathfrak{g}$.

Consequence. Any nilpotent filiform Lie algebra of nonzero rank admits an affine structure (see also [6, 8, 25]).

In case of rank 0, we do not have general results. At the same time, let us consider a $\mathbb{Z}_2$-graded, eight-dimensional characteristically nilpotent Lie algebra $\mathfrak{g}$. Assume that the grading is given in an adapted basis by $\mathfrak{g} = \{e_1, e_3, e_5, e_7\} \bigoplus \{e_0, e_2, e_4, e_6\}$. In this case, from Proposition 2.4, the Lie multiplication μ of $\mathfrak{g}$ is $\mu = \mu_0 + A\psi_{1,4} + B\psi_{1,6} + C\psi_{2,6}$. We consider the bilinear map ∇ defined by $\nabla_{e_0}(e_i) = \mu(e_0, e_i)$,

$$
\nabla_{e_1} = \begin{pmatrix}
0\,0 & 0 & 0 & 0\,0\,0\,0 \\
0\,0 & 0 & 0 & 0\,0\,0\,0 \\
0\,0 & 0 & 0 & 0\,0\,0\,0 \\
0 -A - C + 5p\,0 & 0 & 0\,0\,0\,0 \\
0\,0 & -C + 5/2p\,0 & 0\,0\,0\,0 \\
0\,2m - B & 0 & 3/2p\,0\,0\,0\,0 \\
0\,0 & m & 0 & p\,0\,0\,0 \\
0\,0 & 0 & 0 & 0\,v\,0\,0
\end{pmatrix}
$$

and $\nabla_{e_i} = [\nabla_{e_0}, \nabla_{e_{i-1}}]$. This defines an affine structure on the graded Lie algebra. We can define similar affine structures when the grading is of the second type in Proposition 2.4.

Based on similar examples, we ask the following.
Let $\mathfrak{g}$ be a characteristically nilpotent Lie algebra such that $gr(\mathfrak{g}) = L_{n+1}$. If $\mathfrak{g}$ is $\mathbb{Z}_2$-graded, is it true that $\mathfrak{g}$ admits an affine structure?

The Benoist counterexample is of dimension 11, characteristically nilpotent, with multiplication

$$\mu = \mu_0 + a_{1,4}\psi_{1,4} + a_{3,7}\psi_{3,7} + a_{3,8}\psi_{3,8} + a_{3,9}\psi_{3,9} + a_{4,9}\psi_{4,9} + a_{4,10}\psi_{4,10} + a_{4,11}\psi_{4,11}.$$

It does not admit any $\mathbb{Z}_2$-grading.

Conversely, for a left-symmetric algebra A, $\Phi = (L, \mathrm{id})$ is an étale affine representation of $\mathfrak{g}(A)$ with base 0, where id is the identity transformation on $\mathfrak{g}(A)$. Moreover, if A is a left-symmetric algebra, then $N(A) = \{x \in A | L_x = 0\}$ is an ideal of A which is called a kernel ideal.

Proposition 5.4. *A left-symmetric algebra A has a zero kernel ideal if and only if its corresponding étale affine representation does not contain any nontrivial one-parameter translation subgroups.*

Theorem 5.2. *Let $\mathfrak{g}$ be a Lie algebra. If there exists an étale affine representation $\Phi = (\rho, q)$ of $\mathfrak{g}$ such that ρ is a faithful representation of $\mathfrak{g}$, then the corresponding left-symmetric algebra has the zero kernel ideal. Therefore Φ does not contain any nontrivial one-parameter translation subgroups.*

Proof. Let $x \in N(\mathfrak{g})$. By equation (2.6), we have $ev_v^{-1}[\rho(x)ev_v(y)] = 0$ for every $y \in \mathfrak{g}$. Since ev_v is a linear isomorphism, $\rho(x)z = 0$ for every $z \in \mathfrak{g}$. Then $x \in \mathrm{Ker}\rho$. Since ρ is faithful, $x = 0$. $\square$

5.6 Adapted Structures Associated with $\mathbb{Z}_2^2$-Symmetric Structures

Let G/H be a $\mathbb{Z}_2^k$-symmetric space and $\mathfrak{g} = \mathfrak{h} + \mathfrak{m}$ with $[\mathfrak{h}, \mathfrak{h}] \subset \mathfrak{h}$ and $[\mathfrak{h}, \mathfrak{m}] \subset \mathfrak{m}$ the associated reductive structure of the Lie algebra $\mathfrak{g}$ of G. We have seen that any affine connection on G/H is given by a linear map

$$\bigwedge : \mathfrak{m} \to gl(\mathfrak{m})$$

satisfying

$$\bigwedge[X, Y] = [\bigwedge(X), \lambda(Y)]$$

for all $X \in \mathfrak{m}$ and $Y \in \mathfrak{h}$, where λ is the linear isotropy representation of $\mathfrak{h}$. The corresponding torsion and curvature tensors are given by

$$T(X, Y) = \bigwedge(X)(Y) - \bigwedge(Y)(X) - [X, Y]_\mathfrak{m}$$

and

$$R(X, Y) = [\bigwedge(X), \bigwedge(Y)] - \bigwedge[X, Y] - \lambda([X, Y]_\mathfrak{h})$$

for any $X, Y \in \mathfrak{m}$.

Definition 5.3 ([15]). Consider an affine connection on the $\mathbb{Z}_2^k$-symmetric space G/H defined by the linear map

$$\bigwedge : \mathfrak{m} \to gl(\mathfrak{m}).$$

Then this connection is called *adapted* to the $\mathbb{Z}_2^k$-symmetric structure if any

$$\bigwedge(X_\gamma)(\mathfrak{g}_{\gamma'}) \subset \mathfrak{g}_{\gamma\gamma'}$$

for any $\gamma, \gamma' \in \mathbb{Z}_2^k$, $\gamma, \gamma \neq \varepsilon$. The connection is called homogeneous if any homogeneous component $\mathfrak{g}_\gamma$ of $\mathfrak{m}$ is invariant by $\bigwedge$.

Some example of such connections are described when G is the Heisenberg group in [15].

Proposition 5.5. *Let G/H be a $\mathbb{Z}_2^k$-symmetric space and*

$$\mathfrak{g} = \bigoplus_{\gamma \in \mathbb{Z}_2^k} \mathfrak{g}_\gamma$$

the associated $\mathbb{Z}_2^k$-grading of the Lie algebra $\mathfrak{g}$ of G. If $\mathfrak{g}_\varepsilon = \{0\}$ where ε is the unity of $\mathbb{Z}_2^k$, then any affine connection on G/H with no curvature define an affine structure on $\mathfrak{g}$.

Application. Assume that $\mathfrak{g}$ is a characteristically nilpotent filiform Lie algebra. If $\mathfrak{g}$ admits a $\mathbb{Z}_2^k$-grading with $\mathfrak{g}_\varepsilon = \{0\}$, then $k \geq 2$.

6 Filiform $\mathcal{G}$-Symmetric Spaces

6.1 Symmetric Spaces Associated with Filiform Lie Algebras of Rank 2

A symmetric space is a $\mathbb{Z}_2$-symmetric space G/H. If $\mathcal{G} = \{\varepsilon, \sigma\}$, then σ is an involutive automorphism of G such that H lies between G^σ, the subgroup of G consisting of elements left fixed by σ, and its identity component G_ε^σ. The Lie algebra $\mathfrak{g}$ of G is $\mathbb{Z}_2$-graded, this grading being defined by the automorphism σ of $\mathfrak{g}$ induced by the automorphism σ of G (this explains that we kept the same notation). A $\mathbb{Z}_2$-graded Lie algebra associated with an automorphism σ of $\mathfrak{g}$ is classically called a symmetric Lie algebra and denoted by $(\mathfrak{g}, \mathfrak{h}, \sigma)$. Thus, if G is simply connected, the symmetric Lie algebra $(\mathfrak{g}, \mathfrak{h}, \sigma)$ induces a symmetric space. Thus, in this case, the study of symmetric spaces follows from the study of symmetric Lie algebras. If $(\mathfrak{g}, \mathfrak{h}, \sigma)$ is a symmetric Lie algebra, then the $\mathbb{Z}_2$-grading of $\mathfrak{g}$ can be written as: $\mathfrak{g} = \mathfrak{h} \oplus \mathfrak{m}$ with

$$\begin{cases} [\mathfrak{h}, \mathfrak{h}] \subset \mathfrak{h} \\ [\mathfrak{h}, \mathfrak{m}] \subset \mathfrak{m}, \ [\mathfrak{m}, \mathfrak{m}] \subset \mathfrak{h}. \end{cases}$$

The subspaces $\mathfrak{h}$ and $\mathfrak{m}$ are eigenspaces of σ. Two symmetric Lie algebras $(\mathfrak{g}, \mathfrak{h}, \sigma)$ and $(\mathfrak{g}', \mathfrak{h}', \sigma')$ are isomorphic if there is an isomorphism $\alpha : \mathfrak{g} \to \mathfrak{g}'$ such that $\alpha(\mathfrak{h}) = \mathfrak{h}'$ and $\sigma' \circ \alpha = \alpha \circ \sigma$. We deduce the notion of isomorphism between symmetric spaces if we add some hypothesis on the Lie groups, such as G and H are connected and simply connected.

An important class of symmetric spaces is the class of Riemannian symmetric spaces, that is, Riemannian manifolds M such that the geodesic symmetries are defined at all points of M and are isometries. In this case H is an isotropy group and is compact. From a symmetric space (G, H, σ) with H compact, we can construct a Riemannian symmetric space considering on G/H a G-invariant Riemannian metric. In terms of Lie algebra, we have to consider on a symmetric Lie algebra $(\mathfrak{g}, \mathfrak{h}, \sigma)$ an ad $\mathfrak{h}$-invariant inner product on $\mathfrak{g}$ which admits σ as a linear isometry. The Riemannian symmetric spaces have been classified by E. Cartan. Later on, M. Berger gave the classification of symmetric Lie algebras when $\mathfrak{g}$ is semisimple and irreducible (the adjoint representation of $\mathfrak{h}$ in $\mathfrak{m}$ is irreducible).

The results established in Sect. 2 enable us to describe symmetric Lie algebras when $\mathfrak{g}$ is filiform of rank 2.

6.1.1 $\mathfrak{g} = L_n$

We denote by $\mathcal{L}_n$ the Lie group of matrices

$$
\begin{pmatrix}
1 & 0 & & \cdots & \cdots & & 0 \\
x_2 & 1 & 0 & & & & \vdots \\
x_3 & x_1 & 1 & 0 & & & \vdots \\
x_4 & \frac{(x_1)^2}{2!} & x_1 & \ddots & \ddots & & \vdots \\
\vdots & \vdots & & \ddots & \ddots & \ddots & \vdots \\
\vdots & \vdots & & & \ddots & \ddots & 1 & 0 \\
x_n & \frac{(x_1)^{n-2}}{n-2!} & \cdots & \cdots & \frac{(x_1)^2}{2!} & x_1 & 1
\end{pmatrix}
$$

An element of this group is denoted $[x_1, x_2, \ldots, x_n]$. Its Lie algebra is isomorphic to L_n. In Sect. 2 we classified abelian group gradings on L_n. Their corresponding symmetric spaces are as follows.

(1) $G/H_1 = \mathcal{L}_n/H_1$ where $H_1 = [x_1, 0, \ldots, 0]$.
(2) $G/H_2 = \mathcal{L}_{2p}/H_2$ where $H_2 = [0, x_2, 0, x_4, \ldots, x_{2p}]$, if $n = 2p$.
(3) $G/H_3 = \mathcal{L}_{2p+1}/H_3$ where $H_3 = [0, x_2, 0, x_4, \ldots, x_{2p}, 0]$, if $n = 2p + 1$.
(4) $G/H_4 = \mathcal{L}_{2p+1}/H_4$ where $H_4 = [0, 0, x_3, 0, x_5, \ldots, x_{2p+1}, 0]$, if $n = 2p + 1$.
(5) $G/H_5 = \mathcal{L}_{2p}/H_5$ where $H_5 = [0, 0, x_3, 0, x_5, \ldots, x_{2p-1}, 0]$, if $n = 2p$.

Let us note that in all the cases, the subgroup H_i is an abelian Lie group.

Proposition 6.1. *The nilpotent symmetric spaces $\mathcal{L}_{2p}/H_i$ for $i = 2, 3, 4, 5$ are not Riemannian symmetric spaces.*

Proof. Let us consider the symmetric decomposition of L_n:

$$
L_n = \mathbb{R}\{X_2, X_4, \ldots, X_{2p}\} \oplus \mathbb{R}\{X_1, X_3, \ldots, X_{2p+1}\} = \mathfrak{h} \oplus \mathfrak{m}.
$$

We assume here that $n = 2p + 1$. But the proof is similar if $n = 2p$. Let B be an ad $\mathfrak{h}$-symmetric bilinear form on $\mathfrak{m}$. We have

$$
B([X_2, X_1], X_1) + B(X_1, [X_2, X_1]) = -2B(X_3, X_1) = 0
$$

and, for $k = 1, \ldots, p$,

$$
B([X_{2k}, X_1], X_3) + B(X_1, [X_{2k}, X_3]) = -B(X_{2k+1}, X_3) = 0.
$$

This shows that X_3 is in the kernel of the bilinear form B and B is degenerate. Thus we cannot have a pseudo-Riemannian symmetric metric on this symmetric space. The two other cases can be treated in a similar way. □

6.1.2 $\mathfrak{g} = \mathcal{Q}_n$

We denote by $\mathcal{Q}_n$ the Lie group of matrices

$$
\begin{pmatrix}
1 & 0 & \cdots & \cdots & & & & 0 \\
x_2 & 1 & 0 & & & & & \vdots \\
x_3 & x_1 & 1 & 0 & & & & \vdots \\
x_4 & \frac{(x_1)^2}{2!} & x_1 & \ddots & \ddots & & & \vdots \\
\vdots & \vdots & \ddots & \ddots & \ddots & \ddots & & \vdots \\
\vdots & \vdots & & \ddots & \ddots & 1 & 0 \\
x_{n-1} & \frac{(x_1)^{n-3}}{(n-3)!} & \cdots\cdots & & \frac{(x_1)^2}{2!} & x_1 & 1 \\
x_n & y_{n-1} & \cdots\cdots & & y_3 & x_1 + x_2 & 1
\end{pmatrix}
$$

where y_i are polynomial functions in $x_1, \ldots, x_i$. The set $\{x_1, \ldots, x_n\}$ is a global system of coordinates of $\mathcal{Q}_n$. To simplify we will write $H_i = \{x_1, \ldots, x_{i-1}, 0, \ldots, x_n\}$ for the closed subgroup of $\mathcal{Q}_n$ defined by the equations $x_i = 0$.

A consequence of our results in Sect. 2 is the following.

Proposition 6.2. *Any homogeneous symmetric space* $\mathcal{Q}_n/H$ *is isomorphic to one of the following:*

- $\mathcal{Q}_n/H_1$ *with* $H_1 = \{0, 0, x_3, 0, x_5, \ldots, x_{n-1}, 0\}$;
- $\mathcal{Q}_n/H_2$ *with* $H_2 = \{x_1, -x_1, 0, 0, \ldots, 0, x_n\}$;
- $\mathcal{Q}_n/H_3$ *with* $H_3 = \{0, x_2, 0, x_4, 0, \ldots, 0, x_n\}$.

All the subgroups H_i are abelian.

6.2 Associated Affine Connection

Any symmetric space G/H is an affine space, that is, it can be provided with an affine connection ∇ whose torsion tensor T and curvature tensor R satisfy

$$
T = 0, \quad \nabla R = 0
$$

where

$$
\nabla R(X_1, X_2, X_3, Y) = \nabla(Y, R(X_1, X_2, X_3)) - R(\nabla(Y, X_1), X_2, X_3) \\
- R(X_1, \nabla(Y, X_2), X_3) - R(X_1, X_2, \nabla(Y, X_3))
$$

for any vector fields X_1, X_2, X_3, Y on G/H. This is the only affine connection that is invariant under the symmetries of G/H. When G/H is a Riemannian symmetric space, this connection ∇ coincides with the Levi-Civita connection associated with the Riemannian metric. In all the cases, since G/H is a reductive homogeneous space, that is, $\mathfrak{g}$ admits a decomposition $\mathfrak{g} = \mathfrak{h} \oplus \mathfrak{m}$ with $[\mathfrak{h}, \mathfrak{h}] \subset \mathfrak{h}$ and $[\mathfrak{h}, \mathfrak{m}] \subset \mathfrak{m}$, any connection is given by a linear map

$$\bigwedge : \mathfrak{m} \to gl(\mathfrak{m})$$

satisfying

$$\bigwedge [X, Y] = [\bigwedge(X), \lambda(Y)]$$

for all $X \in \mathfrak{m}$ and $Y \in \mathfrak{h}$, where λ is the linear isotropy representation of $\mathfrak{h}$. The corresponding torsion and curvature tensors are given by

$$T(X, Y) = \bigwedge(X)(Y) - \bigwedge(Y)(X) - [X, Y]_{\mathfrak{m}}$$

and

$$R(X, Y) = [\bigwedge(X), \bigwedge(Y)] - \bigwedge[X, Y] - \lambda([X, Y]_{\mathfrak{h}})$$

for any $X, Y \in \mathfrak{m}$. We have seen that $\mathcal{L}_n/H$ is not a Riemannian symmetric space. Thus the affine connection cannot be computed with a Levi-Civita connection. We shall determine this affine connection in terms of the map $\bigwedge$. As an example, we consider the case $n = 5$, which is not a real restriction. We consider the $\mathbb{Z}_2$-grading of L_5:

$$L_5 = \mathbb{R}\{X_3, X_5\} \oplus \mathbb{R}\{X_1, X_2, X_4\}.$$

Thus $\bigwedge(X_1), \bigwedge(X_2), \bigwedge(X_4)$ are matrices of order 3. If we assume that the torsion T is zero, we obtain

$$\bigwedge(X_1) = \begin{pmatrix} a & 0 & 0 \\ b & 0 & 0 \\ c & d & \frac{a}{2} \end{pmatrix}, \quad \bigwedge(X_2) = \begin{pmatrix} 0 & 0 & 0 \\ 0 & e & 0 \\ d & f & \frac{a}{2} \end{pmatrix}, \quad \bigwedge(X_3) = \begin{pmatrix} 0 & 0 & 0 \\ 0 & 0 & 0 \\ -\frac{a}{2} & 0 & 0 \end{pmatrix}.$$

The linear isotropy representation of H_4 whose Lie algebra is $\mathfrak{h}$ is given by taking the differential of the map $\mathcal{L}_5/H_4 \to \mathcal{L}_5/H_4$ corresponding to the left multiplication $\overline{x} \to h\overline{x}$ with $\overline{x} = xH_4$. We obtain

$$\lambda(X_3) = \begin{pmatrix} 0 & 0 & 0 \\ 0 & 0 & 0 \\ 1 & 0 & 0 \end{pmatrix}, \quad \lambda(X_5) = (0).$$

We deduce that the curvature is always non zero.

6.3 $\mathbb{Z}_2^k$-Symmetric Structures Associated with $\mathcal{L}_n$ and $\mathcal{Q}_n$

Recall that $\mathbb{Z}_2^k$-symmetric structure on G/H is associated with a $\mathbb{Z}_2^k$-grading

$$\mathfrak{g} = \bigoplus_{\gamma \in \mathbb{Z}_2^k} \mathfrak{g}_\gamma$$

of the Lie algebra $\mathfrak{g}$ of G. If ε denotes the unity of $\mathbb{Z}_2^k$, then $\mathfrak{g}_\varepsilon = \mathfrak{h}$ is the Lie algebra of H and the decomposition

$$\mathfrak{g} = \mathfrak{g}_\varepsilon \oplus \mathfrak{m}$$

with

$$\mathfrak{m} = \bigoplus_{\gamma \in \mathbb{Z}_2^k, \gamma \neq \varepsilon} \mathfrak{g}_\gamma.$$

is a reductive decomposition.

Proposition 6.3. *The Lie groups $\mathcal{L}_n$ and $\mathcal{Q}_n$ can be considered as $\mathbb{Z}_2^2$-symmetric spaces by identifying $\mathcal{L}_n$ and $\mathcal{Q}_n$, respectively, with $\mathcal{L}_n/\{1\}$ and $\mathcal{Q}_n/\{1\}$. Moreover there is no $\mathbb{Z}_2^k$-symmetric spaces $(G/H, \mathbb{Z}_2^k)$ for $G = \mathcal{L}_n$ or $G = \mathcal{Q}_n$ if $k > 2$.*

Proof. This is a consequence of the classification of the $\mathbb{Z}_2^k$-gradings of the Lie algebras L_n and Q_n.   □

Since a $\mathbb{Z}_2^k$-symmetric space G/H is reductive, a Riemannian or pseudo-Riemannian metric on G/H is given by a symmetric bilinear form B on $\mathfrak{m}$ which is ad $\mathfrak{h}$-invariant.

Definition 6.1 ([16]). A pseudo-Riemannian $\mathbb{Z}_2^k$-symmetric space is a $\mathbb{Z}_2^k$-symmetric space G/H with a pseudo-Riemannian metric g , which is determined by a nondegenerate bilinear symmetric form B on $\mathfrak{m}$ such that

(1) B is ad $\mathfrak{h}$-invariant.
(2) The homogeneous components $\mathfrak{g}_\gamma$ for $\gamma \neq \varepsilon$ are pairwise orthogonal with respect to B.

Let us note that in case where B is not positive definite, that is, g is not Riemannian, some of the components $\mathfrak{g}_\gamma$ can be degenerate subspaces of $\mathfrak{m}$. In the previous section, we have seen that the symmetric spaces $\mathcal{L}_n/H$ are never Riemannian or pseudo-Riemannian. But if we consider $\mathcal{L}_n$ as a $\mathbb{Z}_2^2$-symmetric space, then $\mathcal{L}_n$ is a pseudo-Riemannian $\mathbb{Z}_2^2$-symmetric space. In fact, let us consider the following $\mathbb{Z}_2^2$-grading of L_n:

$$L_n = \{0\} \oplus \mathbb{R}\{X_2, X_4, \dots, X_{2p}\} \oplus \mathbb{R}\{X_3, X_5, \dots, X_{2p+1}\} \oplus \mathbb{R}\{X_1\}$$

(recall that the brackets of L_n are $[X_1, X_i] = X_{i+1}$, $i = 1, \ldots, n-1$). Then any symmetric bilinear form of type

$$B = B_1(\omega_2, \ldots, \omega_{2p}) + B_2(\omega_3, \ldots, \omega_{2p+1}) + B_3(\omega_1),$$

where $\{\omega_1, \ldots, \omega_n\}$ is the dual basis of $\{X_1, \ldots, X_n\}$ and B_i is a nondegenerate bilinear form on the corresponding homogeneous component of $\mathfrak{m}$, defines a pseudo-Riemannian structure on the $\mathbb{Z}_2^2$-symmetric space $\mathcal{L}_n/\{1\}$. In this case, the corresponding Levi-Civita connection is an affine connection and the $\mathcal{G}$-symmetric space is an affine space. If in the symmetric case we have seen that there was an unique affine connection invariant under the symmetries, it is not the case when we consider $\mathbb{Z}_2^k$-symmetric spaces, with $k \geq 2$. This leads us to considering adapted and homogeneous connections, see Definition 5.3.

Proposition 6.4. *On the $\mathbb{Z}_2^2$-symmetric space $\mathcal{L}_n/\{1\}$ there exists an adapted affine connection whose torsion and curvature are zero.*

Proof. We restrict ourselves to the case where $n = 2p + 1$ is odd. The even case is similar. Consider the grading of L_n given by

$$L_n = \{0\} \oplus \mathbb{R}\{X_3, X_5, \ldots, X_n\} \oplus \mathbb{R}\{X_2, X_4, \ldots, X_{n-1}\} \oplus \mathbb{R}\{X_1\} = \{0\} \oplus \mathfrak{m}.$$

To obtain the connection which is adapted to the grading, we choose the matrices of the linear maps $\bigwedge(X_i)$ in the basis $\{X_3, \ldots, X_n, X_2, \ldots, X_{n-1}, X_1\}$ as follows:

$$\bigwedge(X_i) = \begin{pmatrix} 0\,0\,0 & \ldots\,0 & 0 \\ 0\,0\,0 & \ldots\,0 & 0 \\ 0\,0\,0 & \ldots\,0 & a_1^i \\ \cdot\ \cdot\ \cdot & \ldots\cdot & \cdot \\ 0\,0\,0 & \ldots\,0 & a_p^i \\ 0\,0\,a_{p+1}^i & \ldots\,a_{2p}^i & 0 \end{pmatrix}, \quad i = 3, \ldots, n$$

$$\bigwedge(X_j) = \begin{pmatrix} 0 & \ldots\,0 & 0\,0\,b_1^j \\ \cdot & \ldots\cdot & \cdot\ \cdot\ \cdot \\ 0 & \ldots\,0 & 0\,0\,b_p^j \\ 0 & \ldots\,0 & 0\,0\,0 \\ 0 & \ldots\,0 & 0\,0\,0 \\ b_3^j & \ldots\,b_{2p}^j & 0\,0\,0 \end{pmatrix}, \quad j = 2, \ldots, 2p, \quad \bigwedge(X_1) = \begin{pmatrix} 0_p & C & 0 \\ D & 0_p & 0 \\ 0 & 0 & 0 \end{pmatrix}$$

where C, D, 0_p are square matrices of order p. Because the linear isotropy representation is trivial and

$$R(X, Y) = [\bigwedge(X), \bigwedge(Y)] - \bigwedge([X, Y]),$$

the curvature is zero, so assuming $T = 0$ we deduce, after some computation, that

$$\bigwedge(X_i) = 0, \ i = 2, \ldots, 2p + 1$$

and

$$\bigwedge(X_1) = \begin{pmatrix} 0_p & I_p & 0 \\ D & 0_p & 0 \\ 0 & 0 & 0 \end{pmatrix} \text{ with } D = \begin{pmatrix} 1 & 0 & \ldots & 0 & 0 \\ 0 & 1 & \ldots & 0 & 0 \\ \ldots & \ldots & \ldots & \ldots & \ldots \\ 0 & 0 & \ldots & 1 & 0 \\ 0 & 0 & \ldots & 0 & 0 \end{pmatrix}.$$

The torsion tensor and the curvature tensor of the associated connection are null. □

Remark. Let us consider a homogeneous connection ∇ given by the map $\bigwedge$. This means that the matrices of $\bigwedge(X_i)$ are

$$\begin{pmatrix} A_i & 0_p & 0 \\ 0_p & B_i & 0 \\ 0 & 0 & c_i \end{pmatrix}$$

where A_i and B_i are square matrices of order p. The tensor torsion T of this connection ∇ is not trivial. There exists a connection $\tilde{\nabla}$ which is torsion free and has the same geodesic that ∇. It is defined by

$$\tilde{\nabla} = \nabla - \frac{1}{2}T.$$

This connection $\tilde{\nabla}$ is adapted to the $\mathbb{Z}_2^2$-grading and if we fix $R = 0$, we find again the previous one.

References

1. Ancochea, J.M.; Campoamor Stursberg R., Characteristically nilpotent Lie algebras: a survey. *Extracta Math.* **16** (2001), no. 2, 153–210.
2. Auslander, L.; Markus, L., Holonomy of flat affinely connected manifolds. *Ann. of Math.* (2) 62 (1955), 139–151.
3. Bahturin, Yuri; Goze, Michel., $Z_2 \times Z_2$-symmetric spaces. *Pacific J. Math.* Vol. 2, 615–663, North-Holland, Amsterdam, 2000.
4. Bahturin, Yuri; Goze, Michel; Remm, Elisabeth, Group Gradings on Lie Algebras, with Applications to Geometry. I, in: Developments and Retrospective in Lie theory, Algebraic methods, Developments in Mathematics, Vol. 38, Springer, 2014, 1–50.
5. Bahturin, Yuri; Zaicev, Mikhail, *Involutions on graded matrix algebras*, J. Algebra **315** (2007), 527–540.
6. Benoist, Yves. Une nilvariété non affine. *J. Differential Geom.* **41** (1995), 21–52.

7. Berger, Marcel, Les espaces symétriques noncompacts. (French) Ann. Sci. École Norm. Sup. (3) 74 (1957). 85–177.

8. Burde, Dietrich. Left-symmetric algebras, or pre-Lie algebras in geometry and physics. *Cent. Eur. J. Math.* **4** (2006), no. 3, 323–357 (electronic).

9. Dixmier, J.; Lister, W.G., Derivations of nilpotent Lie algebras. *Proc. AMS.* Vol. 8 (1957), No.1, 155–158.

10. Draper, C; Viruel, A., *Gradings on o*(8, ℂ), arXiv: 0709.0194

11. Elduque, Alberto; Kochetov, Mikhail, Gradings on Simple Lie Algebras, *AMS Mathematical Surveys and Monographs*, 189 (2013), 336 pp.

12. Gómez, J. R.; Jimenéz-Merchán, A.; Khakimdjanov, Y, Low-dimensional filiform Lie algebras, *J. Pure Appl. Algebra* **130** (1998), no. 2, 133–158.

13. Goze, Michel; Khakimdjanov, Yusupdjan, *Nilpotent and solvable Lie algebras.* Handbook of algebra, Vol. 2, 615–663, North-Holland, Amsterdam, 2000.

14. Goze, Michel; Hakimjanov, Yusupdjan, Sur les algèbres de Lie nilpotentes admettant un tore de dérivations. *Manuscripta Math.* **84** (1994), no. 2, 115–124.

15. Goze, Michel; Piu, Paola; Remm Elisabeth, Pseudo-Riemannian Symmetries on Heisenberg groups. To appear in An. Ştiinţ. Univ. Al. I. Cuza Iaşi. Mat. (N.S.)

16. Goze, Michel; Remm, Elisabeth., Riemannian Γ-symmetric spaces, *Differential geometry*, 195–206, World Sci. Publ., Hackensack, NJ, 2009.

17. Goze, Michel; Remm, Elisabeth, *Algèbres de Lie réelles ou complexes.* Preprint, Université de Haute Alsace. 2012

18. Helmstetter, J., Radical d'une algèbre symétrique à gauche, *Ann. Inst. Fourier* 29:4 (1979), 17–35.

19. Kobayashi, Shoshichi; Nomizu, Katsumi, *Foundations of Differential Geometry, Vol. I.* Reprint of the 1963 original. Wiley Classics Library. A Wiley-Interscience Publication. John Wiley and Sons, Inc., New York, 1996. xii+329 pp.

20. Kollross, Andreas, *Exceptional* $\mathbb{Z}_2 \times \mathbb{Z}_2$-*symmetric spaces*, Pacific J. Math. **242** (2009), no. 1, 113–130.

21. Ledger, A. J.; Obata, M., *Affine and Riemannian s-manifolds*, J. Differential Geometry **2** (1968), 451–459.

22. Lutz, Robert, *Sur la géométrie des espaces Γ-symétriques*, C. R. Acad. Sci. Paris Sér. I Math. **293** (1981), no. 1, 55–58.

23. Piu, Paola; Remm, Elisabeth, Riemannian symmetries in flag manifolds. *Arch. Math. (Brno)*, **48** (2012), no. 5, 387–398.

24. Remm, Elisabeth; Goze, Michel, Affine structures on abelian Lie groups. *Linear Algebra Appl.* **360** (2003), 215–230.

25. Remm, Elisabeth, Vinberg algebras associated to some nilpotent Lie algebras, *Non-associative algebra and its applications*, 347–364, Lect. Notes Pure Appl. Math., 246, Chapman - Hall/CRC, Boca Raton, FL, 2006.

26. Goze, Michel; Khakimdjanov, Yusupdjan. Nilpotent Lie Algebras. Mathematics and its Applications. Kluwer Acad. Publ. Group, Dordrecht, 1996. xvi + 336 pp.

Harmonic Analysis on Homogeneous Complex Bounded Domains and Noncommutative Geometry

Pierre Bieliavsky, Victor Gayral, Axel de Goursac, and Florian Spinnler

Abstract We define and study a noncommutative Fourier transform on every homogeneous complex bounded domain. We then give an application in noncommutative differential geometry by defining noncommutative Baumslag–Solitar tori.

Key words Strict deformation quantization • Symmetric spaces • $\star$-representation • $\star$-exponential • noncommutative manifolds

Mathematics Subject Classification (2010): 22E45, 46L87, 53C35, 53D55.

1 Introduction

In [7] the authors developed a tracial symbolic pseudo-differential calculus on every Lie group G whose Lie algebra $\mathfrak{g}$ is a normal j-algebra in the sense of Pyatetskii-Shapiro [16]. The class of such Lie groups is in one-to-one correspondence with the class of homogeneous complex bounded domains. Each of them carries a left-invariant Kähler structure.

This work was supported by the Belgian Interuniversity Attraction Pole (IAP) within the framework "Dynamics, Geometry and Statistical Physics" (DYGEST). Axel de Goursac is supported by the F.R.S.-F.N.R.S. and Florian Spinnler is supported by a F.R.I.A. fellowship (Belgium).

P. Bieliavsky (✉) • A. de Goursac • F. Spinnler
Département de Mathématiques, Université Catholique de Louvain,
Chemin du Cyclotron, 2, 1348 Louvain-la-Neuve, Belgium
e-mail: Pierre.Bieliavsky@uclouvain.be; Axelmg@melix.net; Florian.Spinnler@uclouvain.be

V. Gayral
Laboratoire de Mathématiques, Université de Reims Champagne-Ardenne,
Moulin de la Housse - BP 1039, 51687 Reims cedex 2, France
e-mail: victor.gayral@univ-reims.fr

© Springer International Publishing Switzerland 2014
G. Mason et al. (eds.), *Developments and Retrospectives in Lie Theory*,
Developments in Mathematics 37, DOI 10.1007/978-3-319-09934-7_2

As a by-product, they obtained a G-equivariant continuous linear mapping between the Schwartz space $\mathcal{S}(G)$ of such a Lie group and a subalgebra of Hilbert–Schmidt operators on a Hilbert irreducible unitary G-module. This yields a one-parameter family of noncommutative associative multiplications $\{\star_\theta\}_{\theta \in \mathbb{R}}$ on the Schwartz space, each of them endowing $\mathcal{S}(G)$ with a Fréchet nuclear algebra structure. Moreover, the resulting family of Fréchet algebras $\{(\mathcal{S}(G), \star_\theta)\}_{\theta \in \mathbb{R}}$ deforms the commutative Fréchet algebra structure on $\mathcal{S}(G)$ given by the pointwise multiplication of functions corresponding to the value $\theta = 0$ of the deformation parameter. Note that such a program was achieved in [17] for abelian Lie groups and in [8] for abelian Lie supergroups.

In this article we construct a bijective intertwiner between every noncommutative Fréchet algebra $(\mathcal{S}(G), \star_\theta)$ $(\theta \neq 0)$ and a convolution function algebra on the group G. The intertwiner's kernel consists in a complex-valued smooth function $\mathcal{E}$ on the group $G \times G$ that we call "$\star$-exponential" because of its similar nature with objects defined in [11] and studied in [2] in the context of the Weyl–Moyal quantization of coadjoint orbits of exponential Lie groups.

We then prove that the associated smooth map

$$\mathcal{E} : G \to C^\infty(G)$$

consists in a group-morphism valued in the multiplier (nuclear Fréchet) algebra $\mathcal{M}_{\star_\theta}(G)$ of $(\mathcal{S}(G), \star_\theta)$. The above group-morphism integrates the classical moment mapping

$$\lambda : \mathfrak{g} \to \mathcal{M}_{\star_\theta}(G) \cap C^\infty(G)$$

associated with the (symplectic) action of G on itself by left-translations. Next, we modify the 2-point kernel $\mathcal{E}$ by a power of the modular function of G in such a way that the corresponding Fourier-type transform consists of a unitary operator $\mathcal{F}$ on the Hilbert space of square integrable functions with respect to a left-invariant Haar measure on G.

As an application, we define a class of noncommutative tori associated to generalized Bauslag–Solitar groups in every dimension.

2   Homogeneous Complex Bounded Domains and j-Algebras

The theory of j-algebras was greatly developed by Pyatetskii-Shapiro [16] for studying in a Lie-algebraic way the structure and classification of bounded homogeneous—not necessarily symmetric—domains in $\mathbb{C}^n$. A j-algebra is roughly the Lie algebra $\mathfrak{g}$ of a transitive Lie group of analytic automorphisms of the domain, together with the data of the Lie algebra $\mathfrak{k}$ of the stabilizer of a point in the latter Lie group, an endomorphism j of $\mathfrak{g}$ coming from the complex structure on the domain, and a linear form on $\mathfrak{g}$ whose Chevalley coboundary gives the j-invariant symplectic structure coming from the Kähler structure on the domain. Pyatetskii-

Shapiro realized that among the j-algebras corresponding to a fixed bounded homogeneous domain, there always is at least one whose associated Lie group acts simply transitively on the domain, and which is realizable as upper triangular real matrices. Thoses j-algebras have the structure of *normal j-algebras* which we proceed to describe now.

Definition 2.1. A *normal j-algebra* is a triple $(\mathfrak{g}, \alpha, j)$ where

1. $\mathfrak{g}$ is a solvable Lie algebra which is split over the reals, i.e., ad_X has only real eigenvalues for all $X \in \mathfrak{g}$,
2. j is an endomorphism of $\mathfrak{g}$ such that $j^2 = -Id_{\mathfrak{g}}$ and $[X, Y] + j[jX, Y] + j[X, jY] - [jX, jY] = 0$, $\forall\ X, Y \in \mathfrak{g}$,
3. α is a linear form on $\mathfrak{g}$ such that $\alpha([jX, X]) > 0$ if $X \neq 0$ and $\alpha([jX, jY]) = \alpha([X, Y])$, $\forall\ X, Y \in \mathfrak{g}$.

If $\mathfrak{g}'$ is a subalgebra of $\mathfrak{g}$ which is invariant by j, then $(\mathfrak{g}', \alpha|_{\mathfrak{g}'}, j|_{\mathfrak{g}'})$ is again a normal j-algebra, said to be a j-**subalgebra** of $(\mathfrak{g}, \alpha, j)$. A j-subalgebra whose algebra is at the same time an ideal is called a j-**ideal**.

Remark 2.2. To each simple Lie algebra $\mathfrak{G}$ of Hermitian type (i.e., such that the center of the maximal compact algebra $\mathfrak{k}$ has real dimension one) we can attach a normal j-algebra $(\mathfrak{g}, \alpha, j)$ where

1. $\mathfrak{g}$ is the solvable Lie algebra underlying the Iwasawa factor $\mathfrak{g} = \mathfrak{a} \oplus \mathfrak{n}$ of an Iwasawa decomposition $\mathfrak{k} \oplus \mathfrak{a} \oplus \mathfrak{n}$ of $\mathfrak{G}$.
2. Denoting by $\mathbb{G}/K$ the Hermitian symmetric space associated to the pair $(\mathfrak{G}, \mathfrak{k})$ and by $\mathbb{G} = KAN$ the Iwasawa group decomposition corresponding to $\mathfrak{k} \oplus \mathfrak{a} \oplus \mathfrak{n}$, the global diffeomorphism:

$$G := AN \longrightarrow \mathbb{G}/K : g \mapsto gK\ ,$$

endows the group G with an exact left-invariant symplectic structure as well as a compatible complex structure. The evaluations at the unit element $e \in G$ of these tensor fields define the elements $\mathit{\Omega} = d\alpha$ and j at the Lie algebra level.

It is important to note that not every normal j-algebra arises this way. Indeed, it is with the help of the theory of j-algebras that Pyatetskii-Shapiro discovered the first examples of nonsymmetric bounded homogeneous domains. Nevertheless, they can all be built from these "Hermitian" normal j-algebras by a semidirect product process, as we recall now.

Definition 2.3. A normal j-algebra associated with a rank one Hermitian symmetric space (i.e., $\dim \mathfrak{a} = 1$) is called *elementary*.

Lemma 2.4. *Let (V, ω_0) be a symplectic vector space of dimension $2n$, and let $\mathfrak{h}_V := V \oplus \mathbb{R}E$ be the corresponding Heisenberg algebra : $[x, y] = \omega_0(x, y)E$, $[x, E] = 0\ \forall\ x, y \in V$. Setting $\mathfrak{a} := \mathbb{R}H$, we consider the split extension of Lie algebras:*

$$0 \to \mathfrak{h}_V \to \mathfrak{s} := \mathfrak{a} \ltimes \mathfrak{h}_V \to \mathfrak{a} \to 0,$$

with extension homomorphism $\rho_{\mathfrak{h}} : \mathfrak{a} \to \mathrm{Der}(\mathfrak{h})$ *given by*

$$\rho_{\mathfrak{h}}(H)(x + \ell E) := [H, x + \ell E] := x + 2\ell E, \quad x \in V, \ell \in \mathbb{R}.$$

Then the Lie algebra $\mathfrak{s}$ *underlines an elementary normal* j-*algebra. Moreover, every elementary normal* j-*algebra is of that form.*

The main interest of elementary normal j-algebras is that they are the only building blocks of normal j-algebras, as shown by the following important property [16].

Proposition 2.5. *Let* $(\mathfrak{g}, \alpha, j)$ *be a normal* j-*algebra. Then,*

1. *there exists a one-dimensional ideal* $\mathfrak{z}_1$ *of* $\mathfrak{g}$, *and a vector subspace* V *of* $\mathfrak{g}$, *such that* $\mathfrak{s} = j\mathfrak{z}_1 + V + \mathfrak{z}_1$ *underlies an elementary normal* j-*ideal of* $\mathfrak{g}$. *Moreover, the associated extension sequence*

$$0 \longrightarrow \mathfrak{s} \longrightarrow \mathfrak{g} \longrightarrow \mathfrak{g}' \longrightarrow 0,$$

   is split as a sequence of normal j-*algebras and such that*

   a. $[\mathfrak{g}', \mathfrak{a}_1 \oplus \mathfrak{z}_1] = 0$,
   b. $[\mathfrak{g}', V] \subset V$.

2. *(follows from 1.) every normal* j-*algebra admits a decomposition as a sequence of split extensions of elementary normal* j-*algebras with properties (a) and (b) above.*

2.1 Symplectic Symmetric Space Geometry of Elementary Normal j-Groups

In this section we briefly recall results of [6, 9].

Definition 2.6. The connected simply-connected real Lie group G whose Lie algebra $\mathfrak{g}$ underlies a normal j-algebra is called a *normal* j-*group*. The connected simply connected Lie group $\mathbb{S}$ whose Lie algebra $\mathfrak{s}$ underlies an elementary normal j-algebra is said to be an *elementary normal* j-*group*.

Elementary normal j-groups are exponential (non-nilpotent) solvable Lie groups. As an example, consider the Lie algebra $\mathfrak{s}$ of Definition 2.3 where $V = 0$. It is generated over $\mathbb{R}$ by two elements H and E satisfying $[H, E] = 2E$ and is therefore isomorphic to the Lie algebra of the group of affine transformations of the real line: in this case, $\mathbb{S}$ is the $ax + b$ group.

Now generally, the Iwasawa factor AN of the simple group $SU(1, n)$ (which corresponds to the above example in the case $n = 1$) is an elementary normal j-group.

We realize $\mathbb{S}$ on the product manifold underlying $\mathfrak{s}$:

$$\mathbb{S} = \mathbb{R} \times V \times \mathbb{R} = \{(a, x, \ell)\} .$$

The group law of $\mathbb{S}$ is given by

$$(a, x, \ell) \cdot (a', x', \ell') = \left(a + a', e^{-a'} x + x', e^{-2a'} \ell + \ell' + \frac{1}{2} e^{-a'} \omega_0(x, x') \right) \quad (1.1)$$

and the inverse by

$$(a, x, \ell)^{-1} = (-a, -e^a x, -e^{2a} \ell) .$$

We denote by

$$\mathrm{Ad}^* : \mathbb{S} \times \mathfrak{s}^* : (g, \xi) \mapsto \mathrm{Ad}_g^*(\xi) := \xi \circ \mathrm{Ad}_{g^{-1}}$$

the coadjoint action of $\mathbb{S}$ on the dual space $\mathfrak{s}^*$ of $\mathfrak{s} = \mathbb{R}H \oplus V \oplus \mathbb{R}E$. In the dual $\mathfrak{s}^*$, we consider the elements ^bH and ^bE as well as ^bx $(x \in V)$ defined by

$$^bH|_{V \oplus \mathbb{R}E} \equiv 0, \qquad \langle ^bH, H \rangle = 1,$$

$$^bE|_{\mathbb{R}H \oplus V} \equiv 0, \qquad \langle ^bE, E \rangle = 1,$$

$$^bx|_{\mathbb{R}H \oplus \mathbb{R}E} \equiv 0, \qquad \langle ^bx, y \rangle = \omega_0(x, y) \quad (y \in V) .$$

Proposition 2.7. *Let $\mathcal{O}_\epsilon$ denote the coadjoint orbit through the element $\epsilon \, ^bE$, for $\epsilon = \pm 1$, equipped with its standard Kirillov–Kostant–Souriau symplectic structure (referred to as KKS). Then the map*

$$\mathbb{S} \to \mathcal{O}_\epsilon : (a, x, \ell) \mapsto \mathrm{Ad}_{(a,x,\ell)}^* (\epsilon \, ^bE) = \epsilon(2\ell \, ^bH - e^{-a} \, ^bx + e^{-2a} \, ^bE) \quad (1.2)$$

is a $\mathbb{S}$-equivariant global Darboux chart on $\mathcal{O}_\epsilon$ in which the KKS two-form reads

$$\omega := \omega_{\mathbb{S}} := \epsilon(2da \wedge d\ell + \omega_0) .$$

Within this setting, we consider the moment map of the action of $\mathbb{S}$ on $\mathcal{O}_\epsilon \simeq \mathbb{S}$:

$$\lambda : \mathfrak{s} \to C^\infty(\mathbb{S}) : X \mapsto \lambda_X$$

defined by the relations

$$\lambda_X(g) := \langle \mathrm{Ad}_g^* (\epsilon \, ^bE) , X \rangle .$$

Lemma 2.8. *Denoting for every $X \in \mathfrak{s}$ the associated fundamental vector field by*

$$X_g^* := \frac{d}{dt}|_0 \exp(-tX).g ,$$

one has $(y \in V)$:

$$H^* = -\partial_a , \qquad y^* = -e^{-a}\partial_y + \frac{1}{2}e^{-a}\omega_0(x,y)\partial_\ell , \qquad E^* = -e^{-2a}\partial_\ell .$$

Moreover the moment map reads

$$\lambda_H(a,x,\ell) = 2\epsilon\ell , \qquad \lambda_y(a,x,\ell) = e^{-a}\epsilon\omega_0(y,x) , \qquad \lambda_E(a,x,\ell) = \epsilon e^{-2a} .$$
(1.3)

Proposition 2.9. *The map*

$$s : \mathbb{S} \times \mathbb{S} \to \mathbb{S} : (g,g') \mapsto s_g g'$$

defined by

$$s_{(a,x,\ell)}(a',x',\ell')$$
$$= \Big(2a - a', 2\cosh(a-a')x - x', 2\cosh(2(a-a'))\ell - \ell' + \sinh(a-a')\omega_0(x,x')\Big)$$
(1.4)

endows the Lie group $\mathbb{S}$ *with a left-invariant structure of the symmetric space in the sense of O. Loos (cf. [15]).*

   Moreover the symplectic structure ω *is invariant under the symmetries: for every* $g \in \mathbb{S}$, *one has*

$$s_g^*\omega = \omega .$$

2.2 Normal j-Groups

The above Proposition 2.5 implies that every normal j-group G can be decomposed into a semidirect product

$$G = G_1 \ltimes_\rho \mathbb{S}_2$$
(1.5)

where

$$\mathbb{S}_2 := \mathbb{R}H_2 \times V_2 \times \mathbb{R}E_2$$

is an elementary normal j-group of real dimension $2n_2 + 2$ and G_1 is a normal j-group. This means that the group law of G has the form

$$\forall\, g_1, g_1' \in G_1, \; \forall\, g_2, g_2' \in \mathbb{S}_2 \quad : \quad (g_1, g_2)\cdot(g_1', g_2') = \Big(g_1 \cdot g_1', \; g_2 \cdot(\rho(g_1)g_2')\Big),$$

where $\rho : G_1 \to Sp(V_2, \omega_0)$ denotes the extension homomorphism; and the inverse is given by $(g_1, g_2)^{-1} = (g_1^{-1}, \rho(g_1^{-1})g_2^{-1})$. As a consequence, every normal j-group therefore results in a sequence of semidirect products of a finite number of elementary normal j-groups.

Proposition 2.10. *Consider the decomposition (1.5). Then,*

1. *the Lie group G_1 admits an open coadjoint orbit $\mathcal{O}_1$ through an element $o_1 \in \mathfrak{g}_1^*$ which it acts on in a simply transitive way;*
2. *the coadjoint orbit $\mathcal{O}$ of G through the element $o := o_1 + \epsilon_2{}^\flat E_2$ (same notation as in Sect. 2.1) is open in $\mathfrak{g}^*$;*
3. *denoting by $\mathcal{O}_2$ the coadjoint orbit of $\mathbb{S}_2$ through $\epsilon_2{}^\flat E_2$, the map*

$$\phi : \mathcal{O}_1 \times \mathcal{O}_2 \to \mathcal{O} : (\mathrm{Ad}_{g_1}^* o_1, \epsilon_2 \mathrm{Ad}_{g_2}^*{}^\flat E_2) \mapsto \mathrm{Ad}_{(g_1, g_2)}^*(o) \qquad (1.6)$$

is a symplectomorphism when endowing each orbit with its KKS two-form.

Proof. We proceed by induction on the dimension in proving that $\mathcal{O}$ is acted on by G in a simply transitive way. By induction hypothesis, so is $\mathcal{O}_1$ by G_1. And Proposition 2.7 implies it is the case for $\mathcal{O}_2$ as well. Now denoting $(g_1, e) =: g_1$ and $(e, g_2) =: g_2$, we observe:

$$\mathrm{Ad}_{(g_1, g_2)}^*(o) = \mathrm{Ad}_{g_2 g_1}^*(o) = \mathrm{Ad}_{g_2}^* \left(\mathrm{Ad}_{g_1}^*(o_1) + \epsilon_2{}^\flat E_2 \circ \rho(g_1^{-1})_{*e} \right)$$

where $\rho : G_1 \to \mathrm{Aut}(\mathbb{S}_2)$ denotes the extension homomorphism.

Now for all $\xi_1 \in \mathfrak{g}_1^*$, $X_1 \in \mathfrak{g}_1$, $X_2 \in \mathfrak{s}_2$ and $g_2 \in \mathbb{S}_2$:

$$\langle \mathrm{Ad}_{g_2}^* \xi_1, X_1 + X_2 \rangle = \langle \xi_1, \mathrm{Ad}_{g_2^{-1}} X_1 + \mathrm{Ad}_{g_2^{-1}} X_2 \rangle = \langle \xi_1, \mathrm{Ad}_{g_2^{-1}} X_1 \rangle.$$

But

$$\mathrm{Ad}_{g_2^{-1}} X_1 = \frac{\mathrm{d}}{\mathrm{d}t}|_0 (\exp(tX_1), g_2^{-1})(e, g_2) = \frac{\mathrm{d}}{\mathrm{d}t}|_0 (\exp(tX_1), g_2^{-1} \rho(\exp(tX_1))g_2).$$

Hence

$$\langle \xi_1, \mathrm{Ad}_{g_2^{-1}} X_1 \rangle = \langle \xi_1, X_1 \oplus \left(\frac{\mathrm{d}}{\mathrm{d}t} \log_2^{-1} \rho(\exp(tX_1))g_2 \right) \rangle = \langle \xi_1, X_1 \rangle.$$

Therefore $\mathrm{Ad}_{g_2}^* \xi_1 = \xi_1$ and we get

$$\mathrm{Ad}_{(g_1, g_2)}^*(o) = \mathrm{Ad}_{g_1}^*(o_1) + \epsilon_2 \mathrm{Ad}_{g_2}^* \left({}^\flat E_2 \circ \rho(g_1^{-1})_{*e} \right).$$

The induction hypothesis thus implies that the stabilizer of element o in G is trivial, which shows in particular that the fundamental group of $\mathcal{O}$ is trivial. The map (1.6) being a surjective submersion is therefore a diffeomorphism.

It remains to prove the assertion regarding the symplectic structures. Denoting by $\omega^{\mathcal{O}}$ the KKS form on $\mathcal{O}$, we observe that with obvious notation, for all $Y_1 \in \mathfrak{g}_1$ and $Y_2 \in \mathfrak{s}_2$:

$$\phi^*\omega^{\mathcal{O}}(X_1^* \oplus X_2^*, Y_1^* \oplus Y_2^*) = \omega^{\mathcal{O}}_{\mathrm{Ad}^*_{(g_1,g_2)}(o)}(\phi_* X_1^* + \phi_* X_2^*, \phi_* Y_1^* + \phi_* Y_2^*)$$

$$= \omega^{\mathcal{O}}_{\mathrm{Ad}^*_{(g_1,g_2)}(o)}\left(\left(\mathrm{Ad}_{g_2} X_1\right)^* + X_2^*, \left(\mathrm{Ad}_{g_2} Y_1\right)^* + Y_2^*\right)$$

$$= \left\langle \mathrm{Ad}^*_{(g_1,g_2)}(o), [\mathrm{Ad}_{g_2} X_1 + X_2, \mathrm{Ad}_{g_2} Y_1 + Y_2]\right\rangle$$

$$= \left\langle \mathrm{Ad}^*_{g_1}(o), [X_1 + \mathrm{Ad}_{g_2^{-1}} X_2, Y_1 + \mathrm{Ad}_{g_2^{-1}} Y_2]\right\rangle$$

$$= \left\langle \mathrm{Ad}^*_{g_1}(o), [X_1, Y_1] - \rho(Y_1)\mathrm{Ad}_{g_2^{-1}} X_2 + \rho(X_1)\mathrm{Ad}_{g_2^{-1}} Y_2 + \mathrm{Ad}_{g_2^{-1}}[X_2, Y_2]\right\rangle$$

$$= \omega^{\mathcal{O}_1}_{\mathrm{Ad}^*_{g_1}(o_1)}(X_1^*, Y_1^*) + \epsilon_2 \, \omega^{\mathcal{O}_2}_{\mathrm{Ad}^*_{g_2} {}^\flat E_2}(X_2^*, Y_2^*)$$

$$+ \epsilon_2 \left\langle {}^\flat E_2, \rho(g_1^{-1})_{*e}\left(-\rho(Y_1)\mathrm{Ad}_{g_2^{-1}} X_2 + \rho(X_1)\mathrm{Ad}_{g_2^{-1}} Y_2\right)\right\rangle .$$

The last term in the above expression vanishes identically. Indeed, the specific form of ρ implies that the element $v_2 := -\rho(Y_1)\mathrm{Ad}_{g_2^{-1}} X_2 + \rho(X_1)\mathrm{Ad}_{g_2^{-1}} Y_2$ lives in V_2 as well as in $\rho(g_1^{-1})_{*e} v_2$. $\square$

Remark 2.11. Normal j-groups can be decomposed into elementary normal j-groups $\mathbb{S}_k$ as $G = (\ldots(\mathbb{S}_1 \ltimes_{\rho_1} \mathbb{S}_2) \ltimes_{\rho_2} \ldots) \ltimes_{\rho_{N-1}} \mathbb{S}_N$ and the coadjoint orbits described in Proposition 2.10 are determined by sign choices $\epsilon_k = \pm 1$ for each factor $\mathbb{S}_k$. We will denote by $\mathcal{O}_{(\epsilon)}$ the coadjoint orbit associated to the signs $(\epsilon_k)_{1 \leq k \leq N} \in (\mathbb{Z}_2)^N$.

Example 2.12. Let us describe the following example corresponding to the six-dimensional Siegel domain $\mathrm{Sp}(2, \mathbb{R})/U(2)$. Let $G_1 = \mathbb{S}_1$ be of dimension 2 ($V_1 = 0$, G_1 is the affine group), $\mathbb{S}_2$ of dimension 4, i.e., V_2 is of dimension 2, with basis f_2, f_2' endowed with $\omega_0 = \begin{pmatrix} 0 & 1 \\ -1 & 0 \end{pmatrix}$), and let the action $\rho : \mathbb{S}_1 \to Sp(V_2)$ be given by

$$\rho(a_1, \ell_1) = \begin{pmatrix} e^{a_1} & 0 \\ e^{-a_1} \ell_1 & e^{-a_1} \end{pmatrix}.$$

Then the group law is

$$(a_1, \ell_1, a_2, v_2, w_2, \ell_2) \cdot (a_1', \ell_1', a_2', v_2', w_2' \ell_2') = \left(a_1 + a_1', e^{-2a_1'} \ell_1\right.$$

$$+ \ell_1', a_2 + a_2', e^{-a_2'} v_2 + e^{a_1} v_2', e^{-a_2'} w_2 + e^{-a_1} \ell_1 v_2' + e^{-a_1} w_2',$$

$$\left. e^{-2a_2'} \ell_2 + \ell_2' + \frac{1}{2} e^{-a_2'} (e^{-a_1} \ell_1 v_2 v_2' + e^{-a_1} v_2 w_2' - e^{a_1} w_2 v_2')\right),$$

where $(a_1, \ell_1) \in \mathbb{S}_1$, $(a_2, v_2, w_2, \ell_2) \in \mathbb{S}_2$ and

$$g := (a_1, \ell_1, a_2, v_2, w_2, \ell_2) = e^{a_2 H_2} e^{v_2 f_2 + w_2 f_2'} e^{\ell_2 E_2} e^{a_1 H_1} e^{\ell_1 E_1} .$$

Its Lie algebra is characterized by

$$[H_1, E_1] = 2E_1, \qquad [H_2, f_2] = f_2, \qquad [H_2, f_2'] = f_2', \qquad [f_2, f_2'] = E_2,$$

$$[H_2, E_2] = 2E_2, \qquad [H_1, f_2] = f_2, \qquad [H_1, f_2'] = -f_2', \qquad [E_1, f_2] = f_2',$$

where the other relations vanish. The coadjoint action takes the form

$$\mathrm{Ad}_g^*(\epsilon_1{}^b E_1 + \epsilon_2{}^b E_2) = (2\epsilon_1 \ell_1 + \epsilon_2 v_2 w_2)^b H_1 + (\epsilon_1 e^{-2a_1} - \frac{\epsilon_2}{2} v_2^2)^b E_1$$

$$+ \epsilon_2 (2\ell_2{}^b H_2 - e^{-a_2} v_2{}^b f_2 - e^{-a_2} w_2{}^b f_2' + e^{-2a_2}{}^b E_2) .$$

The moment map can then be extracted from this expression:

$$\lambda_{H_1} = 2\epsilon_1 \ell_1 + \epsilon_2 v_2 w_2, \qquad \lambda_{E_1} = \epsilon_1 e^{-2a_1} - \frac{\epsilon_2}{2} v_2^2, \qquad \lambda_{H_2} = 2\epsilon_2 \ell_2,$$

$$\lambda_{f_2} = \epsilon_2 e^{-a_2} w_2, \qquad \lambda_{f_2'} = -\epsilon_2 e^{-a_2} v_2, \qquad \lambda_{E_2} = \epsilon_2 e^{-2a_2}.$$

3 Determination of the Star-Exponential

3.1 Quantization of Elementary Groups

We follow the analysis developed in [7], where the reader can find all the proofs. In the notation of Sect. 2.1, we choose two Lagrangian subspaces in duality V_0, V_1 of the symplectic vector space (V, ω_0) of dimension $2n$ underlying the elementary group $\mathbb{S}$. We denote the corresponding coordinates $x = (v, w) \in V$ in the global chart, with $v \in V_0$ and $w \in V_1$. Let $\mathfrak{q} = \mathbb{R}H \oplus V_0$ and $Q = \exp(\mathfrak{q})$. The unitary induced representation associated to the coadjoint orbit $\mathcal{O}_\epsilon$ ($\epsilon = \pm 1$) by the method of Kirillov has the form

$$U_{\theta, \epsilon}(a, x, \ell)\varphi(a_0, v_0) = e^{\frac{i\epsilon}{\theta}\left(e^{2(a-a_0)}\ell + \omega_0(\frac{1}{2}e^{a-a_0}v - v_0, e^{a-a_0}w)\right)} \varphi(a_0 - a, v_0 - e^{a-a_0}v) \tag{1.7}$$

for $(a, x, \ell) \in \mathbb{S}$, $\varphi \in L^2(Q)$, $(a_0, v_0) \in Q$ and $\theta \in \mathbb{R}_+^*$. These representations $U_{\theta, \epsilon} : \mathbb{S} \to \mathcal{L}(\mathcal{H})$ are unitary and irreducible, and the unitary dual is described by these two representations. A multiplier $\mathbf{m}$ is a function on Q. There is a particular multiplier:

$$\mathbf{m}_0(a, v) = 2^{n+1} \sqrt{\cosh(2a)} \cosh(a)^n. \tag{1.8}$$

Let us define $\Sigma := (s_{(0,0,0)}|_Q)^*$, where s is the symmetric structure (1.4):

$$\Sigma\varphi(a, v) = \varphi(-a, -v). \tag{1.9}$$

Then, the Weyl-type quantization map is given by

$$
\begin{aligned}
\Omega_{\theta,\epsilon,\mathbf{m}_0}(a, x, \ell)\varphi(a_0, v_0) &:= U_{\theta,\epsilon}(a, x, \ell)\mathbf{m}_0 \Sigma U_{\theta,\epsilon}(a, x, \ell)^{-1}\varphi(a_0, v_0) \\
&= 2^{n+1}\sqrt{\cosh(2(a - a_0))}\cosh(a - a_0)^n \\
&\quad \times e^{\frac{2i\epsilon}{\theta}\left(\sinh(2(a-a_0))\ell + \omega_0(\cosh(a-a_0)v - v_0, \cosh(a-a_0)w)\right)} \\
&\quad \times \varphi(2a - a_0, 2\cosh(a - a_0)v - v_0).
\end{aligned} \tag{1.10}
$$

The operator $\Omega_{\theta,\epsilon,\mathbf{m}_0}(g)$ is a symmetric unbounded operator on $\mathcal{H}$, and $g \in \mathbb{S} \simeq \mathcal{O}_\epsilon$.

On smooth functions with compact support $f \in \mathcal{D}(\mathcal{O}_\epsilon)$, and by denoting $\kappa := \frac{1}{2^n(\pi\theta)^{n+1}}$, one has

$$\Omega_{\theta,\epsilon,\mathbf{m}_0}(f) := \kappa \int_{\mathcal{O}_\epsilon} f(g)\Omega_{\theta,\epsilon,\mathbf{m}_0}(g)d\mu(g)$$

with $d\mu(g) = d^L g$ which corresponds to the Liouville measure of the KKS symplectic form on the coadjoint orbit $\mathcal{O}_\epsilon \simeq \mathbb{S}$. Its extension is continuous and called the quantization map $\Omega_{\theta,\epsilon,\mathbf{m}_0} : L^2(\mathcal{O}_\epsilon) \to \mathcal{L}_{HS}(\mathcal{H})$, with $\mathcal{H} := L^2(Q)$ and $\mathcal{L}_{HS}$ the Hilbert–Schmidt operators. The normalization has been chosen such that $\Omega_{\theta,\epsilon,\mathbf{m}_0}(1) = \mathbb{1}_{\mathcal{H}}$, understood in the distributional sense. Moreover, it is $\mathbb{S}$-equivariant, because of

$$\forall g, g_0 \in \mathbb{S} \quad : \quad \Omega_{\theta,\epsilon,\mathbf{m}_0}(g \cdot g_0) = U_{\theta,\epsilon}(g)\Omega_{\theta,\epsilon,\mathbf{m}_0}(g_0)U_{\theta,\epsilon}(g)^{-1}.$$

The unitary representation $U_{\theta,\epsilon} : \mathbb{S} \to \mathcal{L}(\mathcal{H})$ induces a resolution of the identity.

Proposition 3.1. *By denoting the norm* $\|\varphi\|_w^2 := \int_Q |\varphi(a, v)|^2 e^{2(n+1)a}dadv$ *and* $\varphi_g(q) = U_{\theta,\epsilon}(g)\varphi(q)$ *for* $g \in \mathbb{S}$, $q \in Q$ *and a nonzero* $\varphi \in \mathcal{H}$, *we have*

$$\frac{\kappa}{\|\varphi\|_w^2} \int_{\mathbb{S}} |\varphi_g\rangle\langle\varphi_g|d^L g = \mathbb{1}_{\mathcal{H}}.$$

This resolution of identity shows that the trace has the form

$$\text{Tr}(T) = \frac{\kappa}{\|\varphi\|_w^2} \int_{\mathbb{S}} \langle\varphi_g, T\varphi_g\rangle d^L g \tag{1.11}$$

for any trace-class operator $T \in \mathcal{L}^1(\mathcal{H})$.

Theorem 3.2. *The symbol map, which is the left-inverse of the quantization map* $\Omega_{\theta,\epsilon,m_0}$ *can be obtained via the formula*

$$\forall\, f \in L^2(\mathcal{O}_\epsilon),\; \forall\, g \in \mathcal{O}_\epsilon \quad : \quad \mathrm{Tr}(\Omega_{\theta,\epsilon,m_0}(f)\Omega_{\theta,\epsilon,m_0}(g)) = f(g),$$

where the trace is understood in the distributional sense in the variable $g \in \mathbb{S}$.

Then, the star-product is defined as

$$(f_1 \star_\theta f_2)(g) := \mathrm{Tr}(\Omega_{\theta,\epsilon,m_0}(f_1)\Omega_{\theta,\epsilon,m_0}(f_2)\Omega_{\theta,\epsilon,m_0}(g))$$

for $f_1, f_2 \in L^2(\mathcal{O}_\epsilon)$ and $g \in \mathcal{O}_\epsilon$, where we omitted the subscripts $\epsilon, \mathbf{m}_0$ for the star-product.

Proposition 3.3. *The star-product has the following expression:*

$$(f_1 \star_\theta f_2)(g) = \frac{1}{(\pi\theta)^{2n+2}} \int K_{\mathbb{S}}(g, g_1, g_2) e^{-\frac{2i}{\theta} S_{\mathbb{S}}(g,g_1,g_2)} f_1(g_1) f_2(g_2) d\mu(g_1) d\mu(g_2) \tag{1.12}$$

where the amplitude and the phase are

$$K_{\mathbb{S}}(g, g_1, g_2) = 4\sqrt{\cosh(2(a_1-a_2))\cosh(2(a_1-a))\cosh(2(a-a_2))}\cosh(a_2 - a)^n$$
$$\cosh(a_1 - a)^n \cosh(a_1 - a_2)^n,$$

$$\epsilon S_{\mathbb{S}}(g, g_1, g_2) = -\sinh(2(a_1 - a_2))\ell - \sinh(2(a_2 - a))\ell_1 - \sinh(2(a - a_1))\ell_2$$
$$+ \cosh(a_1 - a)\cosh(a_2 - a)\omega_0(x_1, x_2)$$
$$+ \cosh(a_1 - a)\cosh(a_1 - a_2)\omega_0(x_2, x)$$
$$+ \cosh(a_1 - a_2)\cosh(a_2 - a)\omega_0(x, x_1),$$

with $g_i = (a_i, x_i, \ell_i) \in \mathbb{S}$. *Moreover,* $g \mapsto 1$ *is the unit of this product, is associative,* $\mathbb{S}$-*invariant and satisfies the tracial identity:*

$$\int f_1 \star_\theta f_2 = \int f_1 \cdot f_2. \tag{1.13}$$

Note that this product has first been found [5] by intertwining the Moyal product:

$$(f_1 \star_\theta^0 f_2)(a, x, \ell) = \frac{4}{(\pi\theta)^{2+2n}} \int da_i dx_i d\ell_i \; f_1(a_1 + a, x_1 + x, \ell_1 + \ell)$$
$$f_2(a_2 + a, x_2 + x, \ell_2 + \ell) e^{-\frac{2i\epsilon}{\theta}(2a_1\ell_2 - 2a_2\ell_1 + \omega_0(x_1, x_2))}$$

for $f_1, f_2 \in L^2(\mathcal{O}_\epsilon)$ and $\mathcal{O}_\epsilon \simeq \mathbb{S} \simeq \mathbb{R}^{2n+2}$, which is $\mathfrak{s}$-covariant $([\lambda_X, \lambda_Y]_{\star_\theta^0} = -i\theta\lambda_{[X,Y]})$ but not $\mathbb{S}$-invariant. So for smooth functions with compact support, we have $\hat{f}_1 \star_\theta \hat{f}_2 = T_\theta((T_\theta^{-1} f_1) \star_\theta^0 (T_\theta^{-1} f_2))$ with intertwiners:

$$T_\theta f(a, x, \ell) = \frac{1}{2\pi} \int \sqrt{\cosh(\frac{\theta t}{2})} \cosh(\frac{\theta t}{4})^n e^{\frac{2i}{\theta} \sinh(\frac{\theta t}{2})\ell - i\xi t} f(a, \cosh(\frac{\theta t}{4})x, \xi) dt d\xi$$

$$T_\theta^{-1} f(a, x, \ell) = \frac{1}{2\pi} \int \frac{\sqrt{\cosh(\frac{\theta t}{2})}}{\cosh(\frac{\theta t}{4})^n} e^{-\frac{2i}{\theta} \sinh(\frac{\theta t}{2})\xi + it\ell} f(a, \cosh(\frac{\theta t}{4})^{-1}x, \xi) dt d\xi$$

(1.14)

which will be useful in Sect. 3.4.

3.2 Quantization of Normal j-Groups

Let $G = G_1 \ltimes \mathbb{S}_2$ be a normal j-group, with notation as in Sect. 2.2. Taking into account its structure, the unitary representation U and the quantization map Ω of this group (dependence in $\theta \in \mathbb{R}_+^*$ will be omitted here in the subscripts) can be constructed from the ones U_1 and Ω_1 of G_1 (obtained by recurrence) and the ones U_2 and Ω_2 of $\mathbb{S}_2$, given by (1.7) and (1.10) (without $\mathbf{m}_0$ for the moment).

Let $\mathcal{H}_i$ be the Hilbert space of the representation U_i, associated to a coadjoint orbit $\mathcal{O}_i$ (in the notation of Proposition 2.10). Since U_2 is irreducible and $\rho : G_1 \rightarrow Sp(V_2)$, there exists a unique homomorphism $\mathcal{R} : G_1 \rightarrow \mathcal{L}(\mathcal{H}_2)$ such that for all $g_1 \in G_1$, for all $g_2 \in \mathbb{S}_2$,

$$U_2(\rho(g_1)g_2) = \mathcal{R}(g_1)U_2(g_2)\mathcal{R}(g_1)^{-1}.$$

$\mathcal{R}$ is actually a metaplectic-type representation associated to U_2 and ρ. The matrix $\rho(g_1)$, with smooth coefficients in g_1, is of the form

$$\rho(g_1) = \begin{pmatrix} \rho_+(g_1) & 0 \\ \rho_-(g_1) & (\rho_+(g_1)^T)^{-1} \end{pmatrix}$$

with $\rho_-(g_1)^T \rho_+(g_1) = \rho_+(g_1)^T \rho_-(g_1)$.

Proposition 3.4. *The map* $\mathcal{R} : G_1 \rightarrow \mathcal{L}(\mathcal{H}_2)$ *is given by for all* $g_1 \in G_1$, *for all* $\varphi \in \mathcal{H}_2$ *non-zero,*

$$\mathcal{R}(g_1)\varphi(a_0, v_0) = \frac{1}{|\det(\rho_+(g_1))|^{\frac{1}{2}}} e^{-\frac{i\epsilon_2}{2\theta} v_0 \rho_-(g_1)\rho_+(g_1)^{-1} v_0} \varphi(a_0, \rho_+(g_1)^{-1} v_0)$$

and is unitary, where the sign $\epsilon_2 = \pm 1$ *determines the choice of the coadjoint orbit* $\mathcal{O}_2$ *and the associated irreducible representation* U_2.

The expression

$$U(g)\varphi := U_1(g_1)\varphi_1 \otimes U_2(g_2)\mathcal{R}(g_1)\varphi_2$$

for $g = (g_1, g_2) \in G$, $\varphi = \varphi_1 \otimes \varphi_2 \in \mathcal{H} := \mathcal{H}_1 \otimes \mathcal{H}_2$, defines a unitary representation $U : G \to \mathcal{L}(\mathcal{H})$. Let $\Sigma = \Sigma_1 \otimes \Sigma_2$, with Σ_2 given in (1.9). Then, the quantization map is defined as

$$\Omega(g) := U(g) \circ \Sigma \circ U(g)^{-1}.$$

Using the definition of U and $\mathcal{R}$ together with the property (see Proposition 6.55 in [7]),

$$\mathcal{R}(g_1)\Sigma_2\mathcal{R}(g_1)^{-1} = \Sigma_2,$$

it is easy to check that

$$\Omega((g_1, g_2)) = \Omega_1(g_1) \otimes \Omega_2(g_2),$$

with $(g_1, g_2) \in G$ and $\Omega_i(g_i) = U_i(g_i) \circ \Sigma_i \circ U_i(g_i)^{-1}$. Using the identification $\mathcal{O} \simeq G$ (see Proposition 2.10), we see that Ω is defined on $\mathcal{O}$ and it is again G-equivariant: for $g \in G$ and $g' \in \mathcal{O}$,

$$\Omega(g \cdot g') = U(g)\Omega(g')U(g)^{-1}.$$

In the same way, if $\mathbf{m}_0 := \mathbf{m}_0^1 \otimes \mathbf{m}_0^2$, where $\mathbf{m}_0^2$ is given by (1.8), we also have $\Omega_{\mathbf{m}_0}((g_1, g_2)) = \Omega_{1,\mathbf{m}_0^1}(g_1) \otimes \Omega_{2,\mathbf{m}_0^2}(g_2)$. The quantization map of functions $f \in \mathcal{D}(\mathcal{O})$ has then the form

$$\Omega_{\mathbf{m}_0}(f) := \kappa \int_{\mathcal{O}} f(g)\Omega_{\mathbf{m}_0}(g)d\mu(g)$$

where $d\mu(g) := d\mu_1(g_1)d\mu_2(g_2) = d^L g_1 d^L g_2$ is the Liouville measure of the KKS symplectic form on the coadjoint orbit $\mathcal{O} \simeq G$; $\kappa = \kappa_1\kappa_2$, for $G = G_1 \ltimes \mathbb{S}_2$, is defined recursively with $\kappa_2 = \frac{1}{2^{n_2}(\pi\theta)^{n_2}+1}$ and $\dim(\mathbb{S}_2) = 2n_2 + 2$ in Sect. 3.1. We then have $\Omega_{\mathbf{m}_0}(1) = \mathbb{1}$.

Note that the left-invariant measure for the group $G = G_1 \ltimes \mathbb{S}_2$ has the form

$$d^L g = d^L g_1 d^L g_2$$

which corresponds to the Liouville measure $d\mu(g)$, like the elementary case. As in Sect. 3.1, the unitary representation $U : G \to \mathcal{L}(\mathcal{H})$ induces a resolution of the identity.

Proposition 3.5. *By denoting the norm* $\|\varphi\|_w^2 := \|\varphi_1\|_w^2 \|\varphi_2\|_w^2$ *for* $\varphi = \varphi_1 \otimes \varphi_2 \in \mathcal{H}$ *nonzero, we have*

$$\frac{\kappa}{\|\varphi\|_w^2} \int_G |U(g)\varphi\rangle\langle U(g)\varphi|\mathrm{d}^L g = \mathbb{1}_{\mathcal{H}}.$$

This resolution of identity shows that the trace has the form

$$\mathrm{Tr}(T) = \frac{\kappa}{\|\varphi\|_w^2} \int_G \langle U(g)\varphi, TU(g)\varphi\rangle \mathrm{d}^L g \tag{1.15}$$

for $T \in \mathcal{L}^1(\mathcal{H})$. In particular, for $T = T_1 \otimes T_2$ with $T_i \in \mathcal{L}^1(\mathcal{H}_i)$, one has

$$\mathrm{Tr}(T) = \frac{\kappa}{\|\varphi\|_w^2} \int_G \langle U_1(g_1)\varphi_1, T_1 U_1(g_1)\varphi_1 \rangle \langle U_2(g_2)\mathcal{R}(g_1)\varphi_2, T_2 U_2(g_2)\mathcal{R}(g_1)\varphi_2 \rangle$$

$$\mathrm{d}^L g_1 \mathrm{d}^L g_2 = \mathrm{Tr}(T_1)\,\mathrm{Tr}(T_2). \quad (1.16)$$

Theorem 3.6. *The symbol map, which is the left-inverse of the quantization map* Ω_{m_0} *can be obtained via the formula*

$$\forall f \in L^2(\mathcal{O}), \ \forall g \in \mathcal{O} \quad : \quad \mathrm{Tr}\left(\Omega_{m_0}(f)\Omega_{m_0}(g)\right) = f(g).$$

Proof. Abstractly (in a weak sense), we have

$$\mathrm{Tr}(\Omega_{m_0}(f)\Omega_{m_0}(g))$$

$$= \kappa \int f(g_1', g_2')\,\mathrm{Tr}(\Omega_1(g_1')\Omega_1(g_1))\,\mathrm{Tr}(\Omega_2(g_2')\Omega_2(g_2))\mathrm{d}\mu_1(g_1')\mathrm{d}\mu_2(g_2')$$

$$= f(g_1, g_2).$$

$\square$

The star-product is defined as

$$(f_1 \star_\theta f_2)(g) := \mathrm{Tr}(\Omega_{m_0}(f_1)\Omega_{m_0}(f_2)\Omega_{m_0}(g))$$

for $f_1, f_2 \in L^2(\mathcal{O})$ and $g \in \mathcal{O}$.

Proposition 3.7. *The star-product has the following expression:*

$$(f_1 \star_\theta f_2)(g)$$

$$= \frac{1}{(\pi\theta)^{\dim(G)}} \int_{G \times G} K_G(g, g', g'') e^{-\frac{2i}{\theta} S_G(g, g', g'')} f_1(g') f_2(g'') \mathrm{d}\mu(g') \mathrm{d}\mu(g'')$$

$$\tag{1.17}$$

where the amplitude and the phase are

$$K_G(g, g', g'') = K_{G_1}(g_1, g_1', g_1'') K_{\mathbb{S}_2}(g_2, g_2', g_2''),$$
$$S_G(g, g', g'') = S_{G_1}(g_1, g_1', g_1'') + S_{\mathbb{S}_2}(g_2, g_2', g_2''),$$

with $g = (g_1, g_2) \in \mathcal{O} = \mathcal{O}_1 \times \mathcal{O}_2$ due to (1.6). There is also a tracial identity:

$$\int_{\mathcal{O}} (f_1 \star_\theta f_2)(g) d\mu(g) = \int_{\mathcal{O}} f_1(g) f_2(g) d\mu(g).$$

3.3 Computation of the Star-Exponential

Definition 3.8. We define the **star-exponential** associated to the deformation quantization $(\star, \Omega)$ of Sect. 3.2 as

$$\forall\, g \in G,\ \forall\, g' \in \mathcal{O} \simeq G \quad : \quad \mathcal{E}_g^{\mathcal{O}}(g') = \mathrm{Tr}(U(g)\Omega_{m_0}(g')),$$

where the trace has to be understood in the distributional sense in $(g, g') \in G \times \mathcal{O}$.

By using computation rules of the above sections, we can obtain recursively the number of factors of the normal j-group $G = G_1 \ltimes \mathbb{S}_2$ with the corresponding coadjoint orbit $\mathcal{O} \simeq \mathcal{O}_1 \times \mathcal{O}_2$, the expression of the star-exponential $\mathcal{E}^{\mathcal{O}} \in \mathcal{D}'(G \times \mathcal{O})$.

Theorem 3.9. *We have for all $g, g' \in G$,*

$$\mathcal{E}_g^{\mathcal{O}}(g')$$

$$= \mathcal{E}_{g_1}^{\mathcal{O}_1}(g_1') \frac{2^{n_2} |\det(\rho_+(g_1))|^{\frac{1}{2}} \sqrt{\cosh(a_2)} \cosh(\frac{a_2}{2})^{n_2}}{|\det(1 + \rho_+(g_1))|} \exp\left(\frac{i\epsilon_2}{\theta} \Big[2\sinh(a_2)\ell_2' \right.$$

$$\left. + e^{a_2 - 2a_2'}\ell_2 + e^{\frac{a_2}{2} - a_2'} \cosh(\frac{a_2}{2}) \omega_0(x_2, x_2') + \frac{1}{2}(\tilde{x})^T M_\rho(g_1)\tilde{x} \Big] \right),$$

$$(1.18)$$

where

$$M_\rho(g_1) := \begin{pmatrix} -B_\rho & C_\rho^T \\ C_\rho & 0 \end{pmatrix}, \qquad \tilde{x} := e^{\frac{a_2}{2} - a_2'} x_2 - 2\cosh(\frac{a_2}{2}) x_2'$$

and with $B_\rho = (1 + \rho_+^T(g_1))^{-1} \rho_+^T(g_1)\rho_-(g_1)(1 + \rho_+(g_1))^{-1}$, and $C_\rho = \frac{1}{2}(\rho_+(g_1) - 1)(\rho_+(g_1) + 1)^{-1}$, $g = (g_1, g_2)$, $g_2 = (a_2, x_2, \ell_2) \in \mathbb{S}_2$, and $\mathcal{E}^{\mathcal{O}_1}$ the star-exponential of the normal j-group G_1.

Proof. First, we use Proposition 3.5 and Eq. (1.16):

$$\mathcal{E}_g^{\mathcal{O}}(g') = \mathrm{Tr}(U(g)\Omega_{m_0}(g')) = \mathrm{Tr}(U_1(g_1)\Omega_{1,m_0^1}(g_1'))\,\mathrm{Tr}(U_2(g_2)\mathcal{R}(g_1)\Omega_{2,m_0^2}(g_2')).$$

The second part of the above expression can be computed by using (1.11), and it gives

$$\mathrm{Tr}(U_2(g_2)\mathcal{R}(g_1)\Omega_{2,m_0^2}(g_2')) = \frac{\kappa_2}{\|\varphi\|_w^2} \int_{\mathbb{S}_2} \langle \varphi_{g_2''}, U_2(g_2)\mathcal{R}(g_1)\Omega_{2,m_0^2}(g_2')\varphi_{g_2''}\rangle \mathrm{d}^L g_2''.$$

If we replace U_2, Ω_{2,m_0^2} and $\mathcal{R}$ by their expressions determined previously in (1.7), (1.10) and Proposition 3.4, we find after some integrations and simplifications that for all $g, g' \in G$,

$$\mathcal{E}_g^{\mathcal{O}}(g') = \mathcal{E}_{g_1}^{\mathcal{O}_1}(g_1')\frac{2^{n_2}|\det(\rho_+(g_1))|^{\frac{1}{2}}\sqrt{\cosh(a_2)}\cosh(\frac{a_2}{2})^{n_2}}{|\det(1+\rho_+(g_1))|} \tag{1.19}$$

$$\times \exp(\frac{i\epsilon_2}{\theta}\left[2\sinh(a_2)\ell_2' + e^{a_2 - 2a_2'}\ell_2 + X^T A_\rho X\right]),$$

where

$$A_\rho = \begin{pmatrix} -B_\rho & C_\rho^T & B_\rho & (1+\rho_+^T(g_1))^{-1} \\ C_\rho & 0 & -\rho_+(g_1)(1+\rho_+(g_1))^{-1} & 0 \\ B_\rho & -\rho_+^T(g_1)(1+\rho_+^T(g_1))^{-1} & -B_\rho & C_\rho^T \\ (1+\rho_+(g_1))^{-1} & 0 & C_\rho & 0 \end{pmatrix},$$

$$X = \begin{pmatrix} \frac{1}{\sqrt{2}}e^{\frac{a_2}{2}-a_2'}v_2 \\ \frac{1}{\sqrt{2}}e^{\frac{a_2}{2}-a_2'}w_2 \\ \sqrt{2}\cosh(\frac{a_2}{2})v_2' \\ \sqrt{2}\cosh(\frac{a_2}{2})w_2' \end{pmatrix}$$

and with $x_2 = (v_2, w_2)$. A straightforward computation then gives the result. □

Let us denote by $\mathcal{E}_{(g_1,g_2)}^{\mathcal{O}_2}(g_2')$ the explicit part in the RHS of (1.19) which corresponds to the star-exponential of the group $\mathbb{S}_2$ twisted by the action of $g_1 \in G_1$.

The expression (1.19) seems to be ill-defined when $\det(1+\rho_+^{-1}) = 0$. However, one can obtain in this case a degenerated expression of the star-exponential which is well defined. For example, when $\rho_+(g_1) = -\mathbb{1}_{n_2}$, we have

$$\mathcal{E}^{\mathcal{O}_2}_{(g_1,g_2)}(g_2') = (\pi\theta)^{n_2} \frac{\sqrt{\cosh(a_2)}}{\cosh(\frac{a_2}{2})^{n_2}}$$

$$\times \exp(\frac{i\epsilon_2}{\theta}\Big[2\sinh(a_2)\ell_2' + e^{a_2-2a_2'}\ell_2 + \frac{1}{2}e^{a_2-2a_2'}\omega_0(v_2,w_2)\Big])$$

$$\times \delta\Big(v_2' - \frac{e^{\frac{a_2}{2}-a_2'}}{2\cosh(\frac{a_2}{2})}v_2\Big)\delta\Big(w_2' - \frac{e^{\frac{a_2}{2}-a_2'}}{2\cosh(\frac{a_2}{2})}w_2\Big).$$

In the case where $\rho(g_1) = 1$, i.e., when the action of G_1 on $\mathbb{S}_2$ is trivial in G, we find the second part of the star-exponential

$$\mathcal{E}^{\mathcal{O}_2}_{g_2}(g_2') = \sqrt{\cosh(a_2)}\cosh(\frac{a_2}{2})^{n_2}$$

$$\times \exp(\frac{i\epsilon_2}{\theta}\Big[2\sinh(a_2)\ell_2' + e^{a_2-2a_2'}\ell_2 + e^{\frac{a_2}{2}-a_2'}\cosh(\frac{a_2}{2})\omega_0(x_2,x_2')\Big]). \tag{1.20}$$

which corresponds to the star-exponential of the elementary normal j-group $\mathbb{S}_2$.

By using this characterization in terms of the quantization map, we can derive easily some properties of the star-exponential.

Proposition 3.10. *The star-exponential enjoys the following properties. For all* $g, g' \in G$, *for all* $g_0 \in \mathcal{O}$,

- *hermiticity:* $\overline{\mathcal{E}^{\mathcal{O}}_g(g_0)} = \mathcal{E}^{\mathcal{O}}_{g^{-1}}(g_0)$.
- *covariance:* $\mathcal{E}^{\mathcal{O}}_{g'\cdot g\cdot g'^{-1}}(g'\cdot g_0) = \mathcal{E}^{\mathcal{O}}_g(g_0)$.
- *BCH:* $\mathcal{E}^{\mathcal{O}}_g \star_\theta \mathcal{E}^{\mathcal{O}}_{g'} = \mathcal{E}^{\mathcal{O}}_{g\cdot g'}$.
- *Character formula:* $\int_G \mathcal{E}^{\mathcal{O}}_g(g_0)d\mu(g_0) = \kappa^{-1}\operatorname{Tr}(U(g))$.

Proof. Using Theorem 3.9, we can show that $\overline{\mathcal{E}^{\mathcal{O}}_g(g_0)} = \operatorname{Tr}(U(g^{-1})\Omega_{m_0}(g_0)) = \mathcal{E}^{\mathcal{O}}_{g^{-1}}(g_0)$ since $\Omega_{m_0}(g_0)$ is self-adjoint. In the same way, covariance follows the G-equivariance of Ω_{m_0}. The BCH property is related to the fact that U is a group representation. Finally, we get

$$\int_{\mathcal{O}} \mathcal{E}^{\mathcal{O}}_g(g_0)d\mu(g_0) = \operatorname{Tr}(U(g)\int_{\mathcal{O}} \Omega_{m_0}(g_0)d\mu(g_0)) = \kappa^{-1}\operatorname{Tr}(U(g))$$

using that $\Omega_{m_0}(1) = 1$. □

Note that the BCH property makes sense in a non-formal way only in the functional space $\mathcal{M}_{\star_\theta}(G)$ determined in Sect. 4.2, where we will see that the star-exponential belongs to.

3.4 Other Determination Using PDEs

We give here another way to determine the star-exponential without using the quantization map, but directly by solving the PDE it has to satisfy. We restrict here to the case of an elementary normal j-group $G = \mathbb{S}$ for simplicity.

By using the strong-invariance of the star-product, for any $f \in \mathcal{M}_{\star_\theta}(\mathbb{S})$ (see Sect. 4.2),

$$\forall\, X \in \mathfrak{s} \quad : \quad [\lambda_X, f]_{\star_\theta} = -i\theta X^* f,$$

where λ is the moment map (1.3), and by using also the equivariance of Ω_{m_0}, we deduce that

$$[\Omega_{m_0}(\lambda_X), \Omega_{m_0}(f)] = \Omega_{m_0}([\lambda_X, f]_{\star_\theta}) = -i\theta\Omega_{m_0}(X^* f)$$

$$= -i\theta\frac{d}{dt}|_0\Omega_{m_0}(L^*_{e^{-tX}} f) = -i\theta\frac{d}{dt}|_0 U(e^{tX})\Omega_{m_0}(f)U(e^{-tX})$$

$$= -i\theta[U_*(X), \Omega_{m_0}(f)]$$

Since the center of $\mathcal{M}_{\star_\theta}(\mathbb{S})$ is trivial, this means that there exists a linear map β : $\mathfrak{g} \to \mathbb{C}$ such that

$$\Omega_{m_0}(\lambda_X) = -i\theta U_*(X) + \beta(X)\mathbb{1}.$$

The invariance of the product under Σ (see (1.9)) implies that $\beta(X) = -\beta(X)$ and finally $\beta(X) = 0$. As a consequence, we have the following proposition.

Proposition 3.11. *The star-exponential (see Definition 3.8) of an elementary normal j-group $G = \mathbb{S}$ satisfies the equation*

$$\partial_t \mathcal{E}_{e^{tX}} = \frac{i}{\theta}(\lambda_X \star_\theta \mathcal{E}_{e^{tX}}) \tag{1.21}$$

with initial condition $\lim_{t\to 0}\mathcal{E}_{e^{tX}} = 1.$

Proof. Indeed, by using $\Omega_{m_0}(\lambda_X) = -i\theta U_*(X)$, we derive

$$\partial_t \mathcal{E}_{e^{tX}}(g_0) = \partial_t\,\mathrm{Tr}(U(e^{tX})\Omega_{m_0}(g_0)) = \mathrm{Tr}(U_*(X)U(e^{tX})\Omega_{m_0}(g_0))$$

$$= \frac{i}{\theta}\,\mathrm{Tr}(\Omega_{m_0}(\lambda_X)U(e^{tX})\Omega_{m_0}(g_0)) = \frac{i}{\theta}(\lambda_X \star_\theta \mathcal{E}_{e^{tX}})(g_0)$$

$\square$

Now we can use this equation to find directly the expression of the star-exponential. Let us do it for example for the coadjoint orbit associated to the sign $\epsilon = +1$. Since the equation (1.21) is integro-differential and complicated to solve, we will analyze the following equation:

$$\partial_t f_t = \frac{i}{\theta}(\lambda_X \star_\theta^0 f_t), \qquad \lim_{t \to 0} f_t = 1 \tag{1.22}$$

for the Moyal product $\star_\theta^0$. Indeed, we have the expression of the intertwiner T_θ from $\star_\theta^0$ to $\star_\theta$. We define the partial Fourier transformation as

$$\mathcal{F}f(a, x, \xi) := \hat{f}(a, x, \xi) := \int e^{-i\xi\ell} f(a, x, \ell)d\ell. \tag{1.23}$$

Applying the partial Fourier transformation (1.23), with $X = \alpha H + y + \beta E \in \mathfrak{s}$, on the action of moment maps by the Moyal product, we find

$$\mathcal{F}(\lambda_H \star_\theta^0 f) = \left(2i\partial_\xi + \frac{i\theta}{2}\partial_a\right)\hat{f}$$

$$\mathcal{F}(\lambda_y \star_\theta^0 f) = e^{-a-\frac{\theta\xi}{4}}\left(\omega_0(y, x) + \frac{i\theta}{2}y\partial_x\right)\hat{f}$$

$$\mathcal{F}(\lambda_E \star_\theta^0 f) = e^{-2a-\frac{\theta\xi}{2}}\hat{f},$$

so that Eq. (1.22) can be reformulated as

$$\partial_t \hat{f}_t = \frac{i}{\theta}\left[2i\alpha\partial_\xi + \frac{i\theta\alpha}{2}\partial_a + \beta e^{-2a-\frac{\theta\xi}{2}} + e^{-a-\frac{\theta\xi}{4}}\left(\omega_0(y, x) + \frac{i\theta}{2}y\partial_x\right)\right]\hat{f}_t \tag{1.24}$$

which is a pure PDE. Then, owing to the form of the moment map (1.3), we consider the *ansatz*

$$f_t(a, x, \ell) = v(t) \exp\frac{i}{\theta}\left[2\ell\gamma_1(t) + e^{-2a}\gamma_2(t) + e^{-a}\gamma_3(t)\omega_0(y, x)\right] \tag{1.25}$$

whose partial Fourier transform can be expressed as

$$\hat{f}_t(a, x, \xi) = 4\pi^2\delta\left(\xi - \frac{2\gamma_1(t)}{\theta}\right)v(t)\exp\frac{i}{\theta}\left[e^{-2a}\gamma_2(t) + e^{-a}\gamma_3(t)\omega_0(y, x)\right].$$

Inserting this *ansatz* into Eq. (1.24), it gives

$$\gamma_1'(t) = \alpha, \quad \gamma_2'(t) = \alpha\gamma_2(t) + \beta e^{-\gamma_1(t)}, \quad \gamma_3'(t) = \frac{\alpha}{2}\gamma_3(t) + e^{-\frac{\alpha t}{2}}, \quad v'(t) = 0.$$

We find that the solutions with initial condition $\lim_{t \to 0} f_t = 1$ are

$$\gamma_1 = \alpha t, \qquad \gamma_2 = \frac{\beta}{\alpha}\sinh(\alpha t), \qquad \gamma_3 = \frac{\sinh(\frac{\alpha t}{2})}{\alpha}, \qquad v = 1.$$

Using intertwining operators (1.14), we see that $T_\theta^{-1}\lambda_X = \lambda_X$, and $T_\theta f_t$ is then a solution of (1.21):

$$E_{\star_\theta}(t\lambda_X)(a, x, \ell) := \mathcal{E}_{e^{tX}}(a, x, \ell) = T_\theta f_t(a, x, \ell)$$

$$= \sqrt{\cosh(\alpha t)}\,\cosh(\frac{\alpha t}{2})^n e^{\frac{i}{\theta}\sinh(\alpha t)\left(2\ell + \frac{\beta}{\alpha}e^{-2a} + \frac{e^{-a}}{\alpha}\omega_0(y,x)\right)}.$$

To obtain the star-exponential, we need the expression of the logarithm of the group $\mathbb{S}$: $\mathcal{E}_{g_0} = E_{\star_\theta}(\lambda_{\log(g_0)})$. For $X = \alpha H + y + \beta E \in \mathfrak{s}$, the exponential of the group $\mathbb{S}$ has the expression

$$\exp(\alpha H + y + \beta E) = \left(\alpha, \frac{2e^{-\frac{\alpha}{2}}}{\alpha}\sinh(\frac{\alpha}{2})y, \frac{\beta}{\alpha}e^{-\alpha}\sinh(\alpha)\right),$$

and the logarithm

$$\log(a, x, \ell) = aH + \frac{a}{2}\frac{e^{\frac{a}{2}}}{\sinh(\frac{a}{2})}x + \frac{ae^a}{\sinh(a)}\ell E.$$

Therefore, we obtain

$$\mathcal{E}_{g_0}(g) = \sqrt{\cosh(a_0)}\,\cosh(\frac{a_0}{2})^n e^{\frac{i}{\theta}\left(2\sinh(a_0)\ell + e^{a_0 - 2a}\ell_0 + e^{\frac{a_0}{2}-a}\cosh(\frac{a_0}{2})\omega_0(x_0,x)\right)},$$

which coincides with the expression (1.20) determined by using the quantization map Ω_{m_0}. Note that the BCH property (see Proposition 3.10) can also be checked directly at the level of the Lie algebra $\mathfrak{s}$. From the above expressions of the logarithm and the exponential of the group $\mathbb{S}$, we derive the BCH expression: $\mathrm{BCH}(X_1, X_2) := \log(e^{X_1}e^{X_2})$, i.e.,

$$\mathrm{BCH}(X_1, X_2) = \left(\alpha_1 + \alpha_2, \frac{(\alpha_1 + \alpha_2)}{\sinh(\frac{\alpha_1+\alpha_2}{2})}\left(\frac{e^{-\frac{\alpha_2}{2}}}{\alpha_1}\sinh(\frac{\alpha_1}{2})y_1 + \frac{e^{\frac{\alpha_1}{2}}}{\alpha_2}\sinh(\frac{\alpha_2}{2})y_2\right),\right.$$

$$\frac{(\alpha_1 + \alpha_2)}{\sinh(\alpha_1 + \alpha_2)}\left[\frac{\beta_1}{\alpha_1}e^{-\alpha_2}\sinh(\alpha_1)\right.$$

$$\left.\left. + \frac{\beta_2}{\alpha_2}e^{\alpha_1}\sinh(\alpha_2) + \frac{2}{\alpha_1\alpha_2}e^{\frac{\alpha_1-\alpha_2}{2}}\sinh(\frac{\alpha_1}{2})\sinh(\frac{\alpha_2}{2})\omega_0(y_1, y_2)\right]\right).$$

Then, BCH property $\mathcal{E}_g \star_\theta \mathcal{E}_{g'} = \mathcal{E}_{g\cdot g'}$ is equivalent to

$$\forall\, X_1, X_2 \in \mathfrak{s} \quad : \quad E_{\star_\theta}(\lambda_{X_1}) \star_\theta E_{\star_\theta}(\lambda_{X_2}) = E_{\star_\theta}(\lambda_{\mathrm{BCH}(X_1,X_2)}),$$

which turns out to be true for the star-product (1.12) and the star-exponential determined above.

4 Non-Formal Definition of the Star-Exponential

4.1 Schwartz Spaces

In [7], a Schwartz space adapted to the elementary normal j-group $\mathbb{S}$ has been introduced, which is different from the usual one $\mathcal{S}(\mathbb{R}^{2n+2})$ in the global chart $\{(a, x, \ell)\}$, but related to oscillatory integrals. Let us have a look at the phase (1.12) of the star-product:

$$\epsilon S_{\mathbb{S}}(0, g_1, g_2) = \sinh(2a_1)\ell_2 - \sinh(2a_2)\ell_1 + \cosh(a_1)\cosh(a_2)\omega_0(x_1, x_2)$$

with $g_i = (a_i, x_i, \ell_i) \in \mathbb{S}$. Recall that the left-invariant vector fields of $\mathbb{S}$ are given by

$$\tilde{H} = \partial_a - x\partial_x - 2\ell\partial_\ell, \quad \tilde{y} = y\partial_x + \frac{1}{2}\omega_0(x, y)\partial_\ell, \quad \tilde{E} = \partial_\ell.$$

We define the maps $\tilde{\alpha}$ by $\forall X = (X_1, X_2) \in \mathfrak{s} \oplus \mathfrak{s}$,

$$\tilde{X} \cdot e^{-\frac{2i}{\theta} S_{\mathbb{S}}(0, g_1, g_2)} =: -\frac{2i\epsilon}{\theta}\tilde{\alpha}_X(g_1, g_2)e^{-\frac{2i}{\theta} S_{\mathbb{S}}(0, g_1, g_2)}$$

since it is an oscillatory phase. For example, we have

$$\tilde{\alpha}_{(E,0)}(g_1, g_2) = -\sinh(2a_2), \text{ and}$$
$$\tilde{\alpha}_{(H,0)}(g_1, g_2) = 2\cosh(2a_1)\ell_2 + 2\sinh(2a_2)\ell_1 - e^{-a_1}\cosh(a_2)\omega_0(x_1, x_2).$$

Then we set $\alpha_X(g) := \tilde{\alpha}_{(X,0)}(0, g)$ for any $X \in \mathfrak{s}$ and $g \in G$, whose expressions are

$$\alpha_H(g) = 2\ell, \qquad \alpha_y(g) = \cosh(a)\omega(y, x), \qquad \alpha_E(g) = -\sinh(2a).$$

This leads to the following definition.

Definition 4.1. The *Schwartz space* of $\mathbb{S}$ is defined as

$$\mathcal{S}(\mathbb{S}) = \{f \in C^\infty(\mathbb{S}) \quad \forall j \in \mathbb{N}^{2n+2}, \forall P \in \mathcal{U}(\mathfrak{s}) \text{ such that}$$

$$\|f\|_{j,P} := \sup_{g \in \mathbb{S}} \left|\alpha^j(g)\tilde{P} f(g)\right| < \infty\},$$

where $\alpha^j := \alpha_H^{j_1}\alpha_{e_1}^{j_2} \ldots \alpha_{e_{2n}}^{j_{2n+1}}\alpha_E^{j_{2n+2}}$.

It turns out that the space $\mathcal{S}(\mathbb{S})$ corresponds to the usual Schwartz space in the coordinates (r, x, ℓ) with $r = \sinh(2a)$. It is stable by the action of $\mathbb{S}$:

$$\forall f \in \mathcal{S}(\mathbb{S}), \forall g \in \mathbb{S} \quad : \quad g^* f \in \mathcal{S}(\mathbb{S}).$$

Moreover, $\mathcal{S}(\mathbb{S})$ is a Fréchet nuclear space endowed with the seminorms ($\|f\|_{j,P}$).

For $f, h \in \mathcal{S}(\mathbb{S})$, the product $f \star_\theta h$ is well defined by (1.12). However, to show that it belongs to $\mathcal{S}(\mathbb{S})$, we will use arguments close to oscillatory integral theory. Let us illustrate this concept. One can show that the following operators leave the phase $e^{-\frac{2i}{\theta}S(0,g_1,g_2)}$ invariant:

$$\mathcal{O}_{a_2} := \frac{1}{1 + \tilde{\alpha}^2_{(E,0)}}(1 - \frac{\theta^2}{4}\tilde{E}^2) = \frac{1}{1 + \sinh(2a_2)^2}(1 - \frac{\theta^2}{4}\partial^2_{\ell_1}),$$

$$\mathcal{O}_{a_1} := \frac{1}{1 + \sinh(2a_1)^2}(1 - \frac{\theta^2}{4}\partial^2_{\ell_2}),$$

$$\mathcal{O}_{x_2} := \frac{1}{1 + x_2^2}(1 - \frac{\theta^2}{4\cosh(a_1)^2\cosh(a_2)^2}\partial^2_{x_1}),$$

$$\mathcal{O}_{x_1} := \frac{1}{1 + x_1^2}(1 - \frac{\theta^2}{4\cosh(a_1)^2\cosh(a_2)^2}\partial^2_{x_2}),$$

$$\mathcal{O}_{\ell_2} := \frac{1}{1 + \ell_2^2}(1 - \frac{\theta^2}{4}(\frac{1}{\cosh(2a_1)}(\partial_{a_1} - \tanh(a_1)x_1\partial_{x_1}))^2),$$

$$\mathcal{O}_{\ell_1} := \frac{1}{1 + \ell_1^2}(1 - \frac{\theta^2}{4}(\frac{1}{\cosh(2a_2)}(\partial_{a_2} - \tanh(a_2)x_2\partial_{x_2}))^2).$$

So we can add arbitrary powers of these operators in front of the phase without changing the expression. Then, using integrations by parts, we have for $F \in \mathcal{S}(\mathbb{S}^2)$:

$$\int e^{-\frac{2i}{\theta}S_{\mathbb{S}}(0,g_1,g_2)} F(g_1, g_2)dg_1dg_2$$

$$= \int e^{-\frac{2i}{\theta}S_{\mathbb{S}}(0,g_1,g_2)}(\mathcal{O}^*_{a_1})^{k_1}(\mathcal{O}^*_{a_2})^{k_2}(\mathcal{O}^*_{x_1})^{p_1}(\mathcal{O}^*_{x_2})^{p_2}(\mathcal{O}^*_{\ell_1})^{q_1}(\mathcal{O}^*_{\ell_2})^{q_2} F(g_1, g_2)dg_1dg_2$$

$$= \int e^{-\frac{2i}{\theta}S_{\mathbb{S}}(0,g_1,g_2)}\frac{1}{(1 + \sinh^2(2a_1))^{k_1}(1 + \sinh^2(2a_2))^{k_2}}$$

$$\frac{1}{(1 + x_1^2)^{p_1-q_2}(1 + x_2^2)^{p_2-q_1}(1 + \ell_1^2)^{q_1}(1 + \ell_2^2)^{q_2}}DF(g_1, g_2)dg_1dg_2 \quad (1.26)$$

for any $k_i, q_i, p_i \in \mathbb{N}$ such that $p_1 \geq q_2$ and $p_2 \geq q_1$, and where D is a linear combination of products of bounded functions (with every derivatives bounded) in (g_1, g_2) with powers of ∂_{ℓ_i}, ∂_{x_i} and $\frac{1}{\cosh(2a_i)}\partial_{a_i}$. The first line of (1.26) is not defined for nonintegrable functions F bounded by polynomials in $r_i := \sinh(2a_i)$, x_i and ℓ_i. However, the last two lines of (1.26) are well defined for k_i, p_i, q_i sufficiently large. Therefore it gives a sense to the first line, now understood as an oscillatory integral, i.e., as being equal to the last two lines. This definition of oscillatory integral [7, 9] is unique, in particular unambiguous in the powers

k_i, p_i, q_i. Note that this corresponds to the usual oscillatory integral [13] in the coordinates (r, x, ℓ).

The next theorem, proved in [7], can be showed by using such methods of oscillatory integrals on $\mathcal{S}(\mathbb{S})$.

Theorem 4.2. *Let* $\mathcal{P} : \mathbb{R} \to C^\infty(\mathbb{R})$ *be a smooth map such that* $\mathcal{P}_0 \equiv 1$, *and* $\mathcal{P}_\theta(a)$ *as well as its inverse are bounded by* $C \sinh(2a)^k$, $k \in \mathbb{N}$, $C > 0$. *Then, the expression* (1.12) *yields a* $\mathbb{S}$-*invariant non-formal deformation quantization.*

In particular, $(\mathcal{S}(\mathbb{S}), \star_\theta)$ *is a nuclear Fréchet algebra.*

In what follows we show a factorization property for this Schwartz space. First, by introducing $\gamma(a) = \sinh(2a)$ and $\mathcal{S}(A) := \gamma^* \mathcal{S}(\mathbb{R})$, we note that the group law of $\mathbb{S}$ reads in the coordinates $(r = \gamma(a), x, \ell)$:

$$(r, x, \ell) \cdot (r', x', \ell') = \left(r\sqrt{1 + r'^2} + r'\sqrt{1 + r^2}, (c(r') - s(r'))x + x', \right.$$

$$\left. (\sqrt{1 + r'^2} - r')\ell + \ell' + \frac{1}{2}(c(r') - s(r'))\omega_0(x, x') \right)$$

with the auxiliary functions:

$$c(r) = \frac{\sqrt{2}}{2}(1 + \sqrt{1 + r^2})^{\frac{1}{2}} = \cosh(\frac{1}{2}\operatorname{arcsinh}(r)), \tag{1.27}$$

$$s(r) = \frac{\sqrt{2}}{2}\operatorname{sgn}(r)(-1 + \sqrt{1 + r^2})^{\frac{1}{2}} = \sinh(\frac{1}{2}\operatorname{arcsinh}(r)).$$

Proposition 4.3 (Factorization). *The map* Φ *defined by* $\Phi(f \otimes h) = f \star_\theta h$, *for* $f \in \mathcal{S}(A)$ *and* $h \in \mathcal{S}(\mathbb{R}^{2n+1})$ *realizes a continuous automorphism* $\mathcal{S}(\mathbb{S}) = \mathcal{S}(A) \hat{\otimes} \mathcal{S}(\mathbb{R}^{2n+1}) \to \mathcal{S}(\mathbb{S})$.

Proof. Due to the nuclearity of the Schwartz space, we have indeed $\mathcal{S}(\mathbb{S}) = \mathcal{S}(A) \hat{\otimes} \mathcal{S}(\mathbb{R}^{2n+1})$. For $f \in \mathcal{S}(A)$ (abuse of notation identifying $f(a)$ and $f(r) := f(\gamma^{-1}(r))$) and $h \in \mathcal{S}(\mathbb{R}^{2n+1})$, we reexpress the star-product (1.12) in the coordinates (r, x, ℓ):

$$(f \star_\theta h)(r, x, \ell) = \frac{1}{(\pi\theta)^{2n+2}} \int \left(1 - \frac{r_1 r_2}{\sqrt{(1 + r_1^2)(1 + r_2^2)}}\right)$$

$$\times f(r\sqrt{1 + r_1^2} + r_1\sqrt{1 + r^2})h(\frac{1}{c(r_1)}x_2 + (c(r_2) - \frac{s(r_1)s(r_2)}{c(r_1)})x, \ell_2 + \sqrt{1 + r_2^2}\ell)$$

$$\times \frac{\sqrt{c(r_1)c(r_2)}}{\sqrt{c(r_1\sqrt{1 + r_2^2} - r_2\sqrt{1 + r_1^2})}} e^{-\frac{2i\epsilon}{\theta}(r_1\ell_2 - r_2\ell_1 + \omega_0(x_1, x_2))} dr_i dx_i d\ell_i.$$

By using the partial Fourier transform $\hat{h}(r, \xi) = \int d\ell e^{-i\ell\xi} h(x, \ell)$, and integrating over several variables, we obtain

$$(f \star_\theta h)(r, x, \ell) = \frac{1}{2\pi} \int f(r\sqrt{1 + \frac{\theta^2\xi^2}{4}} + \frac{\epsilon\theta\xi}{2}\sqrt{1 + r^2})\hat{h}(x, \xi)e^{i\ell\xi}d\xi.$$

For $\varphi \in \mathcal{S}(\mathbb{S})$, we have now the following explicit expression for Φ:

$$\Phi(\varphi)(r, x, \ell) = \frac{1}{2\pi} \int \hat{\varphi}(r\sqrt{1 + \frac{\theta^2\xi^2}{4}} + \frac{\epsilon\theta\xi}{2}\sqrt{1 + r^2}, x, \xi)e^{i\ell\xi}d\xi$$

which permits to deduce that Φ is valued in $\mathcal{S}(\mathbb{S})$ and continuous. Then the formula

$$\hat{\varphi}(r, x, \ell) = \int \Phi(\varphi)(r\sqrt{1 + \frac{\theta^2\ell^2}{4}} - \frac{\epsilon\theta\ell}{2}\sqrt{1 + r^2}, x, \xi)e^{-i\ell\xi}d\xi$$

permits to obtain the inverse of Φ which is also continuous. $\square$

For normal j-groups $G = G_1 \ltimes \mathbb{S}_2$, we define the Schwartz space recursively

$$\mathcal{S}(G) = \mathcal{S}(G_1)\hat{\otimes}\mathcal{S}(\mathbb{S}_2)$$

and obtain the same properties as before. In particular, endowed with the star-product (1.17), the Schwartz space $\mathcal{S}(G)$ is a nuclear Fréchet algebra.

4.2   Multipliers

Let us consider the topological dual $\mathcal{S}'(\mathbb{S})$ of $\mathcal{S}(\mathbb{S})$. In the coordinates $(r = \gamma(a), x, \ell)$, it corresponds to tempered distributions. By denoting $\langle -, - \rangle$ the duality bracket between $\mathcal{S}'(\mathbb{S})$ and $\mathcal{S}(\mathbb{S})$, one can extend the product $\star_\theta$ (with tracial identity) as

$$\forall T \in \mathcal{S}'(\mathbb{S}), \forall f, h \in \mathcal{S}(\mathbb{S}) : \langle T \star_\theta f, h \rangle := \langle T, f \star_\theta h \rangle \text{ and } \langle f \star_\theta T, h \rangle := \langle T, h \star_\theta f \rangle,$$

which is compatible with the case $T \in \mathcal{S}(\mathbb{S})$.

Definition 4.4. The *multiplier space* associated to $(\mathcal{S}(\mathbb{S}), \star_\theta)$ is defined as

$$\mathcal{M}_{\star_\theta}(\mathbb{S}) :=$$

$\{T \in \mathcal{S}'(\mathbb{S}), f \mapsto T \star_\theta f \text{ and } f \mapsto f \star_\theta T \text{ are continuous from } \mathcal{S}(\mathbb{S}) \text{ into itself}\}.$

We can endow this space with the topology associated to the seminorms:

$$\|T\|_{B,j,P,L} = \sup_{f \in B} \|T \star f\|_{j,P} \text{ and } \|T\|_{B,j,P,R} = \sup_{f \in B} \|f \star T\|_{j,P}$$

where B is a bounded subset of $\mathcal{S}(\mathbb{S})$, $j \in \mathbb{N}^{2n+2}$, $P \in \mathcal{U}(\mathfrak{s})$ and $\|f\|_{j,P}$ is the Schwartz seminorm introduced in Definition 4.1. Note that B can be described as a set satisfying $\forall j, P, \sup_{f \in B} \|f\|_{j,P}$ exists.

Proposition 4.5. *The star-product can be extended to $\mathcal{M}_{\star_\theta}(\mathbb{S})$ by:*

$$\forall\ S, T \in \mathcal{M}_{\star_\theta}(\mathbb{S}), \ \forall\ f \in \mathcal{S}(\mathbb{S}) \quad : \quad \langle S \star_\theta T, f \rangle := \langle S, T \star_\theta f \rangle = \langle T, f \star_\theta S \rangle.$$

Then $(\mathcal{M}_{\star_\theta}(\mathbb{S}), \star_\theta)$ is an associative Hausdorff locally convex complete and nuclear algebra, with separately continuous product called the multiplier algebra.

Proof. For the extension of the star-product and its associativity, we can show successively for all $S, T \in \mathcal{M}_{\star_\theta}(\mathbb{S})$, for all $f, h \in \mathcal{S}(\mathbb{S})$,

$$(T \star_\theta f) \star_\theta h = T \star_\theta (f \star_\theta h) \quad , \quad (S \star_\theta T) \star_\theta f = S \star_\theta (T \star_\theta f), \text{ and}$$

$$(T_1 \star_\theta T_2) \star_\theta T_3 = T_1 \star_\theta (T_2 \star_\theta T_3),$$

each time by evaluating the distribution on a Schwartz function $\varphi \in \mathcal{S}(\mathbb{S})$ and by using the factorization property (Proposition 4.3).

$\mathcal{M}_{\star_\theta}(\mathbb{S})$ is the intersection of $\mathcal{M}_L$, the left multipliers, and $\mathcal{M}_R$, the right multipliers. By definition, each space $\mathcal{M}_L$ and $\mathcal{M}_R$ is topologically isomorphic to $\mathcal{L}(\mathcal{S}(\mathbb{S}))$ endowed with the strong topology. Since $\mathcal{S}(\mathbb{S})$ is Fréchet and nuclear, so is $\mathcal{L}(\mathcal{S}(\mathbb{S}))$, as well as $\mathcal{M}_L$, $\mathcal{M}_R$ and finally $\mathcal{M}_{\star_\theta}(\mathbb{S})$ (see [18] Propositions 50.1, 50.5 and 50.6). □

Due to the definition of $\mathcal{S}(G)$ for a normal j-group $G = G_1 \ltimes \mathbb{S}_2$ and to the expression of the star-product (1.17), the multiplier space associated to $(\mathcal{S}(G), \star_\theta)$ takes the form

$$\mathcal{M}_{\star_\theta}(G) = \mathcal{M}_{\star_\theta}(G_1) \hat{\otimes} \mathcal{M}_{\star_\theta}(\mathbb{S}_2), \tag{1.28}$$

and is also an associative Hausdorff locally convex complete and nuclear algebra, with separately continuous product. Remember that we have identified coadjoint orbits $\mathcal{O}$ described in Proposition 2.10 with the group G itself, so that we can speak also about the multiplier algebra $\mathcal{M}_{\star_\theta}(\mathcal{O})$.

4.3 Non-Formal Star-Exponential

Theorem 4.6. *Let G be a normal j-group and $\star_\theta$ the star-product (1.17). Then for any $g \in G$, the star-exponential (1.19) $\mathcal{E}_g^{\mathcal{O}}$ lies in the multiplier algebra $\mathcal{M}_{\star_\theta}(\mathcal{O})$.*

Proof. Let us focus for the moment on the case of the elementary group $\mathcal{S}$. The general case can then be obtained recursively due to the structure of the star-exponential (1.19) and of the multiplier algebra (1.28). We use the same notations as before. For $f, h \in \mathcal{S}(A)$, $f \star_\theta h = f \cdot h$. If T belongs to the multiplier space $\mathcal{M}(\mathcal{S}(A))$ of $\mathcal{S}(A)$ for the usual commutative product, we have in particular $T \in \mathcal{S}'(\mathbb{S})$ and by duality $T \star_\theta f = T \cdot f$. Then,

$$\forall f \in \mathcal{S}(A), \ \forall h \in \mathcal{S}(\mathbb{R}^{2n+1}) \quad : \quad T \star_\theta (f \star_\theta h) = (T \cdot f) \star_\theta h.$$

By the factorization property (Proposition 4.3), it means that $T \in \mathcal{M}_{\star_\theta}(\mathbb{S})$, and we have an embedding $\mathcal{M}(\mathcal{S}(A)) \hookrightarrow \mathcal{M}_{\star_\theta}(\mathbb{S})$. If we note as before $\mathbb{R}^{2n+1} = V \oplus \mathbb{R}E$, we can show in the same way that there is another embedding $\mathcal{M}(\mathcal{S}(\mathbb{R}E)) \hookrightarrow \mathcal{M}_{\star_\theta}(\mathbb{S})$. Since $x' \in V \mapsto \mathcal{E}^{\mathcal{O}_2}_{(g_1, g_2)}(0, x', 0)$ is an imaginary exponential of a polynomial of degree less or equal than 2 in x' and since the product $\star_\theta$ coincides with the Moyal product on V, it turns out that $x' \in V \mapsto \mathcal{E}^{\mathcal{O}_2}_{(g_1, g_2)}(0, x', 0)$ is in $\mathcal{M}_{\star_\theta}(\mathcal{S}(V))$. Then, the star-exponential $\mathcal{E}^{\mathcal{O}_2}$ in (1.19) lies in $\mathcal{M}(\mathcal{S}(A)) \hat{\otimes} \mathcal{M}_{\star_\theta}(\mathcal{S}(V)) \hat{\otimes} \mathcal{M}(\mathcal{S}(\mathbb{R}E))$, and it belongs also to $\mathcal{M}_{\star_\theta}(\mathbb{S})$. $\qquad\square$

5 Adapted Fourier Transformation

5.1 Definition

As in the case of the Moyal–Weyl quantization treated in [1,2], we can introduce the notion of adapted Fourier transformation. For normal j-groups $G = G_1 \ltimes \mathbb{S}_2$, which are not unimodular, it is relevant for that to introduce a **modified star-exponential**

$$\tilde{\mathcal{E}}^{\mathcal{O}}_g(g') := \mathrm{Tr}(U(g) d^{\frac{1}{2}} \Omega(g')),$$

where d is the formal dimension operator associated to U (see [10, 12]) and $\mathcal{O}$ is the coadjoint orbit determining the irreducible representation U. Such an operator d is used to regularize the expressions since $\int f(g) U(g) d^{\frac{1}{2}}$ is a Hilbert–Schmidt operator whenever f is in $L^2(G)$. So the trace in the definition of $\tilde{\mathcal{E}}^{\mathcal{O}}$ is understood as a distribution only in the variable $g' \in \mathcal{O}$.

By denoting Δ the modular function, defined by $d^L(g \cdot g') = \Delta(g') d^L g$, whose computation gives

$$\Delta(g) = \Delta_1(g_1) \Delta_2(g_2), \ \text{with} \ \Delta_2(a_2, x_2, \ell_2) = e^{-2(n_2+1)a_2},$$

the operator d is defined (up to a positive constant) by the relation

$$\forall g \in G \quad : \quad U(g) d U(g)^{-1} = \Delta(g)^{-1} d.$$

Since $\mathcal{R}(g_1)d_2\mathcal{R}(g_1)^{-1} = d_2$, it can therefore be expressed as $d = d_1 \otimes d_2$, for d_i the dimension operator associated to U_i, and with for all $\varphi_2 \in \mathcal{H}_2$, for all $(a_0, v_0) \in Q_2$,

$$(d_2\varphi_2)(a_0, v_0) = \kappa_2^2 e^{-2(n_2+1)a_0}\varphi_2(a_0, v_0)$$

where we recall that $\dim(\mathbb{S}_2) = 2(n_2 + 1)$. Note that d_2 is independent here of the choice of the irreducible representation U_2 ($\epsilon_2 = \pm 1$).

Proposition 5.1. *The expression of the modified star-exponential can then be computed the same notation as for Theorem 3.9:*

$$\tilde{\mathcal{E}}_g^{\mathcal{O}}(g') = \tilde{\mathcal{E}}_{g_1}^{\mathcal{O}_1}(g_1')\frac{e^{(n_2+1)(\frac{a_2}{2}-a_2')}}{(\pi\theta)^{n_2+1}}\frac{\sqrt{\cosh(a_2)}\cosh(\frac{a_2}{2})^{n_2}|\det(\rho_+(g_1))|^{\frac{1}{2}}}{|\det(1+\rho_+(g_1))|}$$

$$\exp\left(\frac{i\epsilon_2}{\theta}\left[2\sinh(a_2)\ell_2' + e^{a_2-2a_2'}\ell_2 + X^T A_\rho X\right]\right).$$

Definition 5.2. We can now define the *adapted Fourier transformation*: for $f \in \mathcal{S}(G)$ and $g' \in \mathcal{O}$,

$$\mathcal{F}_{\mathcal{O}}(f)(g') := \int_G f(g)\tilde{\mathcal{E}}_g^{\mathcal{O}}(g')\mathrm{d}^L g.$$

We see that this definition is a generalization of the usual (symplectic) Fourier transformation. For example in the case of the group $\mathbb{R}^2$, the star-exponential associated to the Moyal product is indeed given by $\exp(\frac{2i}{\theta}(a\ell' - a'\ell))$.

5.2 Fourier Analysis

Proposition 5.3. *The modified star-exponential satisfies an orthogonality relation: for $g', g'' \in G$,*

$$\int_G \overline{\tilde{\mathcal{E}}_g^{\mathcal{O}}(g')}\tilde{\mathcal{E}}_g^{\mathcal{O}}(g'')\mathrm{d}^L g = \frac{1}{\Delta(g'')}\delta(g''\cdot(g')^{-1}).$$

Note that $\Delta(g_2'')^{-1}\delta(g_2''\cdot(g_2')^{-1}) = \delta(a_2''-a_2')\delta(x_2''-x_2')\delta(\ell_2''-\ell_2')$. This orthogonality relation does not hold for the unmodified star-exponential.

Proof. We use the expression of Proposition 5.1:

$$\int_G \overline{\tilde{\mathcal{E}}_g^{\mathcal{O}}(g')}\tilde{\mathcal{E}}_g^{\mathcal{O}}(g'')\mathrm{d}^L g = \int_{G_1} \overline{\tilde{\mathcal{E}}_{g_1}^{\mathcal{O}_1}(g_1')}\tilde{\mathcal{E}}_{g_1}^{\mathcal{O}_1}(g_1'')\int_{\mathbb{S}_2}\frac{e^{(n_2+1)(a_2-a_2'-a_2'')}}{(\pi\theta)^{2(n_2+1)}}$$

$$\frac{|\det(\rho_+(g_1))|\cosh(a_2)\cosh(\frac{a_2}{2})^{2n_2}}{|\det(1+\rho_+(g_1))|^2}e^{\frac{i\epsilon_2}{\theta}(2\sinh(a_2)(\ell_2''-\ell_2')+e^{a_2}(e^{-2a_2''}-e^{-2a_2'})\ell_2)}$$

$$e^{\frac{i\epsilon_2}{\theta}((X'')^T A_\rho X''-(X')^T A_\rho X')}\mathrm{d}^L g_2 \mathrm{d}^L g_1$$

with

$$X' = \begin{pmatrix} \frac{1}{\sqrt{2}}e^{\frac{a_2}{2}-a_2'}v_2 \\ \frac{1}{\sqrt{2}}e^{\frac{a_2}{2}-a_2'}w_2 \\ \sqrt{2}\cosh(\frac{a_2}{2})v_2' \\ \sqrt{2}\cosh(\frac{a_2}{2})w_2' \end{pmatrix} \quad \text{and} \quad X'' = \begin{pmatrix} \frac{1}{\sqrt{2}}e^{\frac{a_2}{2}-a_2''}v_2 \\ \frac{1}{\sqrt{2}}e^{\frac{a_2}{2}-a_2''}w_2 \\ \sqrt{2}\cosh(\frac{a_2}{2})v_2'' \\ \sqrt{2}\cosh(\frac{a_2}{2})w_2'' \end{pmatrix}.$$

Integration over ℓ_2 leads to the contribution $\delta(a_2'-a_2'')$. Since A_ρ depends only on g_1, and $a_2' = a_2''$, we see that the gaussian part in (v_2, w_2) disappears and integration over these variables brings $\frac{|\det(1+\rho_+(g_1))|^2}{|\det(\rho_+(g_1))|}\delta(v_2'-v_2'')\delta(w_2'-w_2'')$. Eventually, integration on a_2 can be performed and we find

$$\int_G \overline{\tilde{\mathcal{E}}_g^{\mathcal{O}}(g')}\tilde{\mathcal{E}}_g^{\mathcal{O}}(g'')\mathrm{d}^L g = \left(\int_{G_1}\overline{\tilde{\mathcal{E}}_{g_1}^{\mathcal{O}_1}(g_1')}\tilde{\mathcal{E}}_{g_1}^{\mathcal{O}_1}(g_1'')\mathrm{d}^L g_1\right)\Delta(g_2'')^{-1}\delta(g_2''\cdot(g_2')^{-1})$$

which leads to the result recursively. $\square$

Proposition 5.4. *The adapted Fourier transformation satisfies the following property:* $\forall f_1, f_2 \in \mathcal{S}(G)$,

$$\mathcal{F}_{\mathcal{O}}(f_1 \times f_2) = \frac{\Delta^{\frac{1}{2}}}{\kappa}\left(\Delta^{-\frac{1}{2}}\mathcal{F}_{\mathcal{O}}(f_1)\right)\star_\theta\left(\Delta^{-\frac{1}{2}}\mathcal{F}_{\mathcal{O}}(f_2)\right),$$

with $(f_1 \times f_2)(g) = \int_G f_1(g')f_2((g')^{-1}g)\mathrm{d}^L g'$ *the usual convolution.*

Proof. Due to the BCH property (see Proposition 3.10) and to the computation of the modified star-exponential $\tilde{\mathcal{E}}_g^{\mathcal{O}}(g') = \tilde{\mathcal{E}}_{g_1}^{\mathcal{O}_1}(g_1')\frac{\kappa_2}{\Delta_2(g_2(g_2')^{-2})^{\frac{1}{2}}}\mathcal{E}_g^{\mathcal{O}_2}(g_2')$, we have the modified the BCH property

$$\tilde{\mathcal{E}}_{g\cdot g'}^{\mathcal{O}}(g'') = \frac{\Delta(g'')^{\frac{1}{2}}}{\kappa}\left(\Delta^{-\frac{1}{2}}\tilde{\mathcal{E}}_g^{\mathcal{O}}\right)\star_\theta\left(\Delta^{-\frac{1}{2}}\tilde{\mathcal{E}}_{g'}^{\mathcal{O}}\right)(g'')$$

which leads directly to the result by using the expression of the adapted Fourier transform and the convolution. $\square$

As in Remark 2.11, we consider the coadjoint orbit $\mathcal{O}_{(\epsilon)} = \mathcal{O}_{1,(\epsilon_1)} \times \mathcal{O}_{2,\epsilon_2}$ of the normal j-group $G = G_1 \ltimes \mathbb{S}_2$ determined by the sign choices $(\epsilon) = ((\epsilon_1), \epsilon_2) \in (\mathbb{Z}_2)^N$, with $(\epsilon_1) \in (\mathbb{Z}_2)^{N-1}$ and $\epsilon_2 \in \mathbb{Z}_2$. Due to Proposition 5.1, we can write the modified star-exponential as

$$\tilde{\mathcal{E}}_g^{\mathcal{O}_{(\epsilon)}}(g') = \tilde{\mathcal{E}}_{g_1}^{\mathcal{O}_{1,(\epsilon_1)}}(g_1') \, \tilde{\mathcal{E}}_{(g_1,g_2)}^{\mathcal{O}_{2,\epsilon_2}}(g_2'),$$

with $g = (g_1, g_2) \in G$ and $g' = (g_1', g_2') \in \mathcal{O}_{(\epsilon)}$.

Theorem 5.5. *We have the following inversion formula for the adapted Fourier transformation: for $f \in \mathcal{S}(G)$ and $g \in G$,*

$$f(g) = \sum_{(\epsilon) \in (\mathbb{Z}_2)^N} \int_{\mathcal{O}_{(\epsilon)}} \overline{\tilde{\mathcal{E}}_g^{\mathcal{O}_{(\epsilon)}}(g')} \mathcal{F}_{\mathcal{O}_{(\epsilon)}}(f)(g') d\mu(g').$$

Moreover, the Parseval–Plancherel theorem is true:

$$\int_G |f(g)|^2 d^L g = \sum_{(\epsilon) \in (\mathbb{Z}_2)^N} \int_{\mathcal{O}_{(\epsilon)}} |\mathcal{F}_{\mathcal{O}_{(\epsilon)}}(f)(g')|^2 d\mu(g').$$

Proof. Let us show the dual property to Proposition 5.3, i.e.,

$$\sum_{(\epsilon) \in (\mathbb{Z}_2)^N} \int_{\mathcal{O}_{(\epsilon)}} \overline{\tilde{\mathcal{E}}_{g'}^{\mathcal{O}_{(\epsilon)}}(g)} \tilde{\mathcal{E}}_{g''}^{\mathcal{O}_{(\epsilon)}}(g) d\mu(g) = \frac{1}{\Delta(g'')} \delta(g'' \cdot (g')^{-1}). \qquad (1.29)$$

First, we have

$$\int_{\mathcal{O}_{(\epsilon)}} \overline{\tilde{\mathcal{E}}_{g'}^{\mathcal{O}_{(\epsilon)}}(g)} \tilde{\mathcal{E}}_{g''}^{\mathcal{O}_{(\epsilon)}}(g) d\mu(g)$$

$$= \int_{\mathcal{O}_{1,(\epsilon_1)}} \overline{\tilde{\mathcal{E}}_{g_1'}^{\mathcal{O}_{1,(\epsilon_1)}}(g_1)} \tilde{\mathcal{E}}_{g_1''}^{\mathcal{O}_{1,(\epsilon_1)}}(g_1) \int_{\mathcal{O}_{2,\epsilon_2}} \frac{e^{(n_2+1)(\frac{a_2'+a_2''}{2}-2a_2)}}{(\pi\theta)^{2(n_2+1)}}$$

$$\times \frac{|\det(\rho_+(g_1')) \det(\rho_+(g_1''))|^{\frac{1}{2}} \sqrt{\cosh(a_2') \cosh(a_2'')} \cosh(\frac{a_2'}{2})^{n_2} \cosh(\frac{a_2''}{2})^{n_2}}{|\det(1 + \rho_+(g_1')) \det(1 + \rho_+(g_1''))|}$$

$$\times \exp\left[\frac{i\epsilon_2}{\theta}(-2\sinh(a_2')\ell_2 + 2\sinh(a_2'')\ell_2 - e^{a_2'-2a_2}\ell_2' + e^{a_2''-2a_2}\ell_2'' \right.$$

$$\left. + (X'')^T A_\rho(g_1'')X'' - (X')^T A_\rho(g_1')X')\right] d\mu_2(g_2) d\mu_1(g_1)$$

with

$$X' = \begin{pmatrix} \frac{1}{\sqrt{2}} e^{\frac{a_2'}{2}-a_2} v_2' \\ \frac{1}{\sqrt{2}} e^{\frac{a_2'}{2}-a_2} w_2' \\ \sqrt{2}\cosh(\frac{a_2'}{2})v_2 \\ \sqrt{2}\cosh(\frac{a_2'}{2})w_2 \end{pmatrix} \quad \text{and} \quad X'' = \begin{pmatrix} \frac{1}{\sqrt{2}} e^{\frac{a_2''}{2}-a_2} v_2'' \\ \frac{1}{\sqrt{2}} e^{\frac{a_2''}{2}-a_2} w_2'' \\ \sqrt{2}\cosh(\frac{a_2''}{2})v_2 \\ \sqrt{2}\cosh(\frac{a_2''}{2})w_2 \end{pmatrix}.$$

We want to compute the sum over $(\epsilon) \in (\mathbb{Z}_2)^N$ of such terms. By recurrence, we can suppose that

$$\sum_{(\epsilon_1)\in(\mathbb{Z}_2)^{N-1}} \int_{\mathcal{O}_{1,(\epsilon_1)}} \overline{\tilde{\mathcal{E}}_{g_1'}^{\mathcal{O}_{1,(\epsilon_1)}}(g_1)} \tilde{\mathcal{E}}_{g_1''}^{\mathcal{O}_{1,(\epsilon_1)}}(g_1) d\mu_1(g_1) = \frac{1}{\Delta(g_1'')}\delta(g_1'' \cdot (g_1')^{-1}),$$

which means that $g_1'' = g_1'$ in the following. The integration over ℓ_2 brings a contribution in $\delta(a_2' - a_2'')$. Since $g_1'' = g_1'$ and $a_2' = a_2''$, the gaussian part in (v_2, w_2) disappears and integration over these variables brings $\frac{|\det(1+\rho_+(g_1'))|^2}{|\det(\rho_+(g_1'))|}\delta(v_2' - v_2'')\delta(w_2' - w_2'')$. The remaining term is proportional to

$$\sum_{\epsilon_2=\pm 1} \int_{\mathbb{R}} e^{a_2'-2a_2} e^{\frac{i\epsilon_2}{\theta}} e^{a_2'-2a_2}(\ell_2''-\ell_2') da_2 = \pi\theta\delta(\ell_2'' - \ell_2').$$

The property (1.29) permits showing the inversion formula

$$\sum_{(\epsilon)\in(\mathbb{Z}_2)^N} \int_{\mathcal{O}_{(\epsilon)}} \overline{\tilde{\mathcal{E}}_g^{\mathcal{O}_{(\epsilon)}}(g')} \mathcal{F}_{\mathcal{O}_{(\epsilon)}}(f)(g') d\mu(g')$$

$$= \sum_{(\epsilon)\in(\mathbb{Z}_2)^N} \int \overline{\tilde{\mathcal{E}}_g^{\mathcal{O}_{(\epsilon)}}(g')} f(g'') \tilde{\mathcal{E}}_{g''}^{\mathcal{O}_{(\epsilon)}}(g') d^L g'' d\mu(g') = f(g),$$

as well as the Parseval–Plancherel theorem

$$\sum_{(\epsilon)\in(\mathbb{Z}_2)^N} \int_{\mathcal{O}_{(\epsilon)}} |\mathcal{F}_{\mathcal{O}_{(\epsilon)}}(f)(g')|^2 d\mu(g')$$

$$= \sum_{(\epsilon)\in(\mathbb{Z}_2)^N} \int \overline{f(g)\tilde{\mathcal{E}}_g^{\mathcal{O}_{(\epsilon)}}(g')} f(g'') \tilde{\mathcal{E}}_{g''}^{\mathcal{O}_{(\epsilon)}}(g') d^L g'' d^L g d\mu(g') = \int_G |f(g)|^2 d^L g.$$

$\square$

Corollary 5.6. *The map*

$$\mathcal{F} := \bigoplus_{(\epsilon)\in(\mathbb{Z}_2)^N} \mathcal{F}_{\mathcal{O}_{(\epsilon)}} : L^2(G, d^L g) \to \bigoplus_{(\epsilon)} L^2(\mathcal{O}_{(\epsilon)}, \mu),$$

defined by $\mathcal{F}(f) := \bigoplus_{(\epsilon)}(\mathcal{F}_{\mathcal{O}_{(\epsilon)}} f)$ *realizes an isometric isomorphism.*

Proof. From Proposition 5.3, we deduce that $\forall (\epsilon) \in (\mathbb{Z}_2)^N$, $\mathcal{F}_{\mathcal{O}_{(\epsilon)}} \mathcal{F}_{\mathcal{O}_{(\epsilon)}}^* = \mathbb{1}$. And the Parseval–Plancherel means that $\sum_{(\epsilon)} \mathcal{F}_{\mathcal{O}_{(\epsilon)}}^* \mathcal{F}_{\mathcal{O}_{(\epsilon)}} = \mathbb{1}$. Moreover, we can show that for all $(\epsilon), (\epsilon') \in (\mathbb{Z}_2)^N$, with $(\epsilon) \neq (\epsilon')$, $\mathcal{F}_{\mathcal{O}_{(\epsilon)}} \mathcal{F}_{\mathcal{O}_{(\epsilon')}}^* = 0$. Indeed, if $k \leq N$ is such that $\epsilon_k \neq \epsilon'_k$, then the computation of $\int_G \overline{\tilde{\mathcal{E}}_g^{\mathcal{O}_{(\epsilon')}}(g')}\tilde{\mathcal{E}}_g^{\mathcal{O}_{(\epsilon)}}(g'')\mathrm{d}^L g$ corresponds to having a factor $e^{\frac{i\epsilon_k}{\theta}e^{a_2}(e^{-2a''_2}+e^{-2a'_2})\ell_2}$ in the proof of Proposition 5.3. Integration over ℓ_2 makes this expression vanish.

For each $(\epsilon) \in (\mathbb{Z}_2)^N$ (i.e., for each $(\epsilon) = (\epsilon_1, \cdots \epsilon_N)$ with $\epsilon_j = \pm 1$), we will consider a function $f_{(\epsilon)} \in L^2(\mathcal{O}_{(\epsilon)}, \mu)$. We denote by $\bigoplus_{(\epsilon)} f_{(\epsilon)}$ the 2^N-uplet of these functions on the different orbits. By using the three properties above and the fact that

$$\mathcal{F}^*\left(\bigoplus_{(\epsilon)} f_{(\epsilon)}\right) = \sum_{(\epsilon)} \mathcal{F}^*_{\mathcal{O}_{(\epsilon)}}(f_{(\epsilon)}),$$

we obtain that

$$\mathcal{F}^*\mathcal{F}(f) = \sum_{(\epsilon)} \mathcal{F}^*_{\mathcal{O}_{(\epsilon)}}\mathcal{F}_{\mathcal{O}_{(\epsilon)}}(f) = f, \text{ and}$$

$$\mathcal{F}\mathcal{F}^*\left(\bigoplus_{(\epsilon)} f_{(\epsilon)}\right) = \bigoplus_{(\epsilon)}(\mathcal{F}_{\mathcal{O}_{(\epsilon)}}\mathcal{F}^*_{\mathcal{O}_{(\epsilon)}} f_{(\epsilon)}) = \bigoplus_{(\epsilon)} f_{(\epsilon)}.$$

$\square$

5.3 Fourier Transformation and Schwartz Spaces

Given such an adapted Fourier transformation, we can wonder wether the Schwartz space $\mathcal{S}(G)$ defined in [7] (see Sect. 4.1) is stable by this transformation, as it is true in the flat case: the usual transformation stabilizes the usual Schwartz space on $\mathbb{R}^n$. However, the answer appears to be wrong here. Let us focus on the case of the elementary normal j-group $\mathbb{S}$. The Schwartz space $\mathcal{S}(\mathbb{S})$ of Definition 4.1 corresponds to the usual Schwartz space in the coordinates $(r = \sinh(2a), x, \ell)$. These coordinates are adapted to the phase of the kernel of the star-product (1.12). For the star-exponential of $\mathbb{S}$ given in (1.20), we need also to consider the coordinates corresponding to the moment maps (1.3):

$$\mu : \mathbb{S} \to \mathbb{R}^*_+ \times \mathbb{R}^{2n+1}, \quad (a, x, \ell) \mapsto (e^{-2a}, e^{-a}x, \ell).$$

We will denote the new variables $(s, z, \ell) = \mu(a, x, \ell)$.

Definition 5.7. We define the *moment-Schwartz space* of $\mathbb{S}$ to be

$$\mathcal{S}_\lambda(\mathbb{S}) = \{f \in C^\infty(\mathbb{S}) \quad (\mu^{-1})^* f \in \mathcal{S}(\mathbb{R}^*_+ \times \mathbb{R}^{2n+1})$$

$$\text{and } s^{-\frac{n+1}{2}}(\mu^{-1})^* f(s, z, \ell) \text{ is smooth in } s = 0\}.$$

The space $\mathcal{S}_\lambda(\mathbb{S})$ corresponds to the usual Schwartz space in the coordinates (s, z, ℓ) (for $s > 0$) with some boundary regularity condition in $s = 0$. As before, we identify the group $\mathbb{S}$ with the coadjoint orbit $\mathcal{O}_\epsilon$ ($\epsilon = \pm 1$).

Theorem 5.8. *The adapted Fourier transformation restricted to the Schwartz space induces an isomorphism*

$$\mathcal{F} : \mathcal{S}(\mathbb{S}) \to \mathcal{S}_\lambda(\mathcal{O}_+) \oplus \mathcal{S}_\lambda(\mathcal{O}_-).$$

Proof. Let $f \in \mathcal{S}(\mathbb{S})$. The Fourier transform reads as

$$\mathcal{F}_{\mathcal{O}_\epsilon}(f)(s, z, \ell) = \frac{1}{(\pi\theta)^{n+1}} \int dr' dx' d\ell' \frac{f(r', x', \ell')}{(1 + r'^2)^{\frac{1}{4}}} (\sqrt{1 + r'^2} + r')^{\frac{n+1}{2}} s^{\frac{n+1}{2}} c(r')^n$$
$$e^{\frac{i\epsilon}{\theta}\left(2r'\ell + (\sqrt{1+r'^2}+r')s\ell' + \frac{1}{2}(\sqrt{1+r'^2}+r'+1)\omega_0(x',z)\right)}.$$

Here we use the function $c(r')$ defined in (1.27), the coordinates $s = e^{-2a}$, $z = e^{-a}x$, $r' = \sinh(a')$ and the fact that f is Schwartz in the variable $\sinh(a)$ if and only if it is in the variable $\sinh(2a)$. We denote again by f the function in the new coordinates by a slight abuse of language. We have to check that $h(s, z, \ell) = s^{-\frac{n+1}{2}} \mathcal{F}_{\mathcal{O}_\epsilon}(f)(s, z, \ell)$ is Schwartz in (s, z, ℓ), i.e., we want to estimate expressions of the type

$$\int ds\, dz\, d\ell\, |(1 + s^2)^{k_1}(1 + z^2)^{p_1}(1 + \ell^2)^{q_1} \partial_s^{k_2} \partial_z^{p_2} \partial_\ell^{q_2} h(s, z, \ell)|.$$

Let us provide an analysis in terms of oscillatory integrals.

- Polynomial in ℓ: controlled by an adapted power of the following operator (invariant acting on the phase) $\frac{1}{1+\ell^2}(1 - \frac{\theta^2}{4}(\partial_{r'} - \frac{\ell'}{\sqrt{1+r'^2}}\partial_{\ell'} + \frac{(\sqrt{1+r'^2}+r')}{\sqrt{1+r'^2}(\sqrt{1+r'^2}+r'+1)}x'\partial_{x'})^2)$ (see Sect. 4.1). Indeed, powers and derivatives in the variables r', x', ℓ' are controlled by the Schwartz function f inside the integral.
- Polynomial in z: controlled by an adapted power of the (invariant) operator $\frac{1}{1+z^2}(1 - \frac{4\theta^2}{(\sqrt{1+r'^2}+r'+1)^2}\partial_{x'}^2)$.
- Polynomial in s: controlled by an adapted power of the (invariant) operator $\frac{1}{1+s^2}(1 - \frac{\theta^2}{(\sqrt{1+r'^2}+r')^2}\partial_{\ell'}^2)$. Note that the function $\frac{1}{(\sqrt{1+r'^2}+r')^2}$ is estimated by a polynomial in r' for $r' \to \pm\infty$, as its derivatives.
- Derivations in s: produce terms like powers of $(\sqrt{1 + r'^2} + r')\ell'$ which are controlled.
- Derivations in z: produce terms like powers of $(\sqrt{1 + r'^2} + r' + 1)x'$ which are controlled.
- Derivations in ℓ: produce terms like powers of r'.

This shows that h is Schwartz in (s, z, ℓ), so $\mathcal{F}(f) \in \mathcal{S}_\lambda(\mathbb{S})$.

Conversely, let $f_\epsilon \in \mathcal{S}_\lambda(\mathcal{O}_\epsilon)$. Due to Theorem 5.5, we can write the inverse of the Fourier transform as:

$$\mathcal{F}^{-1}(f_+, f_-)(r, x, \ell)$$
$$= \sum_{\epsilon=\pm 1} \frac{1}{2(\pi\theta)^{n+1}} \int ds' dz' d\ell' \frac{f_\epsilon(s', z', \ell')}{s'^{\frac{n+1}{2}}} (\sqrt{1+r^2}+r)^{\frac{n+1}{2}} \sqrt{1+r^2} c(r)^n$$
$$\times e^{-\frac{i\epsilon}{\theta}\left(2r\ell'+(\sqrt{1+r^2}+r)s'\ell+\frac{1}{2}(\sqrt{1+r^2}+r+1)\omega_0(x,z')\right)}.$$

Here we use now the coordinates $s' = e^{-2a'}$, $z' = e^{-a'}x'$, $r = \sinh(a)$. We want to estimate expressions of the type

$$\int dr\,dx\,d\ell \ |(1+r^2)^{k_1}(1+x^2)^{p_1}(1+\ell^2)^{q_1} \partial_r^{k_2} \partial_x^{p_2} \partial_\ell^{q_2} \mathcal{F}^{-1}(f_+, f_-)(r, x, \ell)|.$$

Let us provide also an analysis in terms of oscillatory integrals.

- <u>Polynomial in ℓ</u>: controlled by an adapted power of the following operator (invariant acting on the phase) $\frac{1}{1+\ell^2}(1 - \frac{\theta^2}{(\sqrt{1+r^2}+r)^2} \partial_{s'}^2)$. As before, powers and derivatives in the variables s', z', ℓ' are controlled by the Schwartz function f inside the integral. Note that $\frac{f_\epsilon(s', z', \ell')}{s'^{\frac{n+1}{2}}}$ is smooth in $s = 0$ so that the integral is well-defined for $s \in \mathbb{R}_+$.
- <u>Polynomial in x</u>: controlled by an adapted power of the (invariant) operator:

$$\frac{1}{1+x^2}\left(1 - \frac{4\theta^2}{(\sqrt{1+r^2}+r+1)^2} \partial_{z'}^2\right).$$

- <u>Polynomial in r</u>: controlled by an adapted power of the (invariant) operator $\frac{1}{1+r^2}(1 - \frac{\theta^2}{4} \partial_{\ell'}^2)$.
- <u>Derivations in r</u>: produce terms like powers of $(\sqrt{1+r^2}+r)$, $\frac{1}{\sqrt{1+r^2}}$, $r, c'(r), \ell', \frac{(\sqrt{1+r^2}+r')}{\sqrt{1+r^2}}s'\ell, \omega_0(x, z'), \ldots$ which are controlled (see just above).
- <u>Derivations in x</u>: produce terms like powers of $(\sqrt{1+r^2}+r+1)z'$ which are controlled.
- <u>Derivations in ℓ</u>: produce terms like powers of $(\sqrt{1+r^2}+r)s'$ which are also controlled.

This shows that $\mathcal{F}^{-1}(f_+, f_-) \in \mathcal{S}(\mathbb{S})$. $\qquad\square$

5.4   Application to Noncommutative Baumslag–Solitar Tori

We consider the decomposition of G into elementary normal j-groups of Sect. 2.2

$$G = \left(\dots (\mathbb{S}_1 \ltimes_{\rho_1} \mathbb{S}_2) \ltimes_{\rho_2} \dots \right) \ltimes_{\rho_{N-1}} \mathbb{S}_N$$

and the associated basis

$$\mathfrak{B} := \left(H_1, (f_1^{(i)})_{1 \le i \le 2n_1}, E_1, \dots, H_N, (f_N^{(i)})_{1 \le i \le 2n_N}, E_N \right)$$

of its Lie algebra $\mathfrak{g}$, where $(f_j^{(i)})_{1 \le i \le 2n_j}$ is a canonical basis of the symplectic space V_j contained in $\mathbb{S}_j$. We note G_{BS} the subgroup of G generated by $\{ e^{\theta X}, \ X \in \mathfrak{B} \}$ and call it the *Baumslag–Solitar subgroup* of G. Indeed, in the case of the "$ax + b$" group (two-dimensional elementary normal j-group), and if $e^{2\theta} \in \mathbb{N}$, this subgroup corresponds to the Baumslag–Solitar group [3]:

$$BS(1, m) := \langle \, e_1, e_2 \ \mid \ e_1 e_2 (e_1)^{-1} = (e_2)^m \, \rangle.$$

We have seen before that the star-exponential associated to a coadjoint orbit $\mathcal{O}$ is a group morphism $\mathcal{E} : G \to M_{\star_\theta}(\mathcal{O}) \simeq M_{\star_\theta}(G)$. Composed with the quantization map Ω, it coincides with the unitary representation $U = \Omega \circ \mathcal{E}$. So, if we now take the subalgebra of $M_{\star_\theta}(G)$ generated by the star-exponential of G_{BS}, i.e., by elements $\{ \mathcal{E}_{e^{\theta X}}, \ X \in \mathfrak{B} \}$, then it is closed for the complex conjugation and it can be completed into a C*-algebra $\mathbf{A}_G$ with norm $\| \Omega(\cdot) \|_{\mathcal{L}(\mathcal{H})}$. This C*-algebra is canonically associated to the group G. Moreover, if $\theta \to 0$, this C*-algebra is commutative and corresponds thus to a certain torus.

Definition 5.9. Let G be a normal j-group. We define the *noncommutative Baumslag–Solitar torus* of G to be the C*-algebra $\mathbf{A}_G$ constructed above.

It turns out that the relation between the generators $\mathcal{E}_{e^{\theta X}}$ ($X \in \mathfrak{B}$) of $\mathbf{A}_G$ can be computed explicitly by using the BCH formula of Proposition 3.10. Let us see some examples.

Example 5.10. In the elementary group case $G = \mathbb{S}$, let

$$U(a, x, \ell) := \mathcal{E}_{(\theta,0,0)}(a, x, \ell) = \sqrt{\cosh(\theta)} \cosh(\tfrac{\theta}{2})^n e^{\frac{2i}{\theta} \sinh(\theta)\ell},$$

$$V(a, x, \ell) := \mathcal{E}_{(0,0,\theta)}(a, x, \ell) = e^{i e^{-2a}},$$

$$W_i(a, x, \ell) := \mathcal{E}_{(0,\theta e_i,0)}(a, x, \ell) = e^{i e^{-a} \omega_0(e_i, x)},$$

where (e_i) is a canonical basis of the symplectic space (V, ω_0) of dimension $2n$ (i.e. $\omega_0(e_i, e_{i+n}) = 1$ if $i \le n$). Then, we can compute relations like

$$U \star_\theta V = V^{e^{2\theta}} \star_\theta U.$$

by using the BCH property of the star-exponential (see Proposition 3.10). We obtain (by omitting the notation $\star$):

$$UV = V^{e^{2\theta}}U \qquad (\text{ and } UV^{\beta} = V^{\beta e^{2\theta}}U),$$

$$UW_i = W_i^{e^{\theta}}U, \qquad W_i W_{i+n} = V^{\theta} W_{i+n} W_i$$

where the other commutation relations are trivial. Note that these relations become trivial at the commutative limit $\theta \to 0$. In the two-dimensional case, where $\mathbb{S}$ is the "$ax+b$ group", the relation $UV = V^{e^{2\theta}}U$ has already been obtained in another way in [14].

Example 5.11. Let us consider the Siegel domain of dimension 6 (see Example 2.12 for definitions and notations). As before, we can define the following generators:

$$U(g) := \mathcal{E}_{(0,0,\theta,0,0,0)}(g) = \sqrt{\cosh(\theta)}\cosh(\frac{\theta}{2})e^{\frac{2i}{\theta}\sinh(\theta)\ell_2},$$

$$V(g) := \mathcal{E}_{(0,0,0,0,0,\theta)}(g) = e^{ie^{-2a_2}},$$

$$W_1(g) := \mathcal{E}_{(0,0,0,\theta,0,0)}(g) = e^{ie^{-a_2}w_2},$$

$$W_2(g) := \mathcal{E}_{(0,0,0,0,\theta,0)}(g) = e^{-ie^{-a_2}v_2},$$

$$R(g) := \mathcal{E}_{(\theta,0,0,0,0,0)}(g) = \frac{e^{\frac{\theta}{2}}\sqrt{\cosh(\theta)}}{\cosh(\frac{\theta}{2})}e^{\frac{2i}{\theta}(\sinh(\theta)\ell_1 + \tanh(\frac{\theta}{2})v_2w_2)},$$

$$S(g) := \mathcal{E}_{(0,\theta,0,0,0,0)}(g) = e^{i(e^{-2a_1}+\frac{1}{2}v_2^2)}.$$

We obtain the relationship:

$$UV = V^{e^{2\theta}}U, \quad UW_1 = (W_1)^{e^{\theta}}U, \quad UW_2 = (W_2)^{e^{\theta}}U, \quad W_1W_2 = V^{\theta}W_2W_1,$$

$$RS = S^{e^{2\theta}}R, \quad RW_1 = (W_1)^{e^{\theta}}R, \quad RW_2 = (W_1)^{e^{-\theta}}R, \quad SW_1 = V^{\frac{\theta^2}{2}}(W_2)^{\theta}W_1S,$$

where the other commutation relations are trivial.

Acknowledgements One of us, Pierre Bieliavsky, spent the academic year 1995–1996 at UC Berkeley as a post-doc in the group of Professor Joseph A. Wolf. It is a great pleasure for Pierre Bieliavsky to warmly thank Professor Wolf for his support not only when a young post-doc but constantly during Pierre Bieliavsky's career. The research presented in this note is closely related to the talk Pierre Bieliavsky gave at the West Coast Lie Theory Seminar in November 1995 when studying some early stage features of the non-formal $\star$-exponential [4].

References

1. D. Arnal, The *-exponential, *in Quantum Theories and Geometry, Math. Phys. Studies* **10** (1988), 23–52.
2. D. Arnal and J.-C. Cortet, Représentations? des groupes exponentiels, *J. Funct. Anal.* **92** (1990) 103–175.
3. G. Baumslag and D. Solitar, Some two-generator one-relator non-Hopfian groups, *Bull. Amer. Math. Soc.* **68** (1962) 199–201.
4. P. Bieliavsky, *Symmetric spaces and star-representations*, talk given at the *West Coast Lie Theory Seminar*, U.C. Berkeley, November 1995.
5. P. Bieliavsky, Strict Quantization of Solvable Symmetric Spaces, *J. Sympl. Geom.* **1** (2002) 269–320.
6. P. Bieliavsky, Non-formal deformation quantizations of solvable Ricci-type symplectic symmetric spaces, Proceedings of the workshop Non-Commutative Geometry and Physics (Orsay 2007), *J. Phys.: Conf. Ser.* **103** (2008).
7. P. Bieliavsky and V. Gayral, Deformation Quantization for Actions of Kahlerian Lie Groups, *to appear in Mem. Amer. Math. Soc.* (2013), arXiv:1109.3419[math.OA].
8. P. Bieliavsky, A. de Goursac and G. Tuynman, Deformation quantization for Heisenberg supergroup, *J. Funct. Anal.* **263** (2012), 549–603.
9. P. Bieliavsky and M. Massar,Oscillatory integral formulae for left-invariant star products on a class of Lie groups, *Lett. Math. Phys.* **58** (2001), 115–128.
10. M. Duflo and C. C. Moore, On the regular representation of a nonunimodular locally compact group, *J. Funct. Anal.* **21** (1976), 209–243.
11. C. Fronsdal, Some ideas about quantization, *Rep. Math. Phys.* **15** (1979), 111–175.
12. V. Gayral, J. M. Gracia-Bondia and J. C. Varilly, Fourier analysis on the affine group, quantization and noncompact Connes geometries, *J. Noncommut. Geom.* **2** (2008), 215–261.
13. L. Hormander, The Weyl calculus of pseudo-differential operators, *Comm. Pure Appl. Math.* **32** (1979), 359.
14. B. Iochum, T. Masson, T. Schucker and A. Sitarz, Compact θ-deformation and spectral triples, *Rep. Math. Phys.* **68** (2011) 37–64.
15. O. Loos, *Symmetric spaces I: General Theory.* Benjamin, 1969.
16. I. I. Pyatetskii-Shapiro, *Automorphic functions and the geometry of classical domains.* Mathematics and Its Applications, Vol. 8, Gordon and Breach Science Publishers, New York, 1969.
17. M. A. Rieffel, Deformation Quantization for actions of $\mathbb{R}^d$, *Mem. Amer. Math. Soc.* **106** (1993) R6.
18. F. Treves, *Topological vector spaces, distributions and kernels.* Academic Press, 1967.

The Radon Transform and Its Dual
for Limits of Symmetric Spaces

Joachim Hilgert and Gestur Ólafsson

Abstract The Radon transform and its dual are central objects in geometric analysis on Riemannian symmetric spaces of the noncompact type. In this article we study algebraic versions of those transforms on inductive limits of symmetric spaces. In particular, we show that normalized versions exists on some spaces of regular functions on the limit. We give a formula for the normalized transform using integral kernels and relate them to limits of double fibration transforms on spheres.

Key words Symmetric spaces • Horospherical spaces • Radon transform • Dual Radon transform • Infinite dimensional analysis • Spherical and conical representations

Mathematics Subject Classification (2010): 43A85, 17B65, 47A67.

1 Introduction

Let G_o be a classical noncompact connected semisimple Lie group and G its complexification. We fix a Cartan involution $\theta : G_o \to G_o$ on G_o and denote the holomorphic extension to G by the same letter. Let $K = G^\theta$ and let $K_o = K \cap G_o$ be the maximal compact subgroup corresponding to θ. Then $\mathbf{X} = G_o/K_o$ is a Riemannian symmetric space of the noncompact type. The space $\mathbf{X}$ is contained in its complexification $\mathbf{Z} = G/K$. The subscript $_o$ will be used to denote subgroups in G_o. Dropping the index will then stand for the corresponding complexification in G.

The research of G. Ólafsson was supported by NSF grant DMS-1101337.

J. Hilgert
Institut für Mathematik, Universität Paderborn, 33098 Paderborn, Germany
e-mail: hilgert@upb.de

G. Ólafsson (✉)
Department of Mathematics, Louisiana State University, Baton Rouge, LA 70803, USA
e-mail: olafsson@math.lsu.edu

© Springer International Publishing Switzerland 2014
G. Mason et al. (eds.), *Developments and Retrospectives in Lie Theory*,
Developments in Mathematics 37, DOI 10.1007/978-3-319-09934-7_3

Let $P_o = M_o A_o N_o$ be a minimal parabolic subgroup of G_o with $A_o \subset \{a \in G_o \mid \theta(a) = a^{-1}\}$ and $M_o = Z_{K_o}(A_o)$. The space $\Xi_o = G_o / M_o N_o$ is the space of horocycles in $\mathbf{X}$. We denote the base point in $\mathbf{X}$ by $x_o = \{K_o\}$ and the base point $\{M_o N_o\}$ in Ξ_o by ξ_o. The (horospherical) Radon transform is the integral transform, initially defined on compactly supported functions on $\mathbf{X}$, given by

$$\mathcal{R}(f)(g \cdot \xi_o) = \int_{N_o} f(gn \cdot x_o) \, dn$$

for a certain normalization of the invariant measure dn on N_o. The dual transform $\mathcal{R}^*$ maps continuous functions on Ξ_o to continuous functions on $\mathbf{X}$ and is given by

$$\mathcal{R}^*(\varphi)(g \cdot x_o) = \int_{K_o} \varphi(gk \cdot \xi_o) \, dk,$$

where dk denotes the invariant probability measure on K_o. If f and φ are compactly supported, then

$$\int_{\Xi} \mathcal{R}(f)(\xi)\varphi(\xi) \, d\xi = \int_{\mathbf{X}} f(x)\mathcal{R}^*(\varphi)(x) \, dx$$

for suitable normalizations of the invariant measures on $\mathbf{X}$, respectively Ξ_o. This explains why $\mathcal{R}^*$ is called the dual Radon transform. For more detailed discussion we refer to Sect. 5.2.

For a complex subgroup $L \subset G$ we call a holomorphic function $f : G/L \to \mathbb{C}$ regular if the orbit $G \cdot f$ with respect to the natural representation spans a finite-dimensional subspace. We denote the G-space of regular functions by $\mathbb{C}[G/L]$. If $L_o \subset G_o$ is a subgroup such that G_o / L_o can be viewed as a real subspace of its complexification G/L, then one calls a smooth function on G_o / L_o regular, if its G_o-orbit spans a finite-dimensional space. Since there is a bijection between regular functions on G_o / L_o and G/L we restrict our attention to $\mathbb{C}[G/L]$. The dual Radon transform can be extended to the space of regular functions on Ξ but the integral defining the Radon transform is in general not defined for regular functions. In fact, a regular function on $\mathbf{Z}$ can be N_o-invariant so the integral is infinite. This problem was first discussed in [HPV02] and then further developed in [HPV03]. Let us describe the main idea from [HPV02] here. We refer to the main body of the article for more details.

Denote the spherical representation of G_o and G with highest weight $\mu \in \mathfrak{a}_o^*$ by (μ_μ, V_μ), and its dual by (π_μ^*, V_μ^*). The duality is written $\langle w, v \rangle$. Note that (π_μ, V_μ) is unitary on a compact real form U which we choose so that $U \cap G_o = K_o$. We fix a highest weight vector $u_\mu \in V_\mu$ of length one and a K-fixed vector $e_\mu^* \in V_\mu^*$ such that $\langle u_\mu, e_\mu^* \rangle = 1$. Fix a highest weight vector u_μ^* in V_μ^* such that $\langle u_\mu, \pi_\mu^*(s_o)u_\mu^* \rangle = 1$, where $s_o \in K$ represents the longest Weyl group element. For $w \in V_\mu$ and $g \in G$ let

$$f_{w,\mu}(g \cdot x_o) := \langle w, \pi_\mu^*(g)e_\mu^* \rangle \quad \text{and} \quad \psi_{w,\mu}(g \cdot \xi_o) := \langle w, \pi_\mu^*(g)u_\mu^* \rangle.$$

Every regular function on $\mathbf{X}$ is a finite linear combination of functions of the form $f_{w,\mu}$ and similarly for Ξ. The *normalized Radon transform* on the space $\mathbb{C}[\mathbf{Z}]$ of regular functions on $\mathbf{Z}$ can now be defined by

$$\Gamma(f_{w,\mu}) := \psi_{w,\mu}.$$

The transform

$$\Gamma^{-1}(\psi_{w,\mu}) := f_{w,\mu}$$

defines a G-equivariant map $\mathbb{C}[\Xi] \to \mathbb{C}[\mathbf{X}]$ which is inverse to Γ. Restricted to each G-type, the transform Γ^{-1} is, up to a normalization given in Lemma 5.5, the dual Radon transform $\mathcal{R}^*$. It is also shown in [HPV03] that the dual Radon transform on $\mathbb{C}[\Xi]$ can be described as a limit of the Radon transform over spheres, see Sect. 5.3 for details.

Our aim in this article is to study the normalized transforms Γ and Γ^{-1} as well as their not normalized counterparts for certain inductive limits of symmetric spaces $\mathbf{X}_j \subset \mathbf{Z}_j$, called propagations of symmetric spaces, introduced in Sect. 3.1. This study is based on results from [ÓW11a, ÓW11b] and [DÓW12] on inductive limits of spherical representations, which we use to study spaces of regular functions on the limit. More precisely, in Sect. 4 we consider two such spaces of regular functions, the projective limit $\varprojlim \mathbb{C}[\mathbf{Z}_j]$ and the inductive limit $\mathbb{C}_i[\mathbf{Z}_\infty] = \varinjlim \mathbb{C}[\mathbf{Z}_j]$. The first main result is Theorem 4.19 which describes how the graded version of Γ extends to the projective limit.

We introduce the Radon transform and its dual in Sect. 5 and in Sect. 5.3 we recall the results from [HPV03] about the Radon transform as a limit of a double fibration transform associated with the spheres in $\mathbf{X}$. In Sect. 5.4 we show that the normalized Radon transform and its dual can be represented as an integral transform against kernel functions. Here the integral is taken over the compact group U. The corresponding result for the direct limit is Theorem 5.16.

Many of the results mentioned so far are valid for propagations of symmetric spaces of arbitrary rank, which means that they apply also to the case of infinite rank. For some results, however, we have to require that the rank of the symmetric space $\varinjlim X_j$ is finite. This is the case in particular in Sect. 5.5, where we define the dual Radon transform $\mathcal{R}^*$ for spaces of finite rank and connect it to the normalized dual Radon transform Γ^{-1}, see Theorem 5.22. Moreover, we define the Radon transform over spheres in this context, and connect it to the dual Radon transform in Theorem 5.25.

2 Finite-Dimensional Geometry

In this section we recall the necessary background from the structure theory of finite-dimensional symmetric spaces and related representation theory. Most of the material is in this section is standard see [H78, H84], but we use this section also to set up the notation for later sections.

2.1   Lie Groups and Symmetric Spaces

Lie group will always be denoted by uppercase Latin letters and their Lie algebra will be denoted by the corresponding lower case German letters. If G and H are Lie groups and $\theta : G \to H$ is a homomorphism, then the derived homomorphism is denoted by $\dot{\theta} : \mathfrak{g} \to \mathfrak{h}$. If $G = H$, then

$$G^\theta = \{a \in G \mid \theta(a) = a\} \quad \text{and} \quad \mathfrak{g}^{\dot{\theta}} = \{X \in \mathfrak{g} \mid \dot{\theta}(X) = X\}.$$

From now on G will stand for a connected simply connected complex semisimple Lie group with Lie algebra $\mathfrak{g}$. Let U be a compact real form of G with Lie algebra $\mathfrak{u}$ and let $\dot{\sigma} : \mathfrak{g} \to \mathfrak{g}$ denote the conjugation on $\mathfrak{g}$ with respect to $\mathfrak{u}$. We denote by $\sigma : G \to G$ the corresponding involution on G. Then, as G is simply connected, $U = G^\sigma$ and U is simply connected.

Let $\theta : U \to U$ be a nontrivial involution and $K_o := U^\theta$. Then K_o is connected and U/K_o is a simply connected symmetric space of the compact type. Extend $\dot{\theta} : \mathfrak{u} \to \mathfrak{u}$ to a complex linear involution, also denoted by $\dot{\theta}$, on $\mathfrak{g}$. Denote by $\theta : G \to G$ the holomorphic involution with derivative $\dot{\theta}$. Write $\mathfrak{u} = \mathfrak{k}_o \oplus \mathfrak{q}_o$ where $\mathfrak{k}_o := \mathfrak{u}^{\dot{\theta}}$ and $\mathfrak{q}_o := \{X \in \mathfrak{u} \mid \dot{\theta}(X) = -X\}$. Let $\mathfrak{s}_o := i\mathfrak{q}_o$ and $\mathfrak{g}_o := \mathfrak{k}_o \oplus \mathfrak{s}_o$. Then $\mathfrak{g}_o$ is a semisimple real Lie algebra. Denote by G_o the analytic subgroup of G with Lie algebra $\mathfrak{g}_o$. Then G_o is θ-stable, $G_o^\theta = K_o$, and G_o/K_o is a symmetric space of the noncompact type. We have $G_o = G^{\theta\sigma}$.

Let $K := G^\theta$. Then $K_o = K \cap U = K \cap G_o$. Let $\mathbf{Z} = G/K$, $\mathbf{X} = G_o/K_o$, and $\mathbf{Y} = U/K_o$. As σ and $\eta := \sigma\theta$ map K into itself, it follows that both involutions define antiholomorphic involutions on $\mathbf{Z}$ and we have

$$\mathbf{X} = \mathbf{Z}^\eta \quad \text{and} \quad \mathbf{Y} = \mathbf{Z}^\sigma.$$

In particular $\mathbf{X}$ and $\mathbf{Y}$ are transversal totally real submanifolds of $\mathbf{Z}$.

If V is a vector space over a field $\mathbb{K}$, then V^* denotes the algebraic dual of V. If V is a topological vector space, then the same notation will be used for the continuous linear forms. If V is finite-dimensional, then each $\alpha \in V^*$ is continuous.

Let $\mathfrak{a}_o$ be a maximal abelian subspace of $\mathfrak{s}_o$ and $\mathfrak{a} = \mathfrak{a}_o^{\mathbb{C}}$. For $\alpha \in \mathfrak{a}_o^* \subset \mathfrak{a}^*$ let

$$\mathfrak{g}_{o\alpha} := \{X \in \mathfrak{g}_o \mid (\forall H \in \mathfrak{a}_o) \, [H, X] = \alpha(H)X\}$$

and

$$\mathfrak{g}_\alpha := \{X \in \mathfrak{g} \mid (\forall H \in \mathfrak{a}) \, [H, X] = \alpha(H)X\}.$$

If $\alpha \neq 0$ and $\mathfrak{g}_\alpha \neq \{0\}$, then $\mathfrak{g}_\alpha = \mathfrak{g}_{o\alpha} \oplus i\mathfrak{g}_{o\alpha}$ and $\mathfrak{g}_\alpha \cap \mathfrak{u} = \{0\}$. The linear form $\alpha \in \mathfrak{a}^* \setminus \{0\}$ is called a (restricted) root if $\mathfrak{g}_\alpha \neq \{0\}$. Denote by $\Sigma := \Sigma(\mathfrak{g}, \mathfrak{a}) \subset \mathfrak{a}^*$ the set of roots. Let $\Sigma_0 := \Sigma_0(\mathfrak{g}, \mathfrak{a}) := \{\alpha \in \Sigma \mid 2\alpha \notin \Sigma\}$, the set of nonmultipliable roots. Then Σ_0 is a root system in the usual sense and the

Weyl group W corresponding to Σ is the same as the Weyl group generated by the reflections s_α, $\alpha \in \Sigma_0$. The Riemannian symmetric spaces $\mathbf{X}$ and $\mathbf{Y}$ are irreducible if and only if the root system Σ_0 is irreducible.

Let $\mathfrak{n} := \bigoplus_{\alpha \in \Sigma^+} \mathfrak{g}_\alpha$, $\mathfrak{m} := \mathfrak{z}_\mathfrak{k}(\mathfrak{a}) = \{X \in \mathfrak{k} \mid [X, \mathfrak{a}] = \{0\}\}$ and $\mathfrak{p} := \mathfrak{m} \oplus \mathfrak{a} \oplus \mathfrak{n}$. All of those algebras are defined over $\mathbb{R}$ and the subscript $_o$ will indicate the intersection of those algebras with $\mathfrak{g}_o$. This intersection can also be described as the η fixed points in the complex Lie algebra.

Define the parabolic subgroups $P_o := N_{G_o}(\mathfrak{p}_o) \subset P := N_G(\mathfrak{p})$. We can write $P_o = M_o A_o N_o$ (semidirect product) where $M_o := Z_K(A_o)$, $A_o := \exp \mathfrak{a}_o$, and $N_o := \exp \mathfrak{n}_o$. Similarly we have $P = MAN$. Let $F := K \cap \exp i\mathfrak{a}_o$. Then each element of F has order two and $M_o = F(M_o)^\circ$ where $^\circ$ denotes the connected component containing the identity element. We let $\Xi_o := G_o/M_o N_o \subset \Xi := G/MN$. As $\theta\sigma$ leaves MN invariant it follows that $\Xi_o = \Xi^{\theta\sigma}$ is a totally real submanifold of Ξ.

Note then $K \cap MN = M$, so we obtain the following double fibration, which is of crucial importance for the Radon transforms:

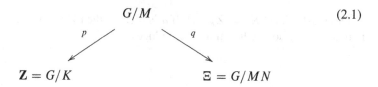

$$G/M \qquad (2.1)$$

$$\mathbf{Z} = G/K \qquad\qquad \Xi = G/MN$$

2.2 Group Spheres

Let $\mathbf{X} = G_o/K_o$ be as in the previous section. Denote by s the symmetry of $\mathbf{X}$ with respect to x_o. Then $\theta(g) = sgs^{-1}$ for $g \in G_o$. Denote by $X(A)$ the (additively written) group $\mathrm{Hom}(A_o, \mathbb{R}_+^\times)$ (where $\mathbb{R}_+^\times$ stands for the multiplicative group of positive numbers). Then $X(A) \simeq \mathfrak{a}^*$ where the isomorphism is given by $\mu \mapsto \chi_\mu$, $\chi_\mu(a) = a^\mu$. We will simply write $\mu(a)$ for $\chi_\mu(a)$.

A *group sphere* in $\mathbf{X}$ is an orbit of a maximal compact subgroup of G_o. Because of the Cartan decomposition $G_o = K_o A_o K_o$, and the fact that all maximal compact subgroups in G_o are G_o-conjugate, any group sphere S is of the form $gK_o a^{-1} \cdot x_o = K_o^g ga^{-1} \cdot x_o$ with $g \in G_o$ and $a \in A_o$. The point $g \cdot x_o$ is called the *center* of S and a is called the *radius* of S. The group sphere of radius a with center at x is denoted by $S_a(x)$.

The group G_o acts transitively on the set $\mathrm{Sph}_a \mathbf{X}$ of group spheres of radius a. The stabilizer of a group sphere $S := S_a(g \cdot x_o)$ is a compact subgroup of G containing the stabilizer K_o^g of the point $g \cdot x_o$ and hence coinciding with it. Since $g \cdot x_o$ is the only fixed point of K_o^g, it is uniquely determined by S. Moreover, $S_{a_1}(x) = S_{a_2}(x)$ if and only if a_1 and a_2 are W-equivalent.

We shall say that $a \in A_o$ tends to infinity, written $a \to \infty$, if $\alpha(a) \to \infty$ for any $\alpha \in \Sigma^+$.

The sphere

$$S_a := S_a(a \cdot x_o) = K^a \cdot x_o$$

of radius a passes through x_o. It is known that it converges to the horosphere $\xi_o = N_o \cdot o$ as $a \to \infty$ (see, e.g., [E73], Proposition 2.6, and [E96], p. 46).

Taking the sphere S_a as the base point for the homogeneous space $\mathrm{Sph}_a \mathbf{X}$, we obtain the representation $\mathrm{Sph}_a \mathbf{X} = G_o/K_o^a$. This gives rise to the double fibration

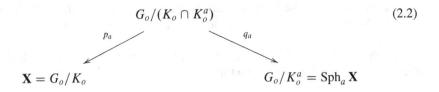

$$(2.2)$$

Obviously, $K_o \cap K_o^a = Z_{K_o}(a)$. If a is regular, then $Z_{K_o}(a) = Z_{K_o}(A_o) = M_o$. In this case the double fibration (2.2) reduces to

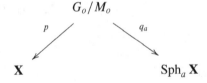

2.3 Spherical Representations

In this subsection we describe the set of spherical representations and the set of fundamental weights. Each irreducible finite-dimensional representation π of U or G_o extends uniquely to a holomorphic irreducible representation π of G and every irreducible holomorphic representation τ of G is a holomorphic extension of an irreducible representation of U and G_o. We will therefore concentrate on irreducible holomorphic representations of G. We will denote by π^U, respectively π^o, the restriction of a holomorphic representation π to U respectively G_o.

For a representation π of a topological group H or a Lie algebra $\mathfrak{h}$ we write V_π for the vector space on which π acts. Let

$$V_\pi^H = \{u \in V_\pi \mid (\forall k \in H) \, \pi(k)u = u\}.$$

Similarly,

$$V_\pi^\mathfrak{h} = \{u \in V_\pi \mid (\forall X \in \mathfrak{h})\ Xu = 0\}.$$

If H is a connected Lie group with Lie algebra $\mathfrak{h}$ and V_π a smooth representation of H, then $\mathfrak{h}$ acts on V_π and $V_\pi^H = V_\pi^\mathfrak{h}$.

Back to our setup, as K_o and K are connected, it follows that if π is a irreducible finite-dimensional holomorphic representation of G (and hence analytic), then

$$V_\pi^K := V_\pi^{K_o} = V_\pi^\mathfrak{k} = V_\pi^{\mathfrak{k}_o}.$$

We say that π is spherical if $V_\pi^K \neq \{0\}$ and that π is conical if $V_\pi^{MN} \neq \{0\}$. Note that even if M_o is not connected, then

$$V_\pi^{MN} = V_\pi^{M_o N_o}.$$

In fact, the inclusion $\subseteq$ is trivial and for the converse it suffices to note that $V_\pi^{M_o N_o} \subseteq V_\pi^{\mathfrak{m}_o + \mathfrak{n}_o} =: V_\pi^{\mathfrak{m}+\mathfrak{n}}$.

Define a representation π^* on $V_\pi^* = V_{\pi*}$ by

$$\langle v, \pi^*(g)v \rangle := \langle \pi(g^{-1})v, v \rangle, \quad g \in G,\ v \in V_\pi,\ v \in V_\pi^*.$$

For the following theorem see [H94], Thm. 4.12 and [H84], Thm. V.1.3 and Thm. V.4.1.

Theorem 2.3. *Let π be an irreducible holomorphic representation of G. Then the following holds:*

(1) *π is spherical if and only if π is conical. In that case*

$$\dim V_\pi^K = \dim V_\pi^{MN} = 1.$$

(2) *π is spherical if and only if π^* is spherical.*

Let

$$\Lambda^+(G, K) := \left\{ \mu \in i\mathfrak{a}_o^* \left| \frac{(\mu, \alpha)}{(\alpha, \alpha)} \in \mathbb{Z}^+ \text{ for all } \alpha \in \Sigma^+ \right. \right\} \tag{2.4}$$

$$= \left\{ \mu \in i\mathfrak{a}_o^* \left| \frac{(\mu, \alpha)}{(\alpha, \alpha)} \in \mathbb{Z}^+ \text{ for all } \alpha \in \Sigma_0^+ \right. \right\}.$$

We mostly write Λ^+ for $\Lambda^+(G, K)$. Let $W = N_{K_o}(\mathfrak{a}_o)/Z_{K_o}(\mathfrak{a}_o)$ denote the Weyl group. The parametrization of the spherical representations is given by the following theorem.

Theorem 2.5. *Let π be a irreducible holomorphic representation of G, and μ its highest weight. Let $w_o \in W$ be such that $w_o \Sigma^+ = -\Sigma^+$. Then the following are equivalent.*

(1) *π is spherical.*
(2) *$\mu \in i\mathfrak{a}_o^*$ and $\mu \in \Lambda^+$.*

Furthermore, if π is spherical with highest weight $\mu \in \Lambda^+$, then π^* has highest weight $\mu^* := -w_o\mu$.

Proof. See [H84, Theorem 4.1, p. 535 and Exer. V.10] for the proof. □

If $\mu \in \Lambda^+$, then π_μ denotes the irreducible spherical representation with highest weight μ.

Denote by $\Psi := \{\alpha_1, \ldots, \alpha_r\}$, $r := \dim_{\mathbb{C}} \mathfrak{a}$, the set of simple roots in Σ_0^+. Define linear functionals $\omega_j \in i\mathfrak{a}_o^*$ by

$$\frac{\langle \omega_i, \alpha_j \rangle}{\langle \alpha_j, \alpha_j \rangle} = \delta_{i,j} \quad \text{for} \quad 1 \leq j \leq r \ . \tag{2.6}$$

Then $\omega_1, \ldots, \omega_r \in \Lambda^+$ and

$$\Lambda^+ = \mathbb{Z}^+\omega_1 + \ldots + \mathbb{Z}^+\omega_r = \left\{ \sum_{j=1}^r n_j \omega_j \ \middle| \ n_j \in \mathbb{Z}^+ \right\} \ .$$

The weights ω_j are called the *spherical fundamental weights for* $(\mathfrak{g}, \mathfrak{k})$. Set $\Omega := \{\omega_1, \ldots, \omega_r\}$.

2.4 Regular Functions

Let L be one of the groups U, G_o and G. Let **M** be a manifold and assume that L acts transitively on **M**. Then L acts on functions on **M** by $a \cdot f(m) = f(a^{-1} \cdot m)$. We say that $f \in C(\mathbf{M})$ is an *L-regular function* if $\{a \cdot f \mid a \in L\}$ spans a finite-dimensional space which we will denote by $\langle L \cdot f \rangle$. We denote by $\mathbb{C}_L[\mathbf{M}]$ the space of L-regular functions on **M**. Coming back to our usual notation we remark that the restriction map defines a G_o-isomorphism $\mathbb{C}_G[\mathbf{Z}] \to \mathbb{C}_{G_o}[\mathbf{X}]$ and a U-isomorphism $\mathbb{C}_G[\mathbf{Z}] \to \mathbb{C}_U[\mathbf{Y}]$. Similarly, restriction defines a G_o-isomorphism $\mathbb{C}_G[\Xi] \to \mathbb{C}_{G_o}[\Xi_o]$. As soon as the acting group is clear from the context we will omit it from the notation.

We will mostly consider regular functions on the two complex spaces **Z** and Ξ. If needed, we will use results only stated or proved for the complex case also for the real cases using the above restriction maps.

For $\mu \in \Lambda^+$ we denote by $\mathbb{C}[\mathbf{Z}]_\mu$, respectively $\mathbb{C}[\Xi]_\mu$, the space of regular functions on **Z**, respectively Ξ, of type π_μ. We recall the following well-known fact (cf. [HPV02]):

Lemma 2.7. *The action of G on $\mathbb{C}[\mathbf{X}]_\mu$ and $\mathbb{C}[\Xi]_\mu$ is irreducible. As a G-module we have*

$$\mathbb{C}[\mathbf{Z}] = \bigoplus_{\mu \in \Lambda^+} \mathbb{C}[\mathbf{Z}]_\mu \quad and \quad \mathbb{C}[\Xi] = \bigoplus_{\mu \in \Lambda^+} \mathbb{C}[\Xi]_\mu .$$

Each representation π_μ occurs with multiplicity one in each of those modules.

Let $f \in \mathbb{C}[\mathbf{X}]_\mu$ be a highest weight vector. Recall that $KAN \subset G$ is open and dense. Let $kan \in KAN$. Then

$$(kan) \cdot f(x_o) = f(n^{-1}a^{-1}k^{-1} \cdot x_o) = a^\mu f(x_o)$$

where $a^\mu = e^{\mu(\log(a))}$. Hence $f(x_o) \neq 0$. We denote by f_μ the unique highest weight vector in $\mathbb{C}[\mathbf{X}]_\mu$ with $f_\mu(e) = 1$.

Let $s_o \in N_{K_o}(\mathfrak{a}_o)$ be a representative of the longest Weyl group element w_o and recall that $N s_o MAN$ is open and dense in G. Let $\psi \in \mathbb{C}[\Xi]_\mu$ be a highest weight vector. Then for $n s_o m a n_1 \in N s_o MAN$ we have

$$(n s_o m a n_1) \cdot \psi(\xi_o) = \psi(n_1^{-1} a^{-1} m^{-1} s_o^{-1} n^{-1} \cdot \xi_o) = a^\mu \psi(s_o^{-1} \cdot \xi_o) .$$

Hence $\psi(s_o^{-1} \cdot \xi_o) \neq 0$. Note that $s_o^2 \in M$. As ψ is M-invariant it follows that $\psi(s_o^{-1} \cdot \xi_o) = \psi(s_o \cdot \xi)$. Let ψ_μ be the unique highest weight vector in $\mathbb{C}[\Xi]_\mu$ with $\psi_\mu(s_o \cdot \xi_o) = 1$. According to Lemma 2.7 there is a unique G-intertwining operator $\Gamma : \mathbb{C}[\mathbf{Z}] \to \mathbb{C}[\Xi]$ such that $\Gamma(f_\mu) = \psi_\mu$ for all $\mu \in \Lambda^+$. For reasons which will become clear in Sect. 5, we call Γ the *normalized Radon transform*. Note that its inverse $\Gamma^{-1} : \mathbb{C}[\Xi] \to \mathbb{C}[\mathbf{Z}]$ is the unique G-isomorphism such that $\psi_\mu \mapsto f_\mu$ for all $\mu \in \Lambda^+$. Let $\Gamma_\mu = \Gamma|_{\mathbb{C}[\mathbf{Z}]_\mu}$. Then $\Gamma_\mu^{-1} = \Gamma^{-1}|_{\mathbb{C}[\Xi]_\mu}$.

The maps Γ_μ and Γ_μ^{-1} have a simple description in terms of the representation (π_μ, V_μ). Fix for all $\nu \in \Lambda^+$ a K-fixed vector $e_\nu \in V_\nu$ and a highest weight vector u_ν in V_ν. Further, choose the highest weight vector $u_\nu^* \in V_\nu^*$ and the spherical vector $e_\mu^* \in V_\nu^*$ according to the normalization

$$\langle u_\nu, \pi_\nu^*(s_o) u_\nu^* \rangle = 1 \quad and \quad \langle u_\mu, e_\mu^* \rangle = 1 .$$

Then, for $v \in V_\mu$

$$f_{v,\mu}(aK) := \langle v, \pi_\mu^*(a) e_\mu^* \rangle \quad and \quad \psi_{v,\mu}(aMN) := \langle v, \pi_\mu^*(a) u_\mu^* \rangle, \qquad (2.8)$$

defines a regular function $f_{v,\mu}$ on $\mathbf{Z}$, respectively $\psi_{v,\mu}$ on Ξ. Furthermore,

$$V_\mu \ni v \mapsto f_{v,\mu} \in \mathbb{C}[\mathbf{Z}]_\mu \quad and \quad V_\mu \ni v \mapsto \psi_{v,\mu} \in \mathbb{C}[\Xi]_\mu$$

are G-isomorphisms. Note that $f_\mu := f_{u_\mu,\mu}$ respectively $\psi_\mu := \psi_{u_\mu,\mu}$ are normalized highest weight vectors.

3   Limits of Symmetric Spaces and Spherical Representations

In this section we introduce the notion of propagation of symmetric spaces and describe the construction of inductive limits of spherical representations from [ÓW11a, ÓW11b]. We then recall the main result from [DÓW12] about the classification of spherical representations in the case where $U_\infty/K_{o\infty}$ has finite rank.

We start with some facts and notations for limits of topological vector spaces, which will always be assumed to be complex, locally convex and Hausdorff. Similar notations for limits will be used for Lie groups and even sets without further comments. Our standard reference is Appendix B in [HY00] and the reference therein.

If $W_1 \subset W_2 \subset \cdots$ is an injective sequence of vector spaces, then we denote the inclusion maps $W_j \hookrightarrow W_k, k \geq j$, by $\iota_{k,j}$. Let

$$W_\infty := \varinjlim W_j = \bigcup_{j=1}^{\infty} W_j \qquad (3.1)$$

and denote by $\iota_{\infty,j}$ the canonical inclusion $W_j \hookrightarrow W_\infty$. If each of the spaces W_j is a topological vector space and each of the maps $\iota_{k,j}$ is continuous, then a set $U \subseteq W_\infty$ is open in the *inductive limit topology* on W_∞ if and only if $U \cap W_j$ is open for all j. Then W_∞ is a (again locally convex and Hausdorff) topological vector space. If $\{W_j\}$ and $\{V_j\}$ are inductive sequences of topological vector spaces and $T_j : W_j \to V_j$ is a family of continuous linear maps such that

$$\iota_{k,j} \circ T_j = T_k \circ \iota_{k,j}$$

where the first inclusion is the one related to the sequence $\{V_j\}$ and the second one is the one associated to $\{W_j\}$, then there exists a unique continuous linear map $T_\infty = \varinjlim T_j : W_\infty \to V_\infty$ such that $\iota_{\infty,j} \circ T_\infty = T_\infty \circ \iota_{\infty,j}$ for all j.

If W is a locally convex Hausdorff complex topological vector space, then W^* will denote the space of continuous linear maps $W \to \mathbb{C}$. We provide it with the weak $*$-topology, i.e., the weakest topology that makes all the maps $W^* \to \mathbb{C}, f \mapsto \langle x, f \rangle := f(x), x \in W$, continuous. Then W^* is also a locally convex Hausdorff topological vector space. If $\{W_j\}$ is a inductive sequence of locally convex Hausdorff topological vector spaces, then $\{W_j^*\}$, with the projections $\mathrm{proj}_{j,k} : W_k^* \to W_j^*, \mathrm{proj}_{j,k}(v) = v|_{W_j}, k \geq j$, is a projective sequence of locally convex Hausdorff topological vector spaces. Denote the projective limit of those spaces by $\varprojlim W_j^* = W_\infty^*$. This notation is justified by the fact that the topological dual of W_∞ is $\varprojlim W_j^*$. We denote by $\mathrm{proj}_{j,\infty} : W_\infty^* \to W_j$ the restriction map. If $\{W_j\}$ and $\{V_j\}$ are injective sequences of topological vector spaces and $T_j : W_j \to V_j$ is as above, then there exists a unique linear map $T_\infty^* = \varprojlim T_j^* : V_\infty^* \to W_\infty^*$ such that $\mathrm{proj}_{j,\infty} \circ T_\infty^* = T_j^* \circ \mathrm{proj}_{j,\infty}$ for all j. In fact, T_∞^* is just the adjoint of T_∞.

We finish the subsection with a simple lemma that connects the inductive limit and the projective limit in case we have a injective sequence of Lie groups G_j and G_j-modules V_j. This will be used several times later on. We leave the simple proof as an exercise for the reader.

Lemma 3.2. *Let $\{G_j\}$ be an injective sequence of Lie groups and let $\{V_j\}$ be a projective sequence of G_j-modules with G_j-equivariant projections $\mathrm{proj}_{j,k} : V_k \to V_j$. Assume that we have G_j-equivariant inclusions $\iota_{k,j} : V_j \to V_k$ making $\{V_j\}$ into an injective sequence and such that $\mathrm{proj}_{j,k} \circ \iota_{k,j} = \mathrm{id}_{V_j}$. For $f \in \varinjlim V_j$, fix j such that $f \in V_j$. Define $\iota_\infty(f) := \{\iota_{k+1,k}(f)\}_{k \geq j}$. Then $\varinjlim V_j$ and $\varprojlim V_j$ are G_∞-modules and ι_∞ is a well-defined G_∞-equivariant embedding $\varinjlim V_j \hookrightarrow \varprojlim V_j$.*

3.1 Propagation of Symmetric Spaces

Assume that $G_1 \subseteq G_2 \subseteq \ldots \subseteq G_k \subseteq G_{k+1} \subseteq \ldots$ is a sequence of connected, simply connected classical complex Lie groups as in the last section. In the following an index k (respectively j) will always indicate objects related to G_k (respectively G_j). We assume that $\theta_k|_{G_j} = \theta_j$ and $\sigma_k|_{G_j} = \sigma_j$ for all $j \leq k$. Then $K_j = G_j \cap K_k$, $U_j = G_j \cap U_k$, and $G_{jo} = G_j \cap G_{ko}$, for $k \geq j$.

This gives rise to an increasing sequence $\{\mathbf{Z}_j = G_j/K_j\}_{j \geq 1}$ of simply connected complex symmetric spaces such that for $k \geq j$ the embedding $\mathbf{Z}_j \hookrightarrow \mathbf{Z}_k$ is a G_j-map. We denote this inclusion by $\iota_{k,j}$ and note that $\{\mathbf{Z}_j\}$ is an injective system.

Similarly, we have a sequence of transversal real forms $\mathbf{X}_j = G_{jo}/K_{jo}$ and $\mathbf{Y}_j = U_j/K_{jo}$. We set

$$G_\infty := \varinjlim G_j, \quad K_\infty := \varinjlim K_j \quad \text{and} \quad \mathbf{Z}_\infty := \varinjlim \mathbf{Z}_j = G_\infty/K_\infty$$

and similarly for other groups and symmetric spaces. Recall that as a set we have $G_\infty = \bigcup G_j$, but the inductive limit comes also with the inductive limit topology and a Lie group structure. The space $\mathbf{Z}_\infty = \bigcup \mathbf{Z}_j$ is a smooth manifold and the action of G_∞ is smooth. Similar comments are valid for the other groups and the corresponding symmetric spaces.

In the following we will always assume that $k \geq j$ and $m \geq n$. As $\theta_k|_{G_j} = \theta_j$ it follows that $\mathfrak{k}_k \cap \mathfrak{g}_j = \mathfrak{k}_j$ and $\mathfrak{s}_k \cap \mathfrak{g}_j = \mathfrak{s}_j$. We choose the sequence $\{\mathfrak{a}_j\}$ of maximal abelian subspaces of $\mathfrak{s}_j$ such that $\mathfrak{a}_k \cap \mathfrak{s}_j = \mathfrak{a}_j$. Then $\Sigma_j \subseteq \Sigma_k|_{\mathfrak{a}_j} \setminus \{0\}$. The ordering in $i\mathfrak{a}_o^*$ is chosen so that $\Sigma_j^+ \subseteq \Sigma_k^+|_{\mathfrak{a}_{jo}} \setminus \{0\}$.

In case each $\mathbf{X}_j$ is irreducible we say that $\mathbf{X}_k$ *propagates* $\mathbf{X}_j$ if (a) $\mathfrak{a}_k = \mathfrak{a}_j$, or (b) the Dynkin diagram for Ψ_k is obtained from the Dynkin diagram for Ψ_j by only adding simple roots at the left end (so the root α_1 stays the same). Note that usually the Dynkin diagram is labeled so that the first simple root is at the *left* end. We have here reversed that labeling. Then, in particular, $\Psi_k = \{\alpha_{k,1}, \ldots, \alpha_{k,r_k}\}$

and $\Psi_j = \{\alpha_{j,1}, \ldots, \alpha_{j,r_j}\}$ are of the same type. Furthermore $\alpha_{k,s}|_{\mathfrak{a}_j} = \alpha_{j,s}$ for $s = 1, \ldots, r_j$, see [ÓW11a]. Furthermore, if $s \geq r_j + 2$, then $\alpha_{k,s}|_{\mathfrak{a}_j} = 0$.

In case of reducible symmetric spaces $\mathbf{X}_t = \mathbf{X}_t^1 \times \cdots \times \mathbf{X}_t^{s_t}$ we say that $\mathbf{X}_k$ propagates $\mathbf{X}_n$, $k \geq n$, $s_k \geq s_n$ and we can arrange the irreducible components so that $\mathbf{X}_k^j$ propagates $\mathbf{X}_n^j$ for $j = 1, \ldots, s_n$. We say that $\mathbf{Z}_k$ propagates $\mathbf{Z}_j$ if $\mathbf{X}_k$ propagates $\mathbf{X}_j$. From now on we will always assume if nothing else is clearly stated that the sequence $\{\mathbf{Z}_j\}$ is such that $\mathbf{Z}_k$ propagates $\mathbf{Z}_j$ for $k \geq j$.

3.2 Inductive Limits of Spherical Representations

In this section we recall the construction of inductive limits of spherical representations from [W09] and [ÓW11b].

As before we assume that $k \geq j$ and that $\mathbf{Z}_k$ propagates $\mathbf{Z}_j$. Moreover, from now on we will always assume that the groups G_j are simple. Denote by $\mathrm{r}_{j,k} : \mathfrak{a}_k^* \to \mathfrak{a}_j^*$ the projection $\mathrm{r}_{j,k}(\mu) = \mu|_{\mathfrak{a}_j}$. As shown in [ÓW11a, ÓW11b] we have

Lemma 3.3. *If $k \geq j$, then $\mathrm{r}_{j,k}(\omega_{k,s}) = \omega_{j,s}$ for $s = 1, \ldots, r_j$.*

This implies that the sets of highest weights $\Lambda_k^+ := \Lambda^+(G_k, K_k)$ form a projective system with restrictions as projections. But those sets also form an injective system as we will now describe. This will allow us to construct an injective system of representations in unique way, starting at a given level j_o.

Let $\mu_j \in \Lambda_j^+$ and write

$$\mu_j = \sum_{s=1}^{r_j} k_s \omega_{j,s}, \qquad k_s \in \mathbb{N}_0.$$

Define $\mu_k \in \Lambda_k^+$ by

$$\mu_k := \sum_{s=1}^{r_j} k_s \omega_{k,s}.$$

The map $\iota_{k,j} : \Lambda_j^+ \to \Lambda_k^+$, $\mu_j \mapsto \mu_k$ is well defined and injective and $\iota_{k,n} \circ \iota_{n,j} = \iota_{k,j}$ for $j \leq n \leq k$. We also have

$$\mathrm{r}_{j,k} \circ \iota_{k,j} = \mathrm{id}. \tag{3.4}$$

Finally, μ_k is the minimal element in $\mathrm{r}_{j,k}^{-1}(\mu_j)$ with respect to the partial ordering $\nu - \mu = \sum_j k_j \omega_{k,j}, k_j \in \mathbb{Z}^+$. In particular we have the following lemma:

Lemma 3.5. *The sequence* $\{\Lambda_j^+\}_j$ *with the maps*

$$\iota_{k,j} : \Lambda_j^+ \to \Lambda_k^+ , \quad \sum_{s=1}^{r_j} k_s \omega_s^j \mapsto \sum_{s=1}^{r_j} k_s \omega_s^k$$

is an injective sequences of sets. Furthermore, there is a canonical inclusion

$$\Lambda_\infty^+ := \varinjlim \Lambda_j^+ \hookrightarrow \varprojlim \Lambda_j^+ .$$

Proof. Most of the proof has been given already. For the last statement the idea is the same as in Lemma 3.2. Given j and $\mu_j \in \Lambda^+$. Then by (3.4) the sequence $(\mu_j, \mu_{j+1}, \ldots)$ is in $\varprojlim \Lambda_j^+$. $\qquad\square$

For $j \in \mathbb{N}$ and $\mu = \mu_j \in \Lambda_j^+$ define $\mu_k = \iota_{k,j}(\mu)$, $k \geq j$, and $\mu_\infty = \varinjlim \mu_j \in \Lambda_\infty^+$. Let (π_{μ_j}, V_{μ_j}) be the spherical representation of G_j with highest weight μ_j. We can and will assume that each V_{μ_j} carries a U_j-invariant inner product such that the embeddings $V_{\mu_j} \hookrightarrow V_{\mu_k}$ are isometric.

Theorem 3.6 (Ó-W). *Assume that $\mathbf{Z}_k$ propagates $\mathbf{Z}_j$. Let $\mu_j \in \Lambda_j^+$ and define $\mu_k \in \Lambda_k^+$ as above. Then the following holds:*

(1) *Let $u_{\mu_k} \in V_{\mu_k}$ be a weight vector chosen as before, let $W_j := \langle \pi_k(G_j)u_{\mu_k} \rangle$ and $\pi_{k,j}(g) := \pi_k(g)|_{W_j}$, $g \in G_j$. Then $\pi_{k,j}$ is equivalent to π_{μ_j} and we can choose the highest weight vector u_{μ_j} in V_{μ_j} so that the linear map generated by $\pi_{\mu_j}(g)u_{\mu_j} \mapsto \pi_{\mu_k}(g)u_{\mu_k}$ is a unitary G_j-isomorphism.*
(2) *The multiplicity of π_{μ_j} in π_{μ_k} is one.*

Remark 3.7. Note that in [ÓW11a] the statement was proved for the compact sequence $\{U_j\}$. But it holds true for the complex groups G_j by holomorphic extension. It is also true for the noncompact groups G_{j_o} by holomorphic extension and then restriction to G_{j_o}. $\qquad\square$

The second half of the above theorem implies that, up to a scalar, the only unitary G_j-isomorphism is the one given in part (1). As a consequence we can and will always think of V_{μ_j} as a subspace of V_{μ_k} such that the highest weight vector u is independent of j, i.e., $u_{\mu_j} = u_{\mu_k}$ for all k and j. We form the inductive limit

$$V_{\mu_\infty} := \varinjlim V_{\mu_j} . \tag{3.8}$$

Starting at a point j_o the highest weight vector u_{μ_j}, $j \geq j_o$ is constant and contained in all V_{μ_j}. In particular, $u_{\mu_j} \in V_{\mu_\infty}$. We also note that $\{\pi_{\mu_j}(g)\}$, $g \in G_j$, forms an injective sequence of continuous linear operators, unitary for $g \in U_j$. Hence $(\pi_{\mu_k})_\infty(g) : V_\infty \to V_\infty$ is a well-defined continuous map; similarly for the Lie algebra. We denote those maps by π_{μ_∞} and $d\pi_{\mu_\infty}$ respectively. Hence the group G_∞ acts continuously, in fact smoothly, on V_{μ_∞}. We denote the corresponding representation of G_∞ by π_{μ_∞}. We have

$$d\pi_{\mu_\infty}(H)u_{\mu_\infty} = \mu_\infty(H)u_{\mu_\infty} \quad \text{for all} \quad H \in \mathfrak{a}_\infty .$$

The representation $(\pi_{\mu\infty}, V_{\mu\infty})$ is (algebraically) irreducible. We can make $\pi_{\mu\infty}|_{U_\infty}$ unitary by completing $V_{\mu\infty}$ to a Hilbert space $\hat{V}_\infty$ as is usually done, see [DÓW12].

The dual of $V_{\mu\infty}$ is given by the corresponding projective limit

$$V^*_{\mu\infty} = \varprojlim V^*_{\mu_j}. \tag{3.9}$$

Note that in this notation $V^*_{\mu\infty} \neq V_{\mu\infty}^*$. We note that the highest weight vector, which we now denote by $u_{\mu\infty}$ is in $V_{\mu\infty}$. If $g \in G_k$, $k \geq j$, then $\pi^*_{\mu_j}(g)$ forms a projective family of operators and hence $\varprojlim \pi^*_{\mu_k}$ is a well-defined continuous representation of G_∞ on $V^*_{\mu\infty}$. We denote this representation by $\pi^*_{\mu\infty}$.

Lemma 3.10. *Let the notation be as above. Then*

$$\dim\left(\varprojlim V^*_{\mu_j}\right)^{K_\infty} = 1.$$

Proof. First fix j_o so that V_{μ_j} is defined for all $j \geq j_o$, i.e., $\{\mu_j\}$ stabilizes from j_o on. As $\dim V^{*K_j}_{\mu_j} = 1$, $j \geq j_o$, there exists a unique, up to scalar, K_{j_o}-fixed element $e^*_{\mu_{j_o}}$. We fix $e^*_{\mu_j}$ now so that $\mathrm{proj}_{j_o,j}(e^*_{\mu_j}) = e^*_{\mu_{j_o}}$, where $\mathrm{proj}_{j_o,j}$ is the dual map of $V_{\mu_{j_o}} \hookrightarrow V_{\mu_j}$. Then $e^*_{\mu\infty} := \{e^*_{\mu_j}\}_{j\geq j_o} \in \left(\varprojlim V^*_{\mu_j}\right)^{K_\infty}$. On the other hand, if $\{e^*_{\mu_j}\}_{j\geq j_o} \in \left(\varprojlim V^*_{\mu_j}\right)^{K_\infty}$, then $e^*_{\mu_{j_o}} \in V^{*K_j}_{\mu_{j_o}}$ is unique up to scalar showing that the dimension is one. $\qquad\square$

From now on we fix $e^*_{\mu\infty}$ so that $\langle u_{\mu\infty}, e^*_{\mu\infty}\rangle = 1$.

Theorem 3.11. $V^*_{\mu\infty}$ *is irreducible.*

Proof. Assume that $W \subset V^*_{\mu\infty}$ is a closed G_∞-invariant subspace. Then $W^\perp = \{u \in V_{\mu\infty} \mid (\forall\varphi \in W)\, \langle u, \varphi\rangle = 0\}$ is closed and G_∞ invariant. Hence $W^\perp = \{0\}$ or $W = V_\infty$, and since all spaces involved are reflexive, this implies that $W = V^*_\infty$ or $W = \{0\}$. $\qquad\square$

The vector $u_{\mu\infty} \in V_{\mu\infty}$ is clearly $M_\infty N_\infty$-invariant. Therefore $V_{\mu\infty}$ is conical (see [DÓ13]). But it is easy to see that with exception of some trivial cases (such as $G_j = G_k$ for all j and k, which we do not consider) the representation $(\pi_{\mu\infty}, V_{\mu\infty})$ is not K_∞-spherical. In fact, suppose that $e \in V_{\mu\infty}$ is a nontrivial, K_∞-invariant vector. Let j be so that $e \in V_{\mu_j}$. Then e has to be fixed for all K_s, $s \geq j$ and hence a multiple of e_s. This is impossible in general as will follow from Lemma 5.5. On the other hand, it was shown in [DÓW12] that the Hilbert space completion $\hat{V}_{\mu\infty}$ is K_∞-spherical if and only if the dimension of $\mathfrak{a}_\infty$ is finite. In this case we can assume that $\mathfrak{a}_j = \mathfrak{a}_\infty$ for all j. Then $\Sigma_j = \Sigma_k = \Sigma_\infty$, $\Sigma^+_j = \Sigma^+_k = \Sigma^+_\infty$ and $\Lambda^+_j = \Lambda^+_k = \Lambda^+_\infty$ for $k \geq j$. But we still use the notation μ_j etc. to indicate what group we are using.

Theorem 3.12 ([DÓW12]). *Let the notation be as above and assume that $\mu \neq 0$. Then $\hat{V}_\infty^{K_\infty} \neq \{0\}$ if and only if the ranks of the compact Riemannian symmetric spaces $\mathbf{X}_k$ are bounded. Thus, in the case where $\mathbf{X}_j$ is an irreducible classical symmetric space, we have $V_\infty^{K_\infty} \neq \{0\}$ only for $\mathrm{SO}(p+\infty)/\mathrm{SO}(p) \times \mathrm{SO}(\infty)$, $\mathrm{SU}(p+\infty)/\mathrm{S}(\mathrm{U}(p) \times \mathrm{U}(\infty))$ and $\mathrm{Sp}(p+\infty)/\mathrm{Sp}(p) \times \mathrm{Sp}(\infty)$ where $0 < p < \infty$.*

Let as usual $\iota_{k,j} : V_{\mu_j} \to V_{\mu_k}$ be the inclusion defined in Theorem 3.6 and $\mathrm{proj}_{j,k} : V_{\mu_k} \to V_{\mu_j}$ the orthogonal projection. Then, as $V_{\mu_j} \simeq < \pi_{\mu_k}(G_j)u_{\mu_k} >$, it follows that $\mathrm{proj}_{j,k} \circ \iota_{k,j} = \mathrm{id}_{V_{\mu_j}}$. By Lemma 3.2 there is a canonical G_∞-inclusion

$$V_{\mu_\infty} \hookrightarrow \varprojlim V_{\mu_j} .$$

Define $u_{\mu_j}^* \in V_{\mu_j}^*$ by

$$\langle \pi_{\mu_j}(s_{o,j})u_{\mu_j}, u_{\mu_j}^* \rangle = 1 \quad \text{and} \quad u_{\mu_j}^*|_{(\pi_\mu(G_j)u_{\mu_j})_{\mu_j}^\perp} = 0$$

where $s_{o,j} \in N_{K_j}(\mathfrak{a}_j)$ is such that $\mathrm{Ad}(s_{*,j})$ maps the set of positive roots into the set of negative roots and $(\pi_\mu(G_j)u_{\mu_j})_{\mu_j}^\perp$ is the orthogonal complement in V_{μ_j}. We use the inner product to fix embeddings $V_{\mu_j}^* \hookrightarrow V_{\mu_k}^*$ for $j \leq k$. Then again, we can take $u_{\mu_j}^*$ independent of j, which defines an $M_\infty N_\infty$-invariant element in $\varinjlim V_{\mu_j}^* \subset \varinjlim V_{\mu_j}^* = V_{\mu_\infty}^*$. As $e_{\mu_\infty}^*$ is K_∞-invariant, it follows that $V_{\mu_\infty}^*$ is both spherical and conical.

We now give another description of the representations $(\pi_{\mu_\infty}, V_{\mu_\infty})$. We say that a representation (π, V) of G_∞ is *holomorphic* if $\pi|_{G_j}$ is holomorphic for all j.

Theorem 3.13 ([DÓ13]). *Assume that X_∞ has finite rank. If $\mu_\infty \in \Lambda_\infty^+ = \Lambda^+$, then $(\pi_{\mu_\infty}, V_{\mu_\infty})$ is irreducible, conical and holomorphic. Conversely, if (π, V) is an irreducible conical and holomorphic representation of G_∞, then there exists a unique $\mu_\infty \in \Lambda_\infty^+$ such that $(\pi, V) \simeq (\pi_{\mu_\infty}, V_{\mu_\infty})$.*

4 Regular Functions on Limit Spaces

In this section we study the spaces $\varinjlim \mathbb{C}[\mathbf{Z}_j]$ and $\varprojlim \mathbb{C}[\mathbf{Z}_j]$ as well as their analogs for Ξ_∞. Our main discussion centers around the injective limits. We only discuss the limits of the complex cases. The corresponding results for the algebras $\mathbb{C}_i[\mathbf{X}_\infty] = \varinjlim \mathbb{C}[\mathbf{X}_j]$ and $\mathbb{C}_i[\mathbf{Y}_\infty] := \varinjlim \mathbb{C}[\mathbf{Y}_\infty]$ can be derived simply by restricting functions from $\mathbf{Z}_\infty$ to the real subspaces $\mathbf{X}_\infty$ respectively $\mathbf{Y}_\infty$.

4.1   Regular Functions on $\mathbf{Z}_\infty$

In this section $\{\mathbf{Z}_j\} = \{G_j/K_j\}$ is a propagated system of symmetric spaces as before. There are two natural ways to extend the notion of a regular function on finite-dimensional symmetric spaces to the inductive limit of those spaces. One is to consider the projective limit $\mathbb{C}[\mathbf{Z}_\infty] := \varprojlim \mathbb{C}[\mathbf{Z}_j]$. The other possible generalization would be to consider the space of functions on $\mathbf{Z}_\infty$ which are algebraic finite sums of algebraically irreducible G_∞-modules and such that each f is locally finite in the sense that for each j the space $\langle G_j \cdot f \rangle$ is finite-dimensional. But as very little is known about those spaces and not all of our previous discussion about the Radon transform and its dual generalize to those spaces we consider first the space

$$\mathbb{C}_i[\mathbf{Z}_\infty] := \varinjlim \mathbb{C}[\mathbf{Z}_j].$$

That this limit in fact exists will be shown in a moment.

Let $x_\infty = \{K_\infty\} \in \mathbf{Z}_\infty$ be the base point in $\mathbf{Z}_\infty$. Then all the spaces $\mathbf{Z}_j$ embed into $\mathbf{Z}_\infty$ via $aK_j \mapsto a \cdot x_\infty$ and in that way $\mathbf{Z}_\infty = \bigcup \mathbf{Z}_j$. Recall from our previous discussion and Lemma 3.5 that the sets Λ_j^+ of highest spherical weights form an injective system and $\Lambda_\infty^+ = \varinjlim \Lambda_j^+$. Each $\mu_\infty = \varinjlim \mu_j$ determines a unique algebraically irreducible (see below for proof) G_∞-module $V_{\mu_\infty} = \varinjlim V_{\mu_k}$ such that the dual space $V_{\mu_\infty}^* = \varprojlim V_{\mu_j}^*$ contains a (normalized) K_∞-fixed vector $e_{\mu_\infty}^*$ normalized by the condition $\langle u_{\mu_\infty}, e_{\mu_\infty}^* \rangle = 1$, as after Lemma 3.10. As before, we denote the G_∞-representation on $V_{\mu_\infty}^*$ by $\pi_{\mu_\infty}^*$ and consider the G_∞-map

$$V_{\mu_\infty} \hookrightarrow \text{space of continuous functions on } G_\infty$$

given by

$$w \mapsto f_{w,\mu_\infty}, \quad \text{where } f_{w,\mu_\infty}(a \cdot x_\infty) := \langle w, \pi_{\mu_\infty}^*(a)e_{\mu_\infty}^* \rangle. \tag{4.1}$$

Denote the image of the map (4.1) by $\mathbb{C}[\mathbf{Z}_\infty]_{\mu_\infty}$. Thus

$$\mathbb{C}[\mathbf{Z}_\infty]_{\mu_\infty} = \{f_{w,\mu_\infty} \mid w \in V_{\mu_\infty}\} \simeq V_{\mu_\infty}. \tag{4.2}$$

Note that the restriction of (4.1) to V_{μ_j} and $\mathbf{Z}_j$ is the G_j-map

$$f_{w,\mu_j}(x) = \langle w, \pi_{\mu_j}^*(a)e_{\mu_j}^* \rangle = \langle w, \pi_{\mu_j}^*(a)e_{\mu_\infty}^* \rangle, \quad x = a \cdot x_\infty,$$

introduced in (2.8). That this is possible follows from the proof of Lemma 3.10. Hence we have a canonical G_j-map $\mathbb{C}[\mathbf{Z}_j]_{\mu_j} \hookrightarrow \mathbb{C}[\mathbf{Z}_k]_{\mu_k}$ for $k \geq j$ such that the following diagram commutes:

$$
\begin{array}{ccc}
V_{\mu_j} & \longrightarrow & V_{\mu_k} & \longrightarrow & \cdots \\
\downarrow & & \downarrow & & \\
\mathbb{C}[\mathbf{Z}_j]_{\mu_j} & \longrightarrow & \mathbb{C}[\mathbf{Z}_k]_{\mu_k} & \longrightarrow & \cdots
\end{array}
$$

As $\mathbb{C}[\mathbf{Z}_j] = \sum^{\oplus} \mathbb{C}[\mathbf{Z}_j]_{\mu_j}$ and $\mathbb{C}[\mathbf{Z}_k] = \sum^{\oplus} \mathbb{C}[\mathbf{Z}_k]_{\mu_k}$ one derives that the spaces $\mathbb{C}[\mathbf{Z}_j]$ form an injective system. Note that $\varinjlim \mathbb{C}[\mathbf{Z}_j]_{\mu_j}$ and $\mathbb{C}_i[\mathbf{Z}_\infty] := \varinjlim \mathbb{C}[\mathbf{Z}_j]$ carry natural G_∞-module structures. This proves part of the following theorem:

Theorem 4.3. *The space* $\mathbb{C}[\mathbf{Z}_\infty]_{\mu_\infty}$ *is an algebraically irreducible* G_∞-*module, and*

$$
\mathbb{C}[\mathbf{Z}_\infty]_{\mu_\infty} = \varinjlim \mathbb{C}[\mathbf{Z}_j]_{\mu_j} \simeq V_{\mu_\infty} .
$$

Furthermore

$$
\mathbb{C}_i[\mathbf{Z}_\infty] = \sum_{\mu_\infty \in \Lambda_\infty^+}^{\oplus} \mathbb{C}[\mathbf{Z}_\infty]_{\mu_\infty}
$$

as a G_∞-*module.*

Proof (See also [KS77, Thm. 1] and [O90, §1.17]). Everything is clear except maybe the irreducibility statement. For that it is enough to show that V_{μ_∞} is algebraically irreducible. So let $W \subset V_{\mu_\infty}$ be G_∞-invariant. If $W \neq \{0\}$, then we must have $W \cap V_{\mu_j} \neq \{0\}$ for some j. But then $W \cap V_{\mu_k} \neq \{0\}$ for all $k \geq j$ and $W \cap V_{\mu_k}$ is G_k-invariant. As V_{μ_k} is algebraically irreducible it follows that $V_{\mu_k} \subset W$ for all $k \geq j$. This finally implies that $W = V_{\mu_\infty}$. □

Remark 4.4. In the case where the real infinite-dimensional space $\mathbf{X}_\infty$ has finite rank the space $\mathbb{C}_i[\mathbf{Z}_\infty]$ has a nice representation-theoretic description. In this case, as mentioned earlier, we may assume $\Lambda_\infty^+ = \Lambda_j^+$. We have also noted that each of the spaces V_{μ_j} is a unitary representation of U_j such that the embedding $V_{\mu_j} \hookrightarrow V_{\mu_k}$ is a G_j-equivariant isometry and the highest weight vector u_{μ_j} gets mapped to the highest weight vector u_{μ_k}. In that way we have $u_{\mu_\infty} = u_{\mu_j}$ for all j. Furthermore this leads to a pre-Hilbert structure on V_{μ_∞} so that V_{μ_∞} can be completed to a unitary irreducible K_∞-spherical representation $\hat{V}_{\mu_\infty}$ of G_∞ (see [DÓW12, Thm. 4.5]). Furthermore, it is shown in [DÓ13] that each unitary K_∞-spherical representation (π, W_π) of G_∞ such that $\pi|_{U_j}$ extends to a holomorphic representation of G_j for each j is locally finite and of the form $\hat{V}_{\mu_\infty}$ for some $\mu_\infty \in \Lambda^+$. Moreover, each of those representations is conical in the sense that $\hat{V}_{\mu_\infty}^{M_\infty N_\infty} \neq \{0\}$. Finally, each irreducible unitary conical representation (π, W_π) of G_∞, whose restriction to U_j extends to a holomorphic representation of G_j is unitarily equivalent to some $\hat{V}_{\mu_\infty}$. □

The inclusions $\iota_{k,j} : \mathbf{Z}_j \hookrightarrow \mathbf{Z}_k$ lead to projections on the spaces of functions given by restriction. In particular, we have the projections

$$\text{proj}_{j,k} : \mathbb{C}[\mathbf{Z}_k] \to \mathbb{C}[\mathbf{Z}_j] \tag{4.5}$$

satisfying $\text{proj}_{j,n} \circ \text{proj}_{n,k} = \text{proj}_{j,k}$. Hence $\{\mathbb{C}[\mathbf{Z}_j]\}$ is a projective sequence and $\varprojlim \mathbb{C}[\mathbf{Z}_j]$ is a G_∞-module, and in fact an algebra, of functions on $\mathbf{Z}_\infty$. In fact, let $f = \{f_j\}_{j \geq j_o} \in \varprojlim \mathbb{C}[\mathbf{Z}_j]$ and $x \in \mathbf{Z}_\infty$. Let j be so that $x \in \mathbf{Z}_j$. Define $f(x) := f_j(x)$. If $k \geq j$, then $\text{proj}_{j,k}(f_k) = f|_{\mathbf{Z}_j}$. In particular $f_k(x) = f_j(x)$. Hence $f(x)$ is well defined.

In general we do not have $\text{proj}_{j,k}(\mathbb{C}[\mathbf{Z}_k]_{\mu_k}) \subset \mathbb{C}[\mathbf{Z}_j]_{\mu_j}$, but

$$\text{proj}_{j,k} \circ \iota_{k,j}|_{\mathbb{C}[\mathbf{Z}_j]_{\mu_j}} = \text{id}_{\mathbb{C}[\mathbf{Z}_j]_{\mu_j}}$$

as this is satisfied on the level of representations $V_{\mu_j} \to V_{\mu_k} \to V_{\mu_j}$ as mentioned before. Hence, by Lemma 3.2, we can view $\varinjlim \mathbb{C}[\mathbf{Z}_j]_{\mu_j}$ as a submodule of $\varprojlim \mathbb{C}[\mathbf{Z}_j]_{\mu_j}$. We record the following lemma which is obvious from the above discussion:

Lemma 4.6. *We have a G_∞-equivariant embedding*

$$\mathbb{C}_i[\mathbf{Z}_\infty] \hookrightarrow \varprojlim \mathbb{C}[\mathbf{Z}_j].$$

4.2 Regular Functions on Ξ_∞

In order to construct regular functions on Ξ_∞ we apply the same construction as above to the horospherical spaces Ξ_j. As the arguments are basically the same, we often just state the results.

The following can easily be proved for at least some examples of infinite rank symmetric spaces like $SL(j, \mathbb{C})/SO(j, \mathbb{C})$, but we only have a general proof in the obvious case of finite rank. In the meantime (see [DÓ13]) Lemma 4.7 has been extended to limits of all irreducible classical symmetric spaces via case by case calculations.

Lemma 4.7. *Assume that the rank of $\mathbf{Z}_j$ is constant. Then for $k \geq j$ we have*

$$M_j = M_k \cap G_j \quad \text{and} \quad N_j = N_k \cap G_j .$$

Definition 4.8. We say that the injective system of propagated symmetric spaces $\mathbf{Z}_j$ is *admissible* if $M_j = M_k \cap G_j$ and $N_j = G_j \cap N_k$ for all $k \geq j$.

From now on we will always assume that the sequence $\{\mathbf{Z}_j\}$ of symmetric spaces is admissible. Let $\xi_o = e M_\infty N_\infty$ be the base point of Ξ_∞ and note that we can view this as the base point in $\Xi_j \simeq G_j \cdot \xi_o \subset \Xi_\infty$.

For $\mu_\infty = \varinjlim \mu_j \in \Lambda_\infty^+$, $w \in V_{\mu_j}$, $\xi = a \cdot \xi_o \in \Xi_j$, $a \in G_j$ we have

$$\psi_{w,\mu_j}(\xi) = \langle w, \pi_{\mu_j}^*(a)u_{\mu_j}^* \rangle = \langle w, \pi_{\mu_j}^*(a)u_{\mu_\infty}^* \rangle =: \psi_{w,\mu_\infty}(\xi).$$

This defines G_j-equivariant inclusions

$$\mathbb{C}[\Xi_j]_{\mu_j} \hookrightarrow \mathbb{C}[\Xi_k]_{\mu_k} \hookrightarrow \mathbb{C}[\Xi_\infty]_{\mu_\infty} := \{\psi_{w,\mu_\infty} \mid w \in V_{\mu_\infty}\} \simeq \varinjlim \mathbb{C}[\Xi_j]_{\mu_j}.$$

We note that $V_{\mu_\infty} \to \mathbb{C}[\Xi_\infty]_{\mu_\infty}$, $w \mapsto \psi_{w,\pi_\infty}$, is a G_∞-isomorphism. With the same argument as above this leads to an injective sequence

$$\mathbb{C}[\Xi_j] \hookrightarrow \mathbb{C}[\Xi_k] \hookrightarrow \varinjlim \mathbb{C}[\Xi_j] =: \mathbb{C}_i[\Xi_\infty].$$

Theorem 4.9. *Assume that the sequence $\{\mathbf{Z}_j\}$ is admissible. Then*

$$\mathbb{C}_i[\Xi_\infty] = \sum_{\mu_\infty \in \Lambda_\infty^+}^{\oplus} \mathbb{C}[\Xi_\infty]_{\mu_\infty}.$$

Furthermore, there exists a G_∞ equivariant inclusion map $\mathbb{C}_i[\Xi_\infty] \hookrightarrow \varprojlim \mathbb{C}[\Xi_j]$.

Denote by Γ_j the normalized Radon transform $\Gamma_j : \mathbb{C}[\mathbf{X}_j] = \mathbb{C}[\mathbf{Z}_j] \to \mathbb{C}[\Xi_j]$ introduced in Sect. 2.4 and set

$$\Gamma_\infty(f_{v,\mu_\infty}) := \psi_{v,\mu_\infty}. \tag{4.10}$$

Then Γ_∞ defines a G_∞-equivariant map $\Gamma_\infty : \mathbb{C}_i[\mathbf{Z}_\infty] \to \mathbb{C}_i[\Xi_\infty]$ and

$$\Gamma_\infty = \varinjlim \Gamma_j.$$

As each Γ_j is invertible it follows that Γ_∞ is also invertible. In fact, the inverse is $\Gamma_\infty^{-1} = \varinjlim \Gamma_j^{-1}$ which maps ψ_{v,μ_∞} to f_{v,μ_∞}. As a consequence we obtain the following theorem:

Theorem 4.11. *Suppose that the sequence $\{\mathbf{Z}_j\}$ is admissible. Let $\mu_\infty = \varinjlim \mu_j \in \Lambda_\infty^+$, $k \geq j$, and $w \in V_{\mu_k}$. Then*

$$\iota_{k,j} \circ \Gamma_j f_{w,\mu_j} = \Gamma_k(\iota_{k,j} \circ f_{w,\mu_j}) = \psi_{w,\mu_\infty}$$

and

$$\iota_{k,j} \circ \Gamma_j^{-1} \psi_{w,\mu_j} = \Gamma_k^{-1}(\iota_{k,j} \circ \psi_{w,\mu_j}) = f_{w,\mu_\infty}.$$

In particular, we have the following commutative diagram:

$$
\begin{array}{ccccccc}
\cdots & \longrightarrow & \mathbb{C}[\mathbf{Z}_j] & \xrightarrow{\;\iota_{k,j}\;} & \mathbb{C}[\mathbf{Z}_k] & \xrightarrow{\;\iota_{k,\infty}\;} & \mathbb{C}_i[\mathbf{Z}_\infty] \\
& & \Gamma_j \big\uparrow \big\downarrow \Gamma_j^{-1} & & \Gamma_k \big\uparrow \big\downarrow \Gamma_k^{-1} & & \Gamma_\infty \big\uparrow \big\downarrow \Gamma_\infty^{-1} \\
\cdots & \longrightarrow & \mathbb{C}[\Xi_j] & \xrightarrow{\;\iota_{k,j}\;} & \mathbb{C}[\Xi_k] & \xrightarrow{\;\iota_{\infty,k}\;} & \mathbb{C}_i[\Xi_\infty]
\end{array}
$$

4.3 The Projective Limit

We discuss the projective limit in more detail. First we need the following notation. For $\mu, \nu \in \mathfrak{a}_o^*$ write

$$
\nu \leq \mu \quad \text{if} \quad \mu - \nu = \sum_{\alpha \in \Sigma^+} n_\alpha \alpha, \quad n_\alpha \in \mathbb{N}_0. \tag{4.12}
$$

If $\nu \leq \mu$ and $\nu \neq \mu$, then we also write $\nu < \mu$. The main problem in studying the projective limit is the decomposition of $V_{\mu_k}|_{G_j}$ for $k \geq j$. It is not clear if these representations decompose into representations with highest weights $\nu_j \leq \mu_j$. In case the rank of $\mathbf{X}_\infty$ is finite, that is correct. We therefore in the remainder of this subsection make the assumption that the rank of $\mathbf{X}_\infty$ is finite. In this case we can—and will—assume that $\mathfrak{a}_j = \mathfrak{a}$ for all j and recall from the earlier discussion that $\Sigma_j = \Sigma$ is constant and so are the sets of positive roots $\Sigma_j^+ = \Sigma^+$ and the sets of highest weights $\Lambda_j^+ = \Lambda^+$.

Write

$$
(\pi_{\mu_k}, V_{\mu_k})|_{G_j} \simeq \bigoplus_{s=0}^{r} (\pi_s, W_s) \tag{4.13}
$$

with $(\pi_0, W_0) \simeq (\pi_{\mu_j}, V_{\mu_j})$ which occurs with multiplicity one.

Lemma 4.14. *Assume that the rank of $\mathbf{X}_\infty$ is finite. Let $\mu = \mu_j = \mu_k \in \Lambda^+$. Then we have the following:*

(1) $e_{\mu_k}^*|_{W_s} = 0$ *if W_s is not spherical.*
(2) *Assume that $W_s \simeq V_\nu$ is spherical. Then $\nu \leq \mu_j$.*

Proof. The first claim is obvious. Let σ be a weight of V_{μ_k}.

$$
\sigma = \mu_k - \sum_{\alpha \in \Psi_k} t_\alpha \alpha
$$

with t_α nonnegative integers. This in particular holds if σ is a highest weight of a spherical representation of G_j, proving the claim. $\square$

Lemma 4.15. *Assume that the rank of* $\mathbf{X}_\infty$ *is finite. Let* $k > j$ *and let* $v \in V_{\mu_k}$. *Then*

$$\psi_{v,\mu_\infty}|_{\Xi_j} = \psi_{v,\mu_k}|_{\Xi_j} = \psi_{\mathrm{proj}_{j,k}(v),\mu_j} \,.$$

Proof. Let $x \in G_j$. Then $\pi_\mu^*(x)u_{\mu_k}^* \in \langle \pi_{\mu_k}(G_j)u_{\mu_k}^* \rangle = V_{\mu_j^*} = V_{\mu_j}^*$. The claim now follows as $u_{\mu_j}^* = u_{\mu_k}^*$. $\square$

We finish this section by recalling the graded version of $\mathbb{C}[\mathbf{Z}_j]$ and Γ_j. Recall that we are assuming that the rank of $\mathbf{X}_j$ is finite. Note that even if $\{\mathbb{C}[\mathbf{Z}_j]\}$ is a projective sequence, the sequence $\{\mathbb{C}[\mathbf{Z}_j]_{\mu_j}\}$ is not projective in general. Consider the ordering on $\mathfrak{a}_o^*$ as above. This defines a filtration on $\mathbb{C}[\mathbf{Z}_j]$ and we denote by $\mathrm{gr}\,\mathbb{C}[\mathbf{Z}_j]$ the corresponding graded module. Thus

$$\mathrm{gr}\,\mathbb{C}[\mathbf{Z}_j]_{\mu_j} = \bigoplus_{\nu_j \le \mu_j} \mathbb{C}[\mathbf{Z}_j]_{\nu_j} / \bigoplus_{\nu_j < \mu_j} \mathbb{C}[\mathbf{Z}_j]_{\nu_j} \qquad (4.16)$$

and

$$\mathrm{gr}\,\mathbb{C}[\mathbf{Z}_j] \simeq_G \bigoplus_{\mu \in \Lambda_j^+} \mathrm{gr}\,\mathbb{C}[\mathbf{Z}_j]_{\mu_j} \,. \qquad (4.17)$$

If $f \in \mathbb{C}[\mathbf{Z}_j]$, then $[f]$ denotes the class of f in $\mathrm{gr}\,\mathbb{C}[\mathbf{Z}_j]$. Let κ_j, respectively κ_{μ_j}, be the G-isomorphism $\mathbb{C}[\mathbf{Z}_j] \simeq \mathrm{gr}\,\mathbb{C}[\mathbf{Z}_j]$, respectively $\mathbb{C}[\mathbf{Z}_j]_{\mu_j} \simeq \mathrm{gr}\,\mathbb{C}[\mathbf{Z}_j]_{\mu_j}$, given by $f \mapsto [f]$. We let $\mathrm{gr}\,\Gamma_j := \Gamma_j \circ \kappa_j^{-1}$ and $\mathrm{gr}\,\Gamma_j\mu_j = \Gamma_j \circ \kappa_{\mu_j}^{-1}$. Note that this construction is also valid for $j = \infty$.

Proposition 4.18 (Prop. 7 [HPV02]). *The* G_j-*map* $\mathrm{gr}\,\Gamma_j$ *is a ring isomorphism* $\mathrm{gr}\,\mathbb{C}[\mathbf{Z}_j] \to \mathbb{C}[\Xi_j]$.

In the following we will not distinguish between Γ_j and $\mathrm{gr}\,\Gamma_j$ except where necessary. Thus we will prove statements for Γ_j and then use it for $\mathrm{gr}\,\Gamma_j$ without any further comments.

Identifying functions on G_k/K_k with right K_k-invariant functions on G_k, the following is clear:

$$\mathrm{proj}_{j,k}(f_{v,\mu_k}) = \langle v, \pi_{\mu_k}^*|_{G_k}e_{\mu_k}^* \rangle \,.$$

Thus, the kernel of $\mathrm{proj}_{j,k}$ is the G_j-module

$$\ker\mathrm{proj}_{j,k} = \{v \in V_k \mid v \perp \pi_k(G_j)e_{\mu_k}^*\} \,.$$

As $\langle v_\mu, e_{\mu_k}^* \rangle \neq 0$ it follows that $\ker\mathrm{proj}_{j,k}$ is a sum of G_j-modules with highest weight $< \mu_j$. Hence

$$\mathrm{gr}\,\mathrm{proj}_{j,k} : \mathrm{gr}\,\mathbb{C}[\mathbf{Z}_k] \to \mathrm{gr}\,\mathbb{C}[\mathbf{Z}_j]$$

is well defined and

$$\operatorname{gr} \operatorname{proj}_{j,k}(\mathbb{C}[\mathbf{Z}_k]_{\mu_k}) = \operatorname{gr} \mathbb{C}[\mathbf{Z}_j]_{\mu_j} .$$

It follows that the sequence $\{\operatorname{gr} \mathbb{C}[\mathbf{Z}_j]_{\mu_j}\}$ is projective.

We can also form the graded algebra $\operatorname{gr} \mathbb{C}_i[\mathbf{Z}_\infty]$ as in (4.16) and (4.17). Again we can view elements in $\operatorname{gr} \mathbb{C}_i[\mathbf{Z}_\infty]$ as functions on $\mathbf{Z}_\infty$ by choosing the unique element in $g \in [f] \in \operatorname{gr} \mathbb{C}_i[\mathbf{Z}_\infty]_\mu$ so that $g \in \mathbb{C}_i[\mathbf{Z}_\infty]_\mu$. The inclusion

$$\operatorname{gr} \iota_{k,j} : \mathbb{C}[\mathbf{Z}_j] = \sum^{\oplus} \operatorname{gr} \mathbb{C}[\mathbf{Z}_j]_{\mu_j} \to \operatorname{gr} \mathbb{C}[\mathbf{Z}_k]_{\mu_k} = \mathbb{C}[\mathbf{Z}_k]$$

given by

$$\sum [f_{v_j, \mu_j}] \mapsto \sum [f_{\iota_{k,j}(v_j), \mu_j}]$$

satisfies the relation $\operatorname{gr} \operatorname{proj}_{j,k} \circ \operatorname{gr} \iota_{k,j} = \operatorname{id}$.

The graded version $\operatorname{gr} \Gamma_\infty$ is also well defined by the requirement that $[f_{v,\mu_\infty}]$ is mapped into ψ_{v,μ_∞} and both $\operatorname{gr} \Gamma_\infty$ and $\operatorname{gr} \Gamma_\infty^{-1}$ are G_j-morphisms of rings.

Theorem 4.19. *Assume that the rank of $\mathbf{X}_\infty$ is finite. Suppose that $k \geq j$, $\mu \in \Lambda^+$, $v \in V_{\mu_k}$ and $w \in V_{\mu_j}$. Denote by $\operatorname{proj}_{V_{\mu_j}}$ the projection $V_{\mu_k} \to V_{\mu_j}$. Then, with $\Gamma_{\mu_s} := \Gamma_s|_{\mathbb{C}[\mathbf{Z}_s]_{\mu_s}}$, $s \in \mathbb{N}$,*

$$\operatorname{proj}_{j,k} \Gamma_{\mu_k}(f_{v,\mu_k}) = \Gamma_{\mu_j}(f_{\operatorname{proj}_{V_{\mu_j}}(v),\mu_j}) = \psi_{\operatorname{proj}_{V_{\mu_j}}(v),\mu_j} \tag{4.20}$$

and

$$\iota_{k,j} \Gamma_{\mu_j}(f_{w,\mu_j}) = \Gamma_{\mu_k}(f_{w,\mu_k}) = \psi_{w,\mu_k} . \tag{4.21}$$

In particular $\operatorname{proj}_{j,k} \circ \iota_{k,j} = \operatorname{id}$ and $\operatorname{proj}_{j,\infty} \circ \iota_{\infty,j}$. Similar statements hold for the inverse maps. Let $\operatorname{gr} \Gamma_\infty = \varprojlim \operatorname{gr} \Gamma_j$ and $(\operatorname{gr} \Gamma_\infty)^{-1} = \varprojlim \operatorname{gr} \Gamma_j^{-1}$. We therefore have a commutative diagram:

$$
\begin{array}{ccccc}
\cdots \longleftarrow & \operatorname{gr} \mathbb{C}[\mathbf{Z}_k] & \overset{}{\longleftarrow} & \operatorname{gr} \mathbb{C}[\mathbf{Z}_j] & \overset{}{\longleftarrow} & \varprojlim \operatorname{gr} \mathbb{C}[\mathbf{Z}_j] \\
 & & \operatorname{proj}_{j,k} & & \operatorname{proj}_{k,\infty} & \\
\operatorname{gr} \Gamma_k \Big\updownarrow \operatorname{gr} \Gamma_k^{-1} & & \operatorname{gr} \Gamma_k \Big\updownarrow \operatorname{gr} \Gamma_j^{-1} & & \operatorname{gr} \Gamma_\infty \Big\updownarrow \operatorname{gr} \Gamma_\infty^{-1} & \\
\cdots \longleftarrow & \mathbb{C}[\Xi_k] & \overset{}{\longleftarrow} & \mathbb{C}[\Xi_j] & \overset{}{\longleftarrow} & \varprojlim \mathbb{C}[\Xi_j] \\
 & & \operatorname{proj}_{k,j} & & \operatorname{proj}_{k,\infty} &
\end{array}
$$

5 The Radon Transform and Its Dual

For the moment we fix the symmetric spaces **X**, **Y**, and **Z** and leave out the index j. The Radon transform or its dual is initially defined on the space of compactly supported function. As the dual Radon transform is an integral over the compact group K_o it is well defined on $\mathbb{C}[\Xi]$ and $\mathbb{C}[\Xi_o]$, the space of regular functions on Ξ. But N_o is noncompact, so the Radon transform cannot be defined on $\mathbb{C}[\mathbf{X}]$ as an integral over N_o. This problem was addressed in [HPV02, HPV03], and we recall the main results here. Then, based on ideas from [G06, GKÓ06], we introduce two integral kernels which allow us to express both the Radon transform and the dual Radon transform as integrals against integral kernels. We start the section by recalling the double fibration transform introduced in [H66, H70].

5.1 The Double Fibration Transform

Assume that G is a Lie group and H and L two closed subgroups. We assume that all of those groups as well as $M = H \cap L$ are unimodular. We have the double fibration

$$(5.1)$$

where π and p are the natural projections. We say that $x = aH$ and $\xi = bL$ are incident if $aH \cap bL \neq \emptyset$. For $x \in G/H$ and $\xi \in G/L$ we set

$$\hat{x} := \{\eta \in G/L \mid x \text{ and } \eta \text{ are incident}\}$$

and similarly

$$\xi^{\vee} := \{y \in G/H \mid \xi \text{ and } y \text{ are incident}\}.$$

Assume that if $a \in L$ and $aH \subset HL$, then $a \in H$ and similarly, if $b \in H$ and $bL \subset LH$, then $b \in L$. Then we can view the points in G/L as subsets of G/H, and similarly points in G/H can be viewed as subsets of G/L. Then $\hat{x}$ is the set of all η such that $x \in \eta$ and $\xi^{\vee}$ is the set of points $y \in G/H$ such that $y \in \xi$. We also have

$$\hat{x} = p(\pi^{-1}(x)) = aH \cdot \xi_0 \simeq H/L \quad \text{and} \quad \xi^{\vee} = \pi(p^{-1}(\xi)) = bL \cdot x_o \simeq K/L.$$

Fix invariant measures on all of the above groups and the homogeneous spaces $G/H, G/L, G/M, H/M$ and L/M such that for $f \in C_c(G)$ we have

$$\int_G f(a)\, da = \int_{G/M} f(am)\, d\mu_{G/M}(aH \cap L)dm$$

$$= \int_{G/H} f(ah)\, d\mu_{G/H}(gH)dh$$

$$= \int_{G/L} f(an)\, d\mu_{G/L}(aL)dn$$

and for $f \in C_c(H)$ and $\varphi \in C_c(L)$

$$\int_H f(h)\, dh = \int_{H/M} \int_M f(am)\, dm d\mu_{H/M}(aM)$$

and

$$\int_L f(a)\, da = \int_{L/M} \int_M f(am)\, dm d\mu_{L/M}(aM).$$

The definition of the *Radon transform* and the *dual Radon transform* is now as follows. Let $x_o = eH$ and $\xi_o = eL$. If $\xi = a \cdot \xi_o \in \Xi$ and $x = b \cdot x_o \in X$, then

$$\hat{f}(\xi) := \int_{L/M} f(al \cdot x_o)\, d\mu_{L/M}(lM), \quad f \in C_c(G/H) \tag{5.2}$$

and

$$\varphi^\vee(x) := \int_{H/M} \varphi(bh \cdot \xi_o)\, d\mu_{H/M}(hM), \quad \varphi \in C_c(H/M). \tag{5.3}$$

Then the following duality holds:

$$\int_{G/L} \hat{f}(\xi)\varphi(\xi)\, d\mu_{G/L}(\xi) = \int_{G/H} f(x)\varphi^\vee(x)\, d\mu_{G/H}(x).$$

5.2   The Horospherical Radon Transform and Its Dual

The example studied most is the case $G = G_o$, $H = K_o$ and $L = M_o N_o$, where we use the notation from the earlier sections. In this case we find the *horospherical* Radon transform which from now on we will simply call the Radon transform and its dual. The corresponding integral transforms are

$$\mathcal{R}f(a \cdot \xi_o) = \hat{f}(a \cdot \xi_o) = \int_{N_o} f(an \cdot x_o) \, dn \, , \quad f \in C_c(\mathbf{X})$$

and

$$\mathcal{R}^* \varphi(b \cdot x_o) = \varphi^\vee(b \cdot x_o) = \int_{K_o} \varphi(bk \cdot z_o) \, dk \, , \quad \varphi \in C_c(\Xi_o) \, .$$

Here dk is the invariant probability measure on K_o.

As mentioned earlier the dual Radon transform

$$\mathcal{R}^* \psi(a \cdot x_o) = \int_{K_o} \psi(ak \cdot \xi_o) \, dk$$

is well defined on $\mathbb{C}[\Xi]$. It is clearly a G-intertwining operator. Thus, there exists $c_\mu \in \mathbb{C}$ such that

$$\mathcal{R}_\mu^* := \mathcal{R}^* |_{\mathbb{C}[\Xi]_\mu} = c_\mu \Gamma_\mu^{-1} \, , \quad \mu \in \Lambda^+. \tag{5.4}$$

To describe the evaluation of c_μ we recall the functions f_μ and ψ_μ from Sect. 2. We find

$$\mathcal{R}^* \psi_\mu(a \cdot x_o) = \int_{K_o} \langle u, \pi_\mu^*(a) \pi_\mu^*(k) u_\mu^* \rangle \, dk$$

$$= \langle \pi_\mu(a)^{-1} u, \int_{K_o} \pi_\mu^*(k) u_\mu^* \, dk \rangle$$

$$= c_\mu f_\mu(a \cdot x_o) \, .$$

Thus c_μ is determined by

$$\int_{K_o} \pi_\mu^*(k) u_\mu^* \, dk = c_\mu e_\mu^* \, .$$

For $g \in G_o$ write $g = k(g)a(g)n(g)$ with $(k(g), a(g), n(g)) \in K_o \times A_o \times N_o$. For $\lambda \in \mathfrak{a}^*$ and $a = \exp X \in A_o$ write $a^\lambda = e^{\lambda(X)}$. Let

$$\rho := \frac{1}{2} \sum_{\alpha \in \Sigma^+} m_\alpha \alpha = \frac{1}{2} \sum_{\alpha \in \Sigma_0^+} \left(\frac{1}{2} m_{\alpha/2} + m_\alpha \right) \alpha$$

where $m_\alpha := \dim_\mathbb{C} \mathfrak{g}_\alpha$. We normalize the Haar measure on $\overline{N}_o = \theta(N_o)$ by

$$\int_{\overline{N}_o} a(\bar{n})^{-2\rho} \, d\bar{n} = 1 \, .$$

Define for $\lambda \in \mathfrak{a}^*(0) = \{\lambda \in \mathfrak{a}^* \mid (\forall \alpha \in \Sigma^+) \, \alpha(\mathrm{Re}\,(\lambda), \alpha) > 0\}$

$$\mathbf{c}(\lambda) := \int_{\bar{N}_o} a(n)^{-\lambda - \rho} \, d\bar{n} \,.$$

Then $\mathbf{c}$ is holomorphic on $\mathfrak{a}^*(0)$. By the Gindikin–Karpelevich formula [GK62] which expresses $\mathbf{c}$ as a rational function in Gamma-functions depending on the multiplicities m_α, the function $\mathbf{c}$ has a meromorphic extension to all of $\mathfrak{a}^*$. The function $\mathbf{c}$, which is called the *Harish–Chandra c-function*, can be used to calculate the constant c_μ from (5.4).

Lemma 5.5. *Let $\mu \in \Lambda^+$. Then $c_\mu = \mathbf{c}(\mu^* + \rho) = \mathbf{c}(\mu + \rho)$.*

Proof. See [DÓW12, Thm. 3.4] or [HPV02, Thm 9]. $\qquad\qquad\qquad\qquad\qquad\qquad$ □

We note that [DÓW12] implies that $c_\mu = \mathbf{c}(\mu^* + \rho)$ and that [HPV02] implies $c_\mu = \mathbf{c}(\mu + \rho)$. But the Gindikin–Karpelevich formula implies that

$$\mathbf{c}(\lambda) = \mathbf{c}(-w_o\lambda) \,.$$

As $-w_o\rho = \rho$ it follows that $\mathbf{c}(\mu^* + \rho) = \mathbf{c}(\mu + \rho)$.

5.3   The Radon Transform as Limit of Integration over Spheres

We have seen that the dual Radon transform and the normalized transform Γ_μ^{-1} are the same up to a normalizing factor that depends on the K-representation μ. No such relation exists for Γ_μ and $\mathcal{R}|_{\mathbb{C}[\mathbf{X}]_\mu}$ because the definition of the Radon transform $\mathcal{R}$ does not make sense for regular functions. However, in [HPV03] a solution was proposed by considering the Radon transform as a limit of Radon transform of a double fibrations transforms with both stabilizers being compact. Hence the corresponding integral transforms are well defined for regular functions. We recall the setup from [HPV03].

Since both K_o and K_o^a are compact, both the Radon transform associated to this double fibration and its dual transform are well defined on regular functions and give G-equivariant linear maps

$$\mathcal{R}_a : \mathbb{C}[\mathbf{X}] \longrightarrow \mathbb{C}[\mathrm{Sph}_a\,\mathbf{X}] \quad \text{and} \quad \mathcal{R}_a^* : \mathbb{C}[\mathrm{Sph}_a\,\mathbf{X}] \longrightarrow \mathbb{C}[\mathbf{X}].$$

One can identify $\mathrm{Sph}_a\,\mathbf{X}$ with $\mathbf{X}$ associating to each sphere its center. Then $\mathcal{R}_a$ and $\mathcal{R}_a^*$ become linear endomorphisms of $\mathbb{C}[\mathbf{X}]$. By definition, $\mathcal{R}_a$ is then obtained by integrating over spheres of radius a, while $\mathcal{R}_a^*$ is obtained by integrating over spheres of radius a^{-1}.

The spaces $\mathbf{X}$, Ξ_o, and $\mathrm{Sph}_a\,\mathbf{X}$ can all be embedded into the algebraic dual space

$$\mathbb{C}[\mathbf{X}]^* = \prod_{\mu \in \Lambda} \mathbb{C}[\mathbf{X}]_\mu^*,$$

which we equip with the product topology. Let $v_\mu^\pm \in \mathbb{C}[\mathbf{X}]_\mu^*$ be the highest (resp. lowest) weight vector uniquely determined by the normalizing condition

$$\langle f_\mu^\pm, v_\mu^\mp \rangle = 1,$$

where $f_\mu^+ := f_\mu$, and $f_\mu^- := s_{x_o} \cdot f_\mu$. Then we have $v_\mu^- = s_{x_o} \cdot v_{\mu^*}^+$ where s_{x_o} is the symmetry around the base point x_o. Similarly, let $v_\mu^0 \in \left(\mathbb{C}[\mathbf{X}]_\mu^* \right)^{K_o}$ be the K_o-invariant vector uniquely determined by the normalizing condition

$$\langle f_\mu^0, v_\mu^0 \rangle = 1,$$

where f_μ^0 is the unique K-invariant function in $\mathbb{C}[\mathbf{X}]_\mu$ such that $f_\mu^0(x_o) = 1$. Note that f_μ^0 is a zonal spherical function. The G_o-equivariant map $\iota_e : \mathbf{X} \to \mathbb{C}[\mathbf{X}]^*$ defined by $\langle f, \iota_e(x) \rangle = f(x)$ for $f \in \mathbb{C}[\mathbf{X}]$ is injective and satisfies $\iota_e(o) = (v_\mu^0)_{\mu \in \Lambda^+}$ (see [HPV03, Sect. 5]). For any $a \in A_o$ obtains an injective G_o-equivariant map $\iota_a : \mathrm{Sph}_a \mathbf{X} \to \mathbb{C}[\mathbf{X}]^*$ by

$$\iota_a(\mathcal{S}_a(x)) = \left(\frac{x_\mu}{a^{\mu^*}} \right)_{\mu \in \Lambda^+}, \qquad \text{if} \quad \iota_e(x) = (x_\mu)_{\mu \in \Lambda^+}.$$

In particular,

$$\iota_a(\mathcal{S}_a) = \left(\frac{a \cdot v_\mu^0}{a^{\mu^*}} \right)_{\mu \in \Lambda^+}.$$

The induced map $\iota_a^* : \mathbb{C}[\mathbf{X}] \to \mathbb{C}[\mathrm{Sph}_a \mathbf{X}]$ is a G_o-module isomorphism. Finally, $\iota(\xi_o) := (v_\mu^+)_{\mu \in \Lambda^+}$ defines a G-equivariant map $\iota : \Xi_o \to \mathbb{C}[\mathbf{X}]^*$. Since the stabilizer of ξ_0, as well the stabilizer of $(v_\mu^+)_{\mu \in \Lambda^+}$, is $M_o N_o$, the map ι is well defined and injective. The induced map $\iota^* : \mathbb{C}[\mathbf{X}] \to \mathbb{C}[\Xi_o]$ coincides with the G_o-module isomorphism Γ, where we note that $\mathbb{C}[\mathbf{X}] = \mathbb{C}[\mathbf{Z}]$ and $\mathbb{C}[\Xi_o] = \mathbb{C}[\Xi]$. We obtain the diagram

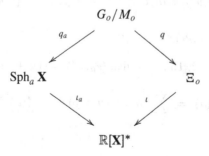

which turns out to be commutative. This is part of the following proposition which is proven in [HPV03, Sect. 6]:

Proposition 5.6.   (i) $\lim_{a\to\infty} \iota_a \circ q_a = \iota \circ q$.

(ii) $\lim_{a\to\infty} q_a^* \circ \iota_a^* = q^* \circ \iota^*$.

(iii) $\lim_{a\to\infty} \mathcal{R}_a^* \circ \iota_a^* = \mathcal{R}^* \circ \iota^*$.

(iv) $\mathcal{R}^*(\iota^*(f)) = \mathbf{c}(\mu + \rho)f$ for all $f \in \mathbb{C}[\mathbf{X}]_\mu$.

We note that those results can be applied to $\mathbb{C}[\mathbf{Z}]$ and $\mathbb{C}[\Xi]$ by restriction and holomorphic extension.

Suppose now as before that we have a propagated sequence of symmetric spaces $\mathbf{Z}_j \to \mathbf{Z}_k, k \geq j$. We also assume that the rank is finite. Then, on each level, we have $\iota_j^* = \Gamma_j$. Therefore, we can define for $f \in \mathbb{C}[\mathbf{Z}_\infty]$

$$\iota_\infty^*(f) = \iota_j^*(f) \quad \text{if} \quad f \in \mathbb{C}[\mathbf{Z}_j].$$

Then

Proposition 5.7.   If the rank of $\mathbf{X}_\infty$ is finite and $f \in \mathbb{C}[\mathbf{Z}_\infty]$, then

$$\Gamma_\infty f = \iota_\infty^*(f).$$

5.4   The Kernels Defining the Normalized Radon Transform and Its Dual

We start by stating the following version of the orthogonality relations which are usually formulated in terms of invariant inner products. The proof for this version is the same as the usual one. The invariant measure on U is always the normalized Haar measure. The following is the usual orthogonality relation stated in form of duality.

Lemma 5.8.   For $\mu, \nu \in \Lambda^+$ let $d(\mu) = \dim V_\mu = \dim V_{\mu^*}$. Then

$$\int_U \langle u, \pi_\mu^*(b)u^* \rangle \langle \pi_\nu(a)v, v^* \rangle \, db = \delta_{\mu,\nu} d(\mu)^{-1} \langle v, u^* \rangle \langle u, v^* \rangle$$

for all $u, v \in V_\mu$ and for all $u^*, v^* \in V_\mu^*$.

It follows by [DÓW12, Thm. 3.4] that $\langle e_\mu, e_\mu^* \rangle = \mathbf{c}(\mu + \rho)$. Define

$$k_{\mathbf{Z}}(a) := \sum_{\mu \in \Lambda^+} d(\mu)\mathbf{c}(\mu + \rho) \langle u_\mu, \pi_\mu^*(a)e_\mu^* \rangle \tag{5.9}$$

and

$$k_\Xi(a) := \sum_{\mu \in \Lambda^+} d(\mu) \langle e_\mu, \pi_\mu^*(a)u_\mu^* \rangle. \tag{5.10}$$

Lemma 5.11. *Let* $\mathcal{O} = \{nak \in NAK \mid (\forall\ j)\ |a^{-\omega_j}| < 1\}$. *Then* $\mathcal{O}$ *is open in* G. *The set* $\mathcal{O}$ *is right* K-*invariant and left* MN-*invariant. Furthermore, the sums defining* $k_{\mathbf{Z}}$ *and* k_{Ξ} *converge uniformly on compact subsets of* $\mathcal{O}$ *and define holomorphic functions on* $\mathcal{O}$.

Proof. Write $\mu = k_1\omega_1 + \ldots + k_r\omega_r$. Let $x \in \mathcal{O}$ and write $b_j = a^{-\omega_j}$. Then $|b_j| < 1$ for $j = 1, \ldots, r$ and

$$d(\mu)\langle u_\mu, \pi_\mu^*(x)u_\mu^*\rangle = d(\mu)\langle \pi_\mu(k^{-1}a^{-1}n^{-1})u_\mu, u_\mu^*\rangle$$
$$= d(\mu)a^{-\mu}$$
$$= d(\mu)\prod_{j=1}^{r} b_j^{k_j}.$$

Similarly, we have

$$d(\mu)\mathbf{c}(\mu + \rho)\langle u_\mu, \pi_\mu^*(x)e_\mu^*\rangle = d(\mu)\mathbf{c}(\mu + \rho)\prod_{j=1}^{r} b_j^{k_j}.$$

The claim follows now because $d(\mu)$ is polynomial in $k_1, \ldots, k_r$ and $\mathbf{c}(\mu + \rho) < 1$.

$\square$

The function $k_{\mathbf{Z}}$ is left MN-invariant and right K-invariant. Hence $k_{\mathbf{Z}}$ can also be viewed as a function on $\Xi \times \mathbf{X}$ given by

$$k_{\mathbf{Z}}(a \cdot \xi_o, b \cdot x_o) := k_{\mathbf{Z}}(a^{-1}b).$$

The function k_{Ξ} is left K-invariant and right MN-invariant and can be viewed as a function k_{Ξ} on $\mathbf{X} \times \Xi$ defined by

$$k_{\Xi}(a \cdot x_o, b \cdot \xi_o) := k_{\Xi}(a^{-1}b).$$

Even if the sums (5.9) and (5.10) do not in general converge for all $a \in G$, they are well defined as linear G-maps $\mathbb{C}[\mathbf{X}] \to \mathbb{C}[\Xi]$, respectively $\mathbb{C}[\Xi] \to \mathbb{C}[\mathbf{X}]$, given by

$$K_{\mathbf{Z}}(f)(\xi) := \int_U f(u \cdot x_o)k_{\mathbf{Z}}(\xi, u \cdot x_o)\, du,$$

respectively

$$K_{\Xi}(f)(x) := \int_U f(u \cdot \xi_o)k_{\Xi}(x, u \cdot \xi_o)\, du,$$

where du is the normalized Haar measure on U. On the right-hand side only finitely many terms are nonzero so the sums converge. Note that the first integral can be written as an integral over the compact symmetric space U/K_o and the second integral is an integral over U/M_o.

Theorem 5.12. *We have* $K_Z = \Gamma$ *and* $K_\Xi = \Gamma^{-1}$.

Proof. It is enough to show that $K_Z|_{\mathbb{C}[X]_\nu} = \Gamma_\nu$ and $K_\Xi|_{\mathbb{C}[\Xi]_\nu} = \Gamma_\nu^{-1}$ for all $\nu \in \Lambda^+$. Thus we have to show that $K_Z(f_\nu) = \psi_\nu$ and $K_\Xi(\psi_\nu) = f_\nu$.

We have

$$K_Z(f_\nu)(a \cdot \xi_o) = \sum_{\nu \in \Lambda^+} \int_U \frac{d(\mu)}{\langle e_\mu, e_\mu^* \rangle} \langle \pi_\mu(a)u_\mu, \pi_\mu^*(b)e_\mu^* \rangle \langle u_\nu, \pi_\nu^*(b)e_\nu^* \rangle \, du$$

$$= \langle \pi_\nu^*(a)u_\nu^*, u_\nu \rangle$$

$$= \psi_\mu(a \cdot \xi_o).$$

The statement for K_Ξ is proved in the same way. $\qquad\square$

Remark 5.13. Above we realized Γ and Γ^{-1} as integral operators. Similar to [G06] one could also consider the integral operator given by the kernel $\tilde{K}(a \cdot \xi_o, b \cdot x_o) = \tilde{k}(a^{-1}b)$ where

$$\tilde{k}(a) = \sum_{\mu \in \Lambda^*} f_\mu(a) = \sum_{\mu \in \Lambda^+} \langle u_\mu, \pi_\mu^*(a)e_\mu^* \rangle. \tag{5.14}$$

Then we have the following theorem, see also [G06]:

Theorem 5.15. *Let* $\mathcal{O}$ *be as in Lemma 5.11 and let* $x = kan \in \mathcal{O}$ *be such that* $|a^{\omega_j}| < 1$ *for* $j = 1, \ldots, r = \mathrm{rank}\, X$. *Then*

$$\tilde{k}(b) = \prod_{j=1}^{r} \frac{1}{1 - a^{\omega_j}}$$

and $\tilde{k}$ *is holomorphic on* $\mathcal{O}$.

Proof. Let $\mu = k_1\omega_1 + \ldots k_r\omega_r \in \Lambda^+$. For $x = nak \in \mathcal{O}$ write as before $b_j = a^{-\omega_j}$. Then

$$\langle u_\mu, \pi_{\mu^*}(x)e_{\mu^*} \rangle = a^{-\mu} = \prod_{j=1}^{r} b_j^{k_j}.$$

It follows that

$$\tilde{k}(x) = \sum_{k_1=0}^{\infty} b_1^{k_1} \ldots \sum_{k_r=0}^{\infty} b_r^{k_r} = \prod_{j=1}^{r} (1 - b_j)^{-1}$$

which finishes the proof. $\qquad\square$

5.5 The Radon Transform and Its Dual on the Injective Limits

In this section we discuss the extension of the Radon transform and its dual on the infinite dimensional spaces.

The kernels defined in (5.9) and (5.10) do not define functions on $\mathbf{Z}_\infty$, respectively Ξ_∞, because of the changes in the dimensions as we move from one space to another. But we still have the following:

Theorem 5.16. *Assume that the sequence $\{\mathbf{Z}_j\}$ is admissible. For $f \in \mathbb{C}_i[\mathbf{Z}_\infty]$ and $\psi \in \mathbb{C}_i[\Xi_\infty]$ the pointwise limits*

$$K_{\mathbf{Z}_\infty} f(\xi) = \lim K_{\mathbf{Z}_j} f(\xi) \text{ and } K_{\Xi_\infty} \psi(x) = \lim K_{\mathbf{Z}_j} \psi(x)$$

are well defined and

$$\Gamma_\infty f = K_{\mathbf{Z}_\infty} f \quad and \quad \Gamma_\infty^{-1} \psi = K_{\Xi_\infty} \psi.$$

Proof. If $v \in V_{\mu_j}$, then $v \in V_{\mu_k}$ for all $k \geq j$. Hence Theorem 5.12 implies that if $s > k > j$ are so that $\xi \in \Xi_k$, we have

$$K_{\mathbf{Z}_s} f(\xi) = K_{\mathbf{Z}_k} f(\xi).$$

Hence the sequence becomes constant and the claim follows. The argument for $\mathbf{K}_{\Xi_\infty} \psi$ is the same. $\qquad\square$

Note that Eqs. (4.20) and (4.21) together with Theorem 5.12 imply that the maps $\mathrm{gr}\, K_{\mathbf{Z}_\infty} = \varprojlim \mathrm{gr}\, K_{\mathbf{Z}_j}$ and $\mathrm{gr}\, K_{\Xi_\infty} = \varprojlim K_{\Xi_\infty}$ are well defined. Here gr stands for $\mathrm{gr}\, K_{\mathbf{Z}_j}([f]) = [K_{\mathbf{Z}_j} f]$ respectively $\mathrm{gr}\, K_{\Xi_j}(f) = [K_{\Xi_j} f]$.

Theorem 5.17. *Assume that the sequence $\mathbf{Z}_j$ is admissible. Then*

$$\mathrm{gr}\, \Gamma_\infty f = \mathrm{gr}\, K_{\mathbf{Z}_\infty} f \quad and \quad (\mathrm{gr}\, \Gamma_\infty)^{-1} f = \mathrm{gr}\, K_{\Xi_\infty} f.$$

In order to make the results of Sect. 5.3 useful for limits of symmetric spaces, we first have to extend the notion of a sphere of radius a. So let $a = \lim a_j \in A_\infty := \varinjlim A_j$. We call A_∞ *regular* if $Z_{K_j}(a_j) = M_j$ for all j. This is a useful notion only in the finite rank case, so we will assume for the remainder of this section that we are in the situation of Remark 4.4.

A simple calculation shows that the diagram

$$
\begin{array}{ccc}
\mathbf{X}_j & \longrightarrow & \mathbf{X}_k \\
\downarrow{\scriptstyle \iota_{j,e}} & & \downarrow{\scriptstyle \iota_{k,e}} \\
\mathbb{C}[\mathbf{X}_j]^* & \longrightarrow & \mathbb{C}[\mathbf{X}_k]^*
\end{array}
$$

of G_j-module morphisms is commutative. The normalizations from Sect. 5.3 are compatible and yield the following commutative diagram

$$
\begin{array}{ccc}
\mathrm{Sph}_{a_j}(\mathbf{X}_j) & \longrightarrow & \mathrm{Sph}_{a_k}(\mathbf{X}_k) \\
\downarrow {\scriptstyle \iota_{j,a_j}} & & \downarrow {\scriptstyle \iota_{k,a_k}} \\
\mathbb{C}[\mathbf{X}_j]^* & \longrightarrow & \mathbb{C}[\mathbf{X}_k]^* \\
\uparrow {\scriptstyle \iota_j} & & \uparrow {\scriptstyle \iota_k} \\
\Xi_{j,o} & \longrightarrow & \Xi_{k,o}
\end{array}
$$

of G_j-module morphisms. In fact, for the commutativity of the upper square one uses the equality $a_j^{\mu_j^*} = a_k^{\mu_k^*} = a^{\mu^*}$, whereas the commutativity of the lower square is a consequence of Theorem 4.11. Thus the $\mathrm{Sph}_{a_j}(\mathbf{X}_j)$ and the $\mathbb{C}[\mathbf{X}_j]^*$ form inductive systems. For the corresponding inductive limits $\mathrm{Sph}_a(\mathbf{X}_\infty)$ and $\mathbb{C}[\mathbf{X}_\infty]^*$ we obtain the commutative diagram

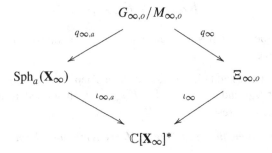

Using this notation Proposition 5.6(i) remains true:

$$
\lim_{a \to \infty} \iota_{\infty,a} \circ q_{\infty,a} = \iota_\infty \circ q_\infty. \tag{5.18}
$$

Due to G-equivariance, it suffices to prove that

$$
\lim_{a \to \infty} \iota_{\infty,a}(q_a(eM)) = \iota_\infty(q_\infty(eM)).
$$

This means that for fixed $\lambda \in \Lambda^+$ and $j \geq j_o$ we have to verify

$$
\lim_{a \to \infty} \frac{a \cdot v_{j,\lambda}^0}{a^{\lambda^*}} = v_{j,\lambda}^+.
$$

Writing $v^0_{j,\lambda} \in \mathbb{C}[\mathbf{X}_j]^*_\lambda$ as a sum of weight vectors (the highest weight being λ^*), we see that

$$\lim_{a \to \infty} \frac{a \cdot v^0_{j,\lambda}}{a^{\lambda^*}} = k_{j,\lambda} v^+_{j,\lambda}$$

for some constant $k_{j,\lambda}$. The calculation

$$\frac{\langle f^-_{j,\lambda}, a \cdot v^0_{j,\lambda} \rangle}{a^{\lambda^*}} = \frac{(a^{-1} f^-_{j,\lambda})(x_o)}{a^{\lambda^*}} = \frac{a^{\lambda^*} f^-_{j,\lambda}(x_o)}{a^{\lambda^*}} = 1, \tag{5.19}$$

shows that $k_{j,\lambda} = 1$. This implies (5.18).

Equation (5.18) yields immediately the convergence of the induced maps of function spaces:

$$\lim_{a \to \infty} q^*_{\infty,a} \circ \iota^*_{\infty,a} = q^*_\infty \circ \iota^*_\infty. \tag{5.20}$$

It was shown in [DÓW12, Thm. 4.7] that the limit

$$\mathbf{c}_\infty(\mu_\infty) := \lim_{j \to \infty} \mathbf{c}(\mu_j + \rho_j)$$

exists and is strictly positive if the rank of $\mathbf{X}_\infty$ is finite. Define

$$\mathcal{R}^*_\infty : \mathbb{C}_i[\Xi_\infty] \to \mathbb{C}_i[\mathbf{Z}_\infty]$$

by

$$\mathcal{R}^*_\infty(f)(x) = \lim_{j \to \infty} \mathcal{R}^*_j f(x). \tag{5.21}$$

Theorem 5.22. *Assume that the rank of* $\mathbf{X}_\infty$ *is finite. Let* $f \in \mathbb{C}_i[\Xi_\infty]$. *Then the pointwise limit (5.21) exists and for* $f \in \mathbb{C}[\Xi_\infty]_{\mu_\infty}$ *we have*

$$\mathcal{R}^*_\infty f = \mathbf{c}_\infty(\mu_\infty)^{1/2} \Gamma^{-1}_\infty f \quad and \quad \mathcal{R}^*_\infty(\iota^*(f)) = \mathbf{c}_\infty(\mu_\infty) f.$$

Proof. As every function in $\mathbb{C}_i[\Xi_\infty]$ is a finite sum of elements in $\mathbb{C}[\mathbf{X}_\infty]_\nu$ we only have to show this for fixed $\mu_\infty \in \Lambda^+_\infty$. But then the claim follows from (5.4), Theorem 5.17, Proposition 5.6, part (iv), and Proposition 5.7. $\square$

As we are assuming that the rank of $\mathbf{X}_\infty$ is finite, it follows from [DÓW12] that $\left(\hat{V}^*_{\mu_\infty}\right)^{K_\infty} \neq \{0\}$. Denote by proj_∞ the orthogonal projection

$$\mathrm{proj}_\infty : \hat{V}^*_{\mu_\infty} \to \left(\hat{V}^*_{\mu_\infty}\right)^{K_\infty}.$$

It follows also from the calculations in [DÓW12] that the sequence $\{e^*_{\mu_j}\}$ converges to $e^*_{\mu_\infty}$ in the Hilbert space $\hat{V}^*_{\mu_\infty} = \hat{V}_{\mu^*_\infty}$. Hence

$$\text{proj}_\infty\left(\hat{V}^*_{\mu_\infty}\right) = \left(\hat{V}^*_{\mu_\infty}\right)^{K_\infty} \subset \varprojlim V^*_{\mu_j}.$$

Finally, a simple calculation shows that

$$\text{proj}_\infty(w) = \lim_{j\to\infty} \int_{K_j} \pi_{\mu_\infty}(k)w\,dk. \tag{5.23}$$

If $f = f_{w,\mu_\infty} \in \mathbb{C}[\mathbf{Z}_\infty]_{\mu_\infty}$, then there exists j_o such that $f|_{\mathbf{Z}_j} = f_{w,\mu_j} \in \mathbb{C}[\mathbf{Z}_j]_{\mu_j}$ for all $j \geq j_o$. We have

$$\mathcal{R}_{a,j}(f|_{\mathbf{Z}_j})(g\cdot\mathcal{S}_a) = \int_{K_j} \langle w, \pi^*_{\mu_j}(g)\pi^*_{\mu_j}(k)\pi^*_{\mu_j}(a)e^*_{\mu_j}\rangle\,dk. \tag{5.24}$$

Theorem 5.25. *Let $f \in \mathbb{C}_i[\mathbf{Z}_\infty]$. Then the pointwise limit*

$$\mathcal{R}_{a,\infty}f(g\cdot\mathcal{S}_a) := \lim_{j\to\infty} \mathcal{R}_{a,j}f(g\cdot\mathcal{S}_a)$$

exists and the following holds:

(1) $\mathcal{R}_{a,\infty}f(g\cdot\mathcal{S}_a) = \langle w, \pi^*_{\mu_\infty}(g)\text{proj}_\infty(\pi^*_{\mu_\infty}(a)e^*_{\mu_\infty})\rangle$ *if* $f \in \mathbb{C}_i[\mathbf{Z}_\infty]_{\mu_\infty}$.
(2) $\lim\limits_{a\to\infty} \mathcal{R}^*_{a,\infty} \circ \iota^*_a = \mathcal{R}^*_\infty \circ \iota^*$.

Proof. This follows from (5.23), (5.24), and Proposition 5.6.   □

Remark 5.26. We note that the following diagrams do not commute

$$
\begin{array}{ccc}
\mathbb{C}[\Xi_j] & \xrightarrow{\iota_{k,j}} & \mathbb{C}[\Xi_k] \\
{\scriptstyle \mathcal{R}^*_j}\downarrow & & \downarrow{\scriptstyle \mathcal{R}^*_k} \\
\mathbb{C}[\mathbf{X}_j] & \xrightarrow{\iota_{k,j}} & \mathbb{C}[\mathbf{X}_k]
\end{array}
\qquad
\begin{array}{ccc}
\mathbb{C}[\text{Sph}_{a_j}(\mathbf{X}_j)] & \xrightarrow{\iota_{k,j}} & \mathbb{C}[\text{Sph}_{a_k}(\mathbf{X}_k)] \\
{\scriptstyle \mathcal{R}^*_{j,a_j}}\downarrow & & \downarrow{\scriptstyle \mathcal{R}^*_{k,a_k}} \\
\mathbb{C}[\mathbf{X}_j] & \xrightarrow{\iota_{k,j}} & \mathbb{C}[\mathbf{X}_k]
\end{array}
$$

This follows from the corresponding commutative diagrams for the normalized dual Radon transforms γ_j^{-1} and the normalizing factor that relates those two transforms. This makes the corresponding theory for infinite rank spaces problematic as in that case $\lim_{j\to\infty} \mathbf{c}(\mu_j + \rho_j) = 0$.   □

Acknowledgements We would like to thank E.B. Vinberg, who suggested to us that dual horospherical Radon transforms may exist also for limits of symmetric spaces.

References

[DÓ13] M. Dawson and G. Ólafsson, *Conical representations for direct limits of symmetric spaces*. In preparation.

[DÓW12] M. Dawson, G. Ólafsson and J. Wolf, *Direct Systems of Spherical Functions and Representations*, Journal of Lie Theory **23** (2013), 711–729.

[E73] P.E. Eberlein, *Geodesic flows on negatively curved manifolds*. II, Trans. Amer. Math. Soc. **178** (1973), 57–82.

[E96] ———, *Geometry of nonpositively curved manifolds,* Chicago Lectures in Mathematics, University of Chicago Press, 1996.

[G06] S. Gindikin, *The horospherical Cauchy-Radon transform on compact symmetric spaces*, Mosc. Math. J. **6** (2006), 299–305.

[GK62] S. Gindikin and F. I. Karpelevich, *Plancherel measure for symmetric Riemannian spaces of non-positive curvaure*, Dokl. Akad. Nauk. SSSR **145** (1962), 252–255, English translation, Soviet Math. Dokl. **3** (1962), 1962–1965.

[GKÓ06] S. Gindikin, B. Krötz, and G. Ólafsson, *Holomorophic horospherical transform on non-compactly causal spaces*, IMRN **2006** (2006), 1–47.

[H66] S. Helgason, *A duality in integral geometry on symmetric spaces*, Proc. U.S.-Japan Seminar in Differential Geometry, Kyoto, Japan, 1965, pp. 37–56. Nippon Hyronsha, Tokyo 1966.

[H70] ———: *A duality for symmetric spaces with applications to group representations*, Advan. Math. **5** (1970), 1–154.

[H78] ———, *Differential Geometry, Lie Groups, and Symmetric Spaces*. Academic Press, 1978.

[H84] ———, *Groups and Geometric Analysis*, Academic Press, 1984.

[H94] ———, *Geometric Analysis on Symmetric Spaces*, Math. Surveys Monogr. **39**, Amer. Math. Soc. Providence, RI 1994.

[HPV02] J. Hilgert, A. Pasquale and E. B. Vinberg, *The dual horospherical Radon transform for polynomials*, Mosc. Math. J. **2** (2002), 113–126, 199.

[HPV03] ———, *The dual horospherical Radon transform as a limit of spherical Radon transforms*, Lie groups and symmetric spaces, 135–143, Amer. Math. Soc. Transl. Ser. 2, 210, Amer. Math. Soc., Providence, RI, 2003.

[HY00] Z. Huang and J. Yan, *Introduction to Infinite Dimensional Stochastic Analysis*, Mathematics and its Applications. Kluwer Academic Press, 2000.

[KS77] V. I. Kolomycev and J. S. Samoilenko, *On irreducible representation of inductive limits of groups*, Ukrainian Math. J. **29** (1977), 402–405.

[ÓW11a] G. Ólafsson and J. Wolf, *Extension of Symmetric Spaces and Restriction of Weyl Groups and Invariant Polynomials*, Contemporary Math. **544** (2011), 85–100.

[ÓW11b] ———, *The Paley-Wiener Theorem and Limits of Symmetric Spaces*. J. Geom. Anal. **24**(1) (2014), 1–31.

[O90] G. Olshanskii, *Unitary representations of infinite dimensional pairs (G, K) and the formalism of R. Howe*. In: Eds. A. M. Vershik and D. P. Zhelobenko: Representation of Lie Groups and Related Topics. Adv. Stud. Contemp. Math. **7**, Gordon and Breach, 1990.

[W09] ———, *Infinite dimensional multiplicity free spaces II: Limits of commutative nilmanifolds*, Contemporary Mathematics, **491** (2009), 179–208.

Cycle Connectivity and Automorphism Groups of Flag Domains

Alan Huckleberry

Dedicated to Arkady Onishchik on the occasion of his 80th birthday.

Abstract A flag domain D is an open orbit of a real form G_0 in a flag manifold $Z = G/P$ of its complexification. If D is holomorphically convex, then, since it is a product of a Hermitian symmetric space of bounded type and a compact flag manifold, $\mathrm{Aut}(D)$ is easily described. If D is not holomorphically convex, then in previous work it was shown that $\mathrm{Aut}(D)$ is a Lie group whose connected component at the identity agrees with G_0, except possibly in situations which arise in Onishchik's list of flag manifolds where $\mathrm{Aut}(Z)^0 = \hat{G}$ is larger than G. In the present work the group $\mathrm{Aut}(D)^0 = \hat{G}_0$ is described as a real form of $\hat{G}$. Using an observation of Kollar, new and much simpler proofs of much of our previous work in the case where D is not holomorphically convex are given.

Key words Flag domains • Automorphism groups • Finiteness theorem

Mathematics Subject Classification (2010): 14M15, 32M05, 57S20.

Research for this project was supported by SFB/TR 12 and SPP 1388 of the Deutsche Forschungsgemeinschaft.

A. Huckleberry (✉)
Institut für Mathematik, Ruhr Universität,
Bochum Universitätsstrasse 150, 44780 Bochum, Germany

Jacobs University Bremen, School of Engineering and Science,
Campus Ring 1, 28759 Bremen, Germany
e-mail: ahuck@gmx.de

© Springer International Publishing Switzerland 2014

G. Mason et al. (eds.), *Developments and Retrospectives in Lie Theory*,
Developments in Mathematics 37, DOI 10.1007/978-3-319-09934-7_4

113

1   Introduction and Statement of Results

Recall that if Z is a compact complex manifold, then its Lie algebra $\mathfrak{g} = \mathrm{Vect}_{\mathcal{O}}(Z)$ of holomorphic vector fields is finite-dimensional and that the fields in $\mathfrak{g}$ can be integrated to define a holomorphic action of the associated simply-connected complex Lie group G. If Z is homogeneous in the sense that this group acts transitively, then we choose a base point $z_0 \in Z$, let $H = G_{z_0}$ denote the isotropy group at that point, and identify Z with the quotient G/H. If Z is projective algebraic with trivial Albanese, i.e., with $b_1(Z) = 0$, then G is semisimple, the isotropy group H is a so-called parabolic subgroup, which from now on we denote by P, and $Z = G/P$ is a G-orbit in the projective space $\mathbb{P}(V)$ of an appropriate G-representation space V. In this case we refer to Z as a flag manifold.

A real form G_0 of G is a real Lie subgroup of G such that the complexification $\mathfrak{g}_0 + i\,\mathfrak{g}_0$ is the Lie algebra $\mathfrak{g}$. If $Z = G/P$ is a flag manifold, then any real form G_0 of G has only finitely many orbits in Z ([W]; see also [FHW] for this as well as other background.). In particular, G_0 always has at least one open orbit D. We refer to such an open orbit as a flag domain. If G_0 is not simple, then, D has product structure corresponding to the factors of G_0. Thus, for our considerations here there is no loss of generality in assuming that G_0 is simple which we do throughout. Note that if G_0 has the abstract structure of a complex Lie group, then its complexification G is, however, not simple. Note also that G_0 could act transitively on Z, e.g., this is always the case for a compact real form. However, from the point of view of this article, in that case all phenomena are well understood and therefore we assume that D is a proper subset of Z.

Since by assumption a flag domain D is noncompact, there is no a priori reason to expect that $\mathrm{Aut}(D)$ or $\mathrm{Vect}_{\mathcal{O}}(D)$ is finite-dimensional. In fact if D possesses non-constant holomorphic functions, the latter is not the case and the former is often not the case as well. Let us begin here by reviewing this situation.

If X is any complex manifold, then the equivalence relation,

$$x \sim y \ \Leftrightarrow \ f(x) = f(y) \text{ for all } f \in \mathcal{O}(X),$$

is equivariant with respect to the full group $\mathrm{Aut}(X)$ of holomorphic automorphisms. If $X = G/H$ is homogeneous with respect to a Lie group of holomorphic transformations, then the reduction $X \to X/\sim$ by this equivalence relation is a G-equivariant holomorphic homogeneous fibration $G/H \to G/I$. If $D = G_0/H_0$ is a flag domain, then this reduction has a particularly simple form ([W, FHW], Sect. 4.4). For this let $D = G_0.z_0$ with H_0 (resp. P) be the G_0-isotropy subgroup (resp. G-isotropy subgroup) at z_0.

Theorem 1.1. *If* $D = G_0.z_0$ *is a flag domain with* $\mathcal{O}(D) \neq \mathbb{C}$, *then the holomorphic reduction* $D = G_0/H_0 \to G_0/I_0 = \tilde{D}$ *is the restriction of a fibration* $Z = G/P \to G/\tilde{P} = \tilde{Z}$ *of the ambient flag manifold with the properties*

1. The fiber of $Z \to \tilde{Z}$, which itself is a flag manifold, agrees with the fiber of $D \to \tilde{D}$.
2. The base $\tilde{D}$ is a G_0-flag domain in $\tilde{Z}$. It is a Hermitian symmetric space of noncompact type embedded in a canonical way in its compact dual $\tilde{Z}$.

Recall that a symmetric space of noncompact type of a simple Lie group is a topological cell and that in the Hermitian case it is a Stein manifold. Thus Grauert's Oka principle implies that the fibration $D \to \tilde{D}$ is a (holomorphically) trivial bundle. As a consequence we have the following more refined version of the above result.

Corollary 1.2. A flag domain D with $\mathcal{O}(D) \neq \mathbb{C}$ is the product $\tilde{D} \times F$ of a Hermitian symmetric space $\tilde{D}$ of noncompact type and a compact flag manifold F. In particular, D is holomorphically convex and $D \to \tilde{D}$ is its Remmert reduction.

As indicated above our goal here is to describe the connected component at the identity $\text{Aut}(D)^0$ of the group of holomorphic automorphisms of any given flag domain D. With certain exceptions which we cover in detail below, we carried out this project in [H1] by studying the associated action of $\text{Aut}(D)$ on a certain space (described below) $\mathcal{C}_q(D)$ of holomorphic cycles. If $D = \tilde{D}$ is a Hermitian symmetric space of noncompact type, such cycles are just isolated points and $\mathcal{C}_q(D) = D$. Thus the cycle space gives us no additional information. However, in this case D possesses the invariant Bergman metric and as a result $\text{Aut}(D)$ is well-understood.

If $D = \tilde{D} \times F$ is a product with nontrivial base and fiber, then, although it is infinite-dimensional, $\text{Aut}(D)$ is in a certain sense easy to describe: The fibration $D \to \tilde{D}$ induces a surjective homomorphism $\text{Aut}(D) \to \text{Aut}(\tilde{D})$. The kernel is the space $\text{Hol}(\tilde{D}, \text{Aut}(F))$ of holomorphic maps from the base to the complex Lie group $\text{Aut}(F)$ and as a result $\text{Aut}(D) = \text{Hol}(\tilde{D}, \text{Aut}(F)) \rtimes \text{Aut}(D)$ has semidirect product structure.

Having settled the case where $\mathcal{O}(D) \neq \mathbb{C}$, or equivalently where D is holomorphically convex, we turn to the situation where $\mathcal{O}(D) = \mathbb{C}$. In [H1] we showed that $\text{Aut}(D)$ is a (finite-dimensional) Lie group which, with certain exceptions that are handled below, $\text{Aut}(D)^0 = G_0$. Other than taking care of these exceptional cases, where in fact $\text{Aut}(D)^0$ contains G_0 as a proper Lie subgroup, here we also make use of an observation of Kollar ([K]) which leads to a simple proof of $\text{Aut}(D)^0 = G_0$ with the possible exceptions. This proof is given in Sect. 2.

Before going into the details of proofs, let us state the main result of the paper. For this the following classification theorem of A. Onishchik ([O1, O2]) is the key first step for handling the exceptional cases mentioned above.

Theorem 1.3. The following is a list of the flag manifolds Z and (connected) complex simple Lie groups G and $\hat{G}$ so that $Z = G/P$ and $\hat{G} := \text{Aut}(Z)^0$ properly contains G.

1. The manifold Z is the odd-dimensional projective space $\mathbb{P}(\mathbb{C}^{2n})$ where, after lifting to simply-connected coverings, $G = \text{Sp}_{2n}(\mathbb{C})$ and $\hat{G} = \text{SL}_{2n}(\mathbb{C})$.

2. *The five-dimensional complex quadric Z is equipped with the standard action of $\hat{G} = \mathrm{SO}_7(\mathbb{C})$ and G is the exceptional complex Lie group G_2 embedded in $\hat{G}$ as the automorphism group of the octonions.*

3. *Equipping $\mathbb{C}^{2n}$ with a nondegenerate complex bilinear form b, Z is the space of n-dimensional b-isotropic subspaces, $\hat{G}$ is the b-orthogonal group $\mathrm{SO}_{2n}(\mathbb{C})$ and G is the complex orthogonal group $\mathrm{SO}_{2n-1}(\mathbb{C})$ which is embedded in $\hat{G}$ as the connected component at the identity of the isotropy group of the $\hat{G}$-action at some nonzero point in $\mathbb{C}^{2n}$.*

Referring to the above list of exceptions as Onishchik's list, our main result can be stated as follows.

Theorem 1.4. *If D is a G_0-flag domain in $Z = G/Q$, then $\mathrm{Aut}(D)$ can be described as follows:*

1. *If $\mathcal{O}(D) \neq \mathbb{C}$, or equivalently if it is holomorphically convex, D is a product $\tilde{D} \times F$ of a Hermitian symmetric space $\tilde{D}$ of noncompact type and a compact flag manifold F, and $\mathrm{Aut}(D)$ is correspondingly a semidirect product $\mathrm{Aut}(D) = \mathrm{Hol}(D, \mathrm{Aut}(F)) \rtimes \mathrm{Aut}(\tilde{D})$.*

2. *If $\mathcal{O}(D) = \mathbb{C}$, then $\mathrm{Aut}(D)$ is a finite-dimensional Lie group of holomorphic transformations on D and, if the complexification G is the full group $\mathrm{Aut}(Z)^0$, then $\mathrm{Aut}(D)^0 = G_0$.*

3. *If $\mathcal{O}(D) = \mathbb{C}$ and G is a proper subgroup of $\hat{G} = \mathrm{Aut}(Z)^0$, then in each case of Onishchik's list $\mathrm{Aut}(D)^0 = \hat{G}_0$ is a uniquely determined real form of $\hat{G}$ which contains G_0 as a proper subgroup.*

It should be remarked that the simple proof given here of the fact that if $\mathcal{O}(D) = \mathbb{C}$, then $\mathrm{Aut}(D)^0$ is a Lie group acting on Z does not yield a proof that in this case full group $\mathrm{Aut}(D)$ is a Lie group. At the present time we have no other proof of this fact other than that in [H1].

2   Cycle Connectivity

In [H2] we used chains of cycles to study the pseudo-convexity and pseudoconcavity of flag domains . We continued the use of these chains in our study of $\mathrm{Aut}(D)$ in [H1]. Here, in particular, compared to the chains in [K], it is sufficient to consider chains of a very special type which we now introduce.

A basic fact, which is the tip of the iceberg of Matsuki duality, is that for a flag domain D, any given maximal compact subgroup K_0 of G_0 has exactly one orbit $C_0 = K_0.z_0$ in D which is a complex submanifold. In fact it is the (unique) orbit of minimal dimension. If K is the complexification of K_0, then since C_0 is complex, K stabilizes it. Denoting $q := \dim_{\mathbb{C}} C_0$, we usually regard C_0 as a point in the Barlet cycle space $\mathcal{C}_q(D)$, but for our purposes here we may regard it as a point in the full Chow space $\mathcal{C}_q(Z)$ where G is acting algebraically. The group theoretical cycle space of D is then defined as the connected open subset

$$\mathcal{M}(D) = \{g(C_0) : g \in G, \ g(C_0) \subset D\}^0$$

of the orbit of the base cycle C_0. One can show that $\mathcal{M}(D)$ is a closed submanifold of $\mathcal{C}_q(D)$ (See [FHW] for background and a systematic study of these cycle spaces.). For the purposes of this paper a chain of cycles is a finite connected union of (supports of) cycles in $\mathcal{M}(D)$. We often write such a chain as $(C_1, \ldots, C_m)$ to indicate that $C_i \cap C_{i+1} \neq \emptyset$. Using such chains we have the *cycle connection* equivalence relation

$$x \sim y \ \Leftrightarrow \ x \text{ and } y \text{ are contained in a chain}.$$

Note that this relation is G_0-equivariant. In particular, if $D = G_0/H_0$, then there is a (possibly not closed) subgroup I_0 of G_0 which contains H_0 so that the quotient of D by this equivalence relation is given by $G_0/H_0 \to G_0/I_0$. Now if $z_0 \in D$ is the base point where $H_0 := G_{z_0}$ and $K_0.z_0 = K.z_0 = C_0$ is the base cycle, then, since C_0 is by definition contained in the equivalence class of z_0, it is immediate that $I_0 \supset K_0$. Since K_0 is a maximal subgroup of G_0, i.e., any (not necessarily closed) subgroup of G_0 which contains K_0 is either K_0 or G_0, the following is immediate (see also [H1] and [H2] for the same proof).

Proposition 2.1. *The following are equivalent:*

1. $\mathcal{O}(D) = \mathbb{C}$
2. *D is not holomorphically convex.*
3. *There is no nontrivial G_0-equivariant holomorphic map of D to a Hermitian symmetric space $\tilde{D}$ of noncompact type.*
4. *D is cycle connected.*

Proof. The equivalence of the first three conditions follows from the discussion in Sect. 1. If D is cycle connected, then, since $\tilde{D}$ is Stein and therefore every holomorphic map to $\tilde{D}$ is constant along every chain, (4) $\Rightarrow$ (3). Conversely, if D is not cycle connected, then the equivalence class containing the base point z_0 is just the cycle C_0 which is therefore stabilized by the G-isotropy P as well as K. Since the cycle connection reduction is given by $G_0/H_0 \to G_0/K_0$, it follows that this fibration is the restriction of the fibration $G/P \to G/\tilde{P}$ of Z where $\tilde{P} = KP$ and therefore the base G_0/K_0 is the Hermitian symmetric space $\tilde{D}$. In other words, the cycle connected reduction is just the holomorphic reduction and in particular $\mathcal{O}(D) \neq \mathbb{C}$.   □

Remark. For applications in another context, Griffiths, Robles and Toledo recently gave another proof a result which is essentially equivalent to Proposition 2.1. (see [GRT]).

Although it is well-known that K_0 is a maximal subgroup of G_0, for the convenience of the reader we would like to give the following nice proof of J. Brun which was pointed out to us by Keivan Mallahi Karai (see the Appendix of [B]).

Theorem 2.2. *If G is a connected simple Lie group, K is a maximal compact subgroup and L is an abstract group which contains K, then L is either G or K.*

Proof. Standard results in the theory of symmetric spaces show that K is connected and the adjoint representation of K on $\mathfrak{g}/\mathfrak{k}$ is irreducible. Thus if ℓ is a Lie subalgebra of $\mathfrak{g}$ which properly contains $\mathfrak{k}$, then $\ell = \mathfrak{g}$. Thus, if L is closed, then the result is immediate. Furthermore, if L is not closed and properly contains K, then its closure $c\ell(L)$ is the full group G. In that case we let $\mathfrak{k}'$ be the vector subspace of the Lie algebra $\mathfrak{g}$ of G which is generated by $\mathrm{Ad}(x)(\mathfrak{k})$ for all $x \in L$. Since $c\ell(L) = G$, it follows by continuity that $\mathfrak{k}'$ is G-invariant and since G is simple, it is immediate that $\mathfrak{k}' = \mathfrak{g}$. Therefore there are finitely many elements $x_i \in L$ so that

$$\mathfrak{g} = \sum_1^m \mathrm{Ad}(x_i)(\mathfrak{k})$$

and as a result the map

$$K^m \to G, \ K(k_1, \ldots, k_m) \mapsto \prod(x_i k_i x_i^{-1})$$

has maximal rank at the origin. Thus L contains a neighborhood of the origin and therefore, contrary to assumption $L = G$. □

3 Finiteness Theorem

Our original goal in this setting was to show that a flag domain D is either pseudo-convex or pseudoconcave ([H2]). More precisely, we had hoped to show that if D is not holomorphically convex, then C_0 has a pseudoconcave neighborhood which is filled out by cycles. If this would be possible, then using Andreotti's finiteness theorem ([A]) we would be able to conclude that the space of sections of any holomorphic vector bundle, in particular the space $\mathrm{Vect}_{\mathcal{O}}(D)$, is finite-dimensional. Although we have been successful in constructing such a neighborhood in a number of cases ([H2]), we have failed do this in general. Recently, in a substantially more general setting, Kollar proved the desired finiteness theorem along with a number of equivalent properties which would follow from the pseudoconcavity of D ([K]). Here we make use of Kollar's result, leaving the question of existence of the pseudoconcave neighborhood open.

Formulated in our setting, Kollar's finiteness result can be stated as follows.

Theorem 3.1. *The space $\Gamma(D, E)$ of sections of any holomorphic vector bundle on a cycle connected flag domain is finite-dimensional.*

This is an immediate consequence of the same result for line bundles which in turn is proved using the following lemma (Lemma 15 in [K]), again formulated in our restricted context.

Lemma 3.2. *Let L be a holomorphic line bundle on D. Then, given $d \in \mathbb{N}$, there exists $d_0 \in \mathbb{N}$ so that for every $C \in \mathcal{M}_D$ and any $z_0 \in C$ every section $s \in \Gamma(D, L)$ which vanishes of order d_0 at z_0 vanishes of order d along C.*

The proof is given by classical methods which are reminiscent of Siegel's Schwarz Lemma. One key point is that C can be filled out by rational curves which in our case are closures of orbits of one-parameter groups.

Now, given a chain of cycles $(C_1, \ldots, C_m)$ with $z_i \in C_i \cap C_{i+1}$, and given $d_m \in \mathbb{N}$ we apply the lemma to obtain $d_{m-1} \in \mathbb{N}$ so that if s vanishes of order d_{m_1} at z_{m-1}, then it vanishes of order d_m along C_m. Working backwards to the first cycle in the chain, we see that the lemma holds for chains.

Corollary 3.3. *Given $d \in \mathbb{N}$ there exists $d_1 \in \mathbb{N}$ so that for any chain $(C_1, \ldots, C_m)$ of length m and any $z_1 \in C_1$ if s vanishes of order d_1 at z_1, then it vanishes of order d along C_m.*

It should be emphasized that for a fixed d, the required vanishing order d_1 depends on m. Thus to apply this result we need some sort of uniform estimate for the length of a chain connecting two given points. This can be given as follows.

For example, let C_1 be a base cycle for a given maximal compact subgroup K_0. Recall that the complexification K has only finitely many orbits in Z and therefore has a (unique) open dense orbit Ω. Take $z_1 \in C_1$ and any point $z \in \Omega$ and let $(C_1, \ldots, C_m)$ be a chain connecting z_0 to z. For $k \in K$ sufficiently close to the identity, the chain $(k(C_1), \ldots, k(C_m))$ is still contained in D. Thus, since $k(C_1) = C_1$ and $k(z)$ can be an arbitrary point in a sufficiently small neighborhood U of z, we have the desired vanishing theorem.

Corollary 3.4. *If $s \in \Gamma(D, L)$ vanishes of sufficiently high order at a given point $z_1 \in C_1$, then it vanishes identically. In particular, $\Gamma(D, L)$ is finite-dimensional.*

Proof. Since the required vanishing order d_1 only depends on the number d_m and the length m, Corollary 3.3 implies that if s vanishes of order d_1 at z_1, then it vanishes at every point of the set U which was constructed above. The desired result then follows from the identity principle. $\square$

As we remarked above, the finiteness theorem for vector bundles is an immediate consequence of this corollary (see [K], p. 8).

4 Integrability of Vector Fields

The following is the main result of this section.

Theorem 4.1. *Let $Z = G/Q$ be a complex flag manifold and $\hat{\mathfrak{g}}$ a finite-dimensional complex Lie algebra which contains $\mathfrak{g} := \mathrm{Lie}(G)$. Let $\hat{G}$ be a complex Lie group* *which contains G and is associated to* $\hat{\mathfrak{g}}$. *If* $\hat{\mathfrak{q}}$ *is a complex subalgebra of* $\hat{\mathfrak{g}}$ *so that the quotient map* $\hat{\mathfrak{g}} \to \hat{\mathfrak{g}}/\hat{\mathfrak{q}}$ *induces an isomorphism*

$$\hat{\mathfrak{g}}/\hat{\mathfrak{q}} = \mathfrak{g}/\mathfrak{q},$$

then $\hat{G}$ *acts holomorphically on* Z *with*

$$Z = \hat{G}/\hat{Q} = G/Q.$$

Proof. We apply a basic idea of Tits. For this regard $x_0 := \hat{\mathfrak{q}}$ as a point in the Grassmannian $X := \mathrm{Gr}_k(\hat{\mathfrak{g}})$ of subspaces of dimension $k = \dim_{\mathbb{C}}\hat{\mathfrak{q}}$ in $\hat{\mathfrak{g}}$. The isotropy group at x_0 of the $\hat{G}$-action on X is the normalizer

$$\hat{N} = \{\hat{g} \in \hat{G} : \mathrm{Ad}(\hat{g})(\hat{\mathfrak{q}}) = \hat{\mathfrak{q}}\}.$$

Denote by $N = \hat{N} \cap G$ the G-isotropy at x_0 and note that if $g \in N$ and $\xi \in \mathfrak{q}$, it follows that $\mathrm{Ad}(g)(\xi) \in \hat{\mathfrak{q}} \cap \mathfrak{g} = \mathfrak{q}$. In other words N is contained in the normalizer of $\mathfrak{q}$ in $\mathfrak{g}$. Since the parabolic group Q is self-normalizing in G, it follows that $N \subset Q$. But $\hat{\mathfrak{n}} \supset \hat{\mathfrak{q}}$ and $\hat{\mathfrak{q}} \cap \mathfrak{g} = \mathfrak{q}$. Therefore $\mathfrak{n} \supset \mathfrak{q}$. Consequently $N = Q$ and the G-orbit of $\hat{\mathfrak{q}}$ is the compact manifold $Z = G/Q$. Since $\hat{\mathfrak{g}}/\hat{\mathfrak{q}} = \mathfrak{g}/\mathfrak{q}$, the $\hat{G}$-orbit of $\hat{\mathfrak{q}}$ has dimension at most that of G/Q. But on the other hand $\hat{G} \supset G$ and therefore the $\hat{G}$-orbit has the same dimension as the G-orbit. Consequently $G.x_0$ is open in $\hat{G}.x_0$ and the compactness of $G.x_0$ implies that these orbits agree.   □

Applying the Finiteness Theorem, the following is now immediate.

Corollary 4.2. *Let* G_0 *be a simple real form of a complex semisimple Lie group* G *and let* D *be a cycle connected* G_0-*flag domain in a* G-*flag manifold* $Z = G/Q$. *Let* $\hat{\mathfrak{g}}$ *be the Lie algebra of holomorphic vector fields on* D. *Then the restriction mapping* $R : \mathrm{aut}(Z) \to \hat{\mathfrak{g}}$ *is an isomorphism and the action of* $\hat{\mathfrak{g}}$ *can be integrated to the action of a connected complex Lie group* $\hat{G}$ *which is thereby identified with* $\mathrm{Aut}(Z)^0$.

As a consequence we have the description of $\mathrm{Aut}(D)$ which was proved by other means in [H1].

Corollary 4.3. *If* D *is a* G_0-*flag domain* $Z = G/Q$, *then one of the following holds:*

1. *If* D *is holomorphically convex, it is a product of a compact flag manifold and Hermitian symmetric space of noncompact type and* $\mathrm{Aut}(D)^0$ *can be described as in Sect. 1.*
2. *If* D *is not holomorphically convex or equivalently it is cycle connected, then* $\mathrm{Aut}(D)^0$ *is a finite-dimensional Lie group which is acting on* Z *and agrees with* G_0 *with the possible exceptions in the situations classified by Onishchik where* G *is a proper subgroup of* $\hat{G} = \mathrm{Aut}(Z)^0$.

5 Exceptional Cases

To complete our project of understanding the automorphism groups of flag domains, we must analyze the exceptional cases indicated in the above corollary. We do this here, proving the following result.

Theorem 5.1. *Suppose that $D \subset Z = G/Q$ is a G_0-flag domain which is not holomorphically convex and that G is properly contained in complex Lie group $\hat{G} = \mathrm{Aut}(Z)^0$. Then there is a uniquely determined real form $\hat{G}_0 = \mathrm{Aut}(D)^0$ of $\hat{G}$ which properly contains G_0 and which stabilizes D.*

Our proof of this fact amounts to a concrete discussion for each of the three classes of exceptions in Onishchik's list which was given in Sect. 1. Below we show that these cases not only occur but also occur at the level of real forms. This is the content of (3) in Theorems 1.4 and 5.1 above.

5.1 Projective Space

Here we consider the case where $Z = \mathbb{P}(V)$ is the projective space of an even-dimensional complex vector space $V = \mathbb{C}^{2n}$. Define the complex bilinear form b by $b(z, w) = z^t w$. In the standard basis $(e_1, \ldots, e_{2n})$ define $J : V \to V$ by $J(e_i) = e_{n+i}, i \leq n$, and $J(e_i) = -e_{i-n}, i > n$. Note J is b-orthogonal with $J^2 = -\mathrm{Id}$ and define a (complex, bilinear) symplectic form by $\omega(z, w) = z^t J w$. Define $V_+ := \mathrm{Span}\{e_1, \ldots, e_n\}$ and $V_- = \mathrm{Span}\{e_{n+1}, \ldots, e_{2n}\}$ and correspondingly $E := +\mathrm{Id} \oplus -\mathrm{Id}$. If $C : V \to V$ denotes the standard complex conjugation given by $z \mapsto \bar{z}$, define a nondegenerate (mixed-signature) Hermitian structure on V by $h(z, w) = z^t E C(w)$. Finally, if the antilinear map $\varphi : V \to V$ is defined by $z \mapsto -JECw$, it follows that $h(z, w) = \omega(z, \varphi(w))$ and, since $\varphi^2 = -\mathrm{Id}$, that φ is an h-isometry. Observe that if P is a φ-invariant subspace of V, then $P^{\perp_h} = P^{\perp_\omega}$. In particular, P is symplectic if and only if it is h-nondegenerate and in either of these cases $V = P \oplus P_h^\perp$ is a decomposition of V into h-nondegenerate, symplectic subspaces.

The complex symplectic group $G = \mathrm{Sp}_{2n}(\mathbb{C})$ defined by ω has two types of real forms. The first case to be considered is where $\hat{G}_0$ is the real form $\mathrm{SU}(n, n)$ of $\hat{G} = \mathrm{SL}_{2n}(\mathbb{C})$ which is defined as the group of h-isometries. In this case the real form $G_0 = \mathrm{Sp}_{2n}(\mathbb{R})$ of $G = \mathrm{Sp}_{2n}(\mathbb{C})$ is defined as the intersection $\hat{G}_0 \cap \mathrm{Sp}_{2n}(\mathbb{C})$. Considering the orbits of these groups on $\mathbb{P}(V)$ we let D_+ (resp. D_-) be the open sets in $\mathbb{P}(V)$ of h-positive (reps. h-negative) lines.

Proposition 5.2. *The open sets D_+ and D_- are both orbits of G_0 and $\hat{G}_0$.*

Proof. It is clear that D_+ and D_- are G_0- and $\hat{G}_0$-invariant and that $D_+ \cup D_-$ is dense in Z. Since $G_0 \subset \hat{G}_0$, it is therefore enough to show that G_0 acts transitively on both sets. The proof for D_+ is exactly the same as for D_- and therefore we only give it for D_+. For this, given positive lines $L = \mathbb{C}.z$ and $\tilde{L} = \mathbb{C}.\tilde{z}$, we define

$P = \text{Span}\{z, \varphi(z)\}$ and $\tilde{P} = \text{Span}\{\tilde{z}, \varphi(\tilde{z})\}$. These planes are h-nondegenerate and symplectic. We normalize z and $\tilde{z}$ so that $\|z\|_h^2 = \|\tilde{z}\|^2 = 1$ and, since $\varphi : E_+ \to E_-$, $\|\varphi(z)\|^2 = \|\varphi(\tilde{z})\|^2 = -1$. Applying this procedure to $P^\perp$ and $\tilde{P}^\perp$ we have h- and ω-orthogonal decompositions

$$V = P_1 \oplus \ldots \oplus P_n = \tilde{P}_1 \oplus \ldots \oplus \tilde{P}_n$$

of V. Furthermore, every P_i (resp. $\tilde{P}_i$) comes equipped with a basis $(z_i, \varphi(z_i))$ (resp. $(\tilde{z}_i, \varphi(\tilde{z}_i))$) such that the mapping $T_i : P_i \to \tilde{P}_i$ defined by $z_i \to \tilde{z}_i$ and $\varphi(z_i) \mapsto \varphi(\tilde{z}_i)$ is both symplectic and an h-isometry. It follows that $T = T_1 \oplus \ldots \oplus T_n$ is both a symplectic isomorphism and h-isometry of V, i.e., $T \in G_0$. Since $T(L) = \tilde{L}$, the proof is complete. $\qquad\square$

Now let us turn to the real form $G_0 = \text{Sp}(2p, 2q)$ of $G = \text{Sp}_{2n}(\mathbb{C})$. In this case we line up J and E in a different way. The decomposition $V := V_+ \oplus V_-$ and J are the same, but now h has signature (p, q) on both spaces, being defined by the block diagonal matrix $E_{p,q} = (\text{Id}_p, -\text{Id}_q)$. Then $\hat{G}_0 = \text{SU}(2p, 2q)$ is defined as above by the Hermitian form h and $G_0 = G \cap \hat{G}_0$. The proof of the following fact is exactly the same as that of Proposition 5.2 above.

Zusatz. Proposition 5.2 also holds for $G_0 = \text{Sp}(2p, 2q)$ and $\hat{G}_0 = \text{SU}(2p, 2q)$. $\qquad\square$

5.2   Five-Dimensional Quadric

Here we consider $V = \mathbb{C}^7$ equipped with the complex bilinear form b defined by $\|z\|_b^2 = (z_1^2 + z_2^2 + z_3^2) - (z_4^2 + \ldots + z_7^2)$ and Hermitian form h defined by $\|z\|_h^2 = (|z_1|^2 + |z_2|^2 + |z_3^2|^2) - (|z_3|^2 + \ldots + |z_7|^2)$. Denote by $\hat{G} = \text{SO}_7(\mathbb{C})$ the associated complex orthogonal group and by $\hat{G}_0 := \text{SO}(3, 4)$ the associated group of Hermitian isometries.

We regard the exceptional complex Lie group $G = G_2$ as being embedded in $\hat{G}$ as the automorphism group $\text{Aut}(\mathbb{O})$ of the octonions. It has a unique noncompact real form $G_0 = \text{Aut}(\tilde{\mathbb{O}})$, the automorphism group of the split octonions $\tilde{\mathbb{O}}$. In this way G_0 is the intersection $G \cap \hat{G}_0$ of G with the real from $\hat{G}_0 = \text{SO}(3, 4)$ (see, e.g., [Ha] for details). Note that $\hat{G}_0$ is invariant by the standard complex conjugation $z \mapsto \bar{z}$.

The remainder of this section is devoted to the proof of the following fact.

Proposition 5.3. *For every $z \in Z$ it follows that $G_0.z = \hat{G}_0.z$. In particular the open orbits of G_0 and $\hat{G}_0$ coincide.*

We should note that as indicated below, the open orbits of $\hat{G}_0$ are the spaces D_+ and D_- of positive and negative lines, respectively.

For the proof of Proposition 5.3 we use Matsuki duality (see, e.g., Chap. 8 in Part II of [FHW]) which states that there is a $1 - 1$ correspondence between the G_0-orbits and K-orbits in Z. This can be given as follows: For every G_0-orbit there is a unique K-orbit which intersects it in the unique K_0-orbit of minimal dimension and vice versa, i.e., given a K-orbit there is a unique G_0-orbit which intersects it in the unique K_0-orbit of minimal dimension. Due to our interest in the open G_0-orbits (resp. $\hat{G}_0$-orbits) in Z, we have stated the above result on that side of the duality. However, we have found it more convenient to prove the corresponding dual statement.

Let us fix the maximal compact subgroup $K_0 \cong (\mathrm{SU}_2 \times \mathrm{SU}_2)/(-\mathrm{Id}, -\mathrm{Id})$ of G_0 being diagonally embedded in the maximal compact subgroup $\hat{K}_0 = S(O(3) \times O(4))$ of $\hat{G}_0$. If $E_+ := \mathrm{Span}\{e_1, e_2, e_3\}$ and $E_- := \mathrm{Span}\{e_4, \ldots, e_7\}$, then we define $z_+ := e_1 + i e_2, z_- := e_4 + i e_5$ and observe that the base cycles C_+ and C_- for the open orbits of the $\hat{G}_0$-action are the quadrics of b-isotropic lines in E_+ and E_-, respectively. The corresponding open orbits are the spaces $D_+ = \hat{G}_0.z_+$ of positive lines in Z and $D_- = \hat{G}_0.z_-$ of negative lines, respectively. The complement of $D_+ \cup D_-$, which is the space of lines that are both b- and h-isotropic, consists of two $\hat{G}_0$-orbits, the real points $Z_{\mathbb{R}}$ and its complement.

The $\hat{K}$-orbits that correspond via Matsuki duality to the four $\hat{G}_0$-orbits are the two base cycles C_+ and C_-, the open $\hat{K}$-orbit of any point on $Z_{\mathbb{R}}$ and a fourth orbit $\mathcal{O}$ which has two ends, i.e., that has the two base cycles on its boundary. In fact this fourth orbit is a $\mathbb{C}^*$-principal bundle over the two-dimensional cycle C_- (see [FHW], Sect. 16.4 for a detailed discussion in the case of the $K3$-period domain which can be transferred verbatim to the case at hand). To prove the above Proposition 5.3 we show that K acts transitively on each of these four $\hat{K}$-orbits.

Now the second factor of $\hat{K}$ acts trivially on C_+ and the first factor acts trivially on C_- and vice versa. Since K is diagonally embedded in $\hat{K}$ and projects onto both factors, it is immediate that it acts transitively on both C_+ and C_- as well. Since $\mathcal{O}$ is a $\mathbb{C}^*$-bundle over C_-, K acts transitively on the base of this bundle and has an open orbit in the bundle space $\mathcal{O}$, it is immediate that it acts transitively on $\mathcal{O}$.

It remains to show that K acts transitively on the open $\hat{K}$-orbit. For this we first note that since $e_3 + e_4 \in Z_{\mathbb{R}}$ and the connected component at the identity of $\hat{K}_0$ is the product of the special orthogonal groups of E_+ and E_-, it follows that up to finite group quotients $Z_{\mathbb{R}}$ is the corresponding product $S^2 \times S^3$ of spheres. One immediately observes that G_0 acts transitively on $Z_{\mathbb{R}}$, because every G_0-orbit is at least half-dimensional over $\mathbb{R}$. Thus K_0 acts transitively on $Z_{\mathbb{R}}$ and if $z_{\mathbb{R}}$ is an arbitrary point of $Z_{\mathbb{R}}$, it follows that $K.z_{\mathbb{R}}$ is open in $\hat{K}.z_{\mathbb{R}}$.

To complete the proof of Proposition 5.3 we must show that $K.z_{\mathbb{R}} = \hat{K}.z_{\mathbb{R}}$. For this we let $\hat{K}_1$ be the first factor of the product decomposition of the connected component of $\hat{K}$ and consider the homogeneous fibration

$$\hat{K}.z_{\mathbb{R}} = \hat{K}/\hat{L} \to \hat{K}/\hat{K}_1\hat{L} = \hat{K}_2/\hat{L}_2 = B.$$

Since $Z_{\mathbb{R}}$ is essentially a product $S^2 \times S^3$ corresponding to the decomposition of the connected component $\hat{K}_0$, it follows that up to finite group quotients the base B is the complexification of S^3, i.e., the affine quadric $Q_{(3)} = \mathrm{SO}_4(\mathbb{C})/\mathrm{SO}_3(\mathbb{C})$. Since K

projects surjectively onto both factors of $\hat{K}$, it is immediate that K acts transitively on $B = K/M$. Now the induced fibration of $K.z_\mathbb{R}$ is a homogeneous bundle $K/L \to K/M$ where the fiber M/L is an open M-orbit in the corresponding fiber $\hat{F}$ of the $\hat{K}$-bundle $\hat{K}/\hat{L} \to \hat{K}/\hat{M}$. But $\hat{M}$ acts on this fiber as $SO_3(\mathbb{C})$ so that $\hat{F}$ is the affine quadric $Q_{(2)}$. Since K/M is affine, M is reductive. But the only reductive subgroup of $SO_3(\mathbb{C})$ with an open orbit in $Q_{(2)}$ is $SO_3(\mathbb{C})$ itself. Consequently K does indeed act transitively on the open $\hat{K}$-orbit and the proof of Proposition 5.3 is complete.   □

5.3   Space of Isotropic n-Planes in $\mathbb{C}^{2n}$

Now let $\hat{V} = \mathbb{C}^{2n}$ be equipped with its standard basis $(e_1, \ldots, e_{2n})$ and complex bilinear form defined by $b(z, w) = z^t w$. The complex orthogonal group $SO_{2n}(\mathbb{C})$ of b-isometries is denoted by $\hat{G}$. We let $G := \text{Fix}_{\hat{G}}(e_{2n})$. In this way $G \cong SO_{2n-1}(\mathbb{C})$ is the orthogonal group of the restriction of b to $V := \text{Span}\{e_1, \ldots, e_{2n-1}\}$. We consider the action of these groups on the flag manifold Z of n-dimensional b-isotropic subspaces of $\hat{V}$.

Proposition 5.4. *The groups G and $\hat{G}$ act transitively on Z.*

Proof. Note that the intersection $W := \hat{W} \cap V$ of an isotropic n-plane in $\hat{V}$ is an isotropic $(n-1)$-plane in V. It follows that $\hat{W} = W \oplus \mathbb{C}.(v + ie_{2n})$ for some $v \in V$. Applying an appropriate element of G, we may assume that $v = e_{2n-1}$ and it then follows that $W \subset \text{Span}\{e_1, \ldots, e_{2n-2}\}$. We then apply the induction assumption to obtain a transformation in the corresponding $SO_{2n-2}(\mathbb{C})$ to bring W to the normal form with basis $(e_1 + ie_{n+1}, \ldots, e_{n-1} + ie_{2n-2})$ so that altogether we have found a transformation in G which brings W to the normal form with the basis $(e_1 + ie_{n+1}, \ldots, e_{2n-1} + ie_{2n})$.   □

Recall that up to conjugation the only real forms of $SO_{2n-1}(\mathbb{C})$ are the isometry groups $G_0 = SO(p, q)$ for the mixed signature Hermitian form defined by $h(z, w) = z^t EC(w)$ on V where $E = E_{p,q}$ is defined in the same way as in Sect. 5.1. Without loss of generality we may choose h to be this form and note that an appropriately chosen arbitrarily small perturbation of an isotropic n-plane $\hat{W}$ will result in the intersection $W = \hat{W} \cap V$ being h-nondegenerate. Thus, if $G_0.z =: D$ is an open orbit in Z, the $(n-1)$-plane W associated to z is h-nondegenerate.

Note that if p is even, then q is odd and vice versa. To make the notation more explicit, we assume that p is even. Now the space of h-positive b-isotropic lines in V is an open G_0-orbit. Thus, given W as above, we may apply an element $g \in G_0$ so that after replacing W by $g(W)$ we have $L = \mathbb{C}(e_1 + ie_2) \subset W$. Notice that the subspace of V of vectors which are both h- and b-orthogonal to L is simply $\text{Span}\{e_3, \ldots, e_{2n-1}\}$. Thus, after going to this smaller space, we have the same situation as before. Hence we may continue on by induction to obtain a maximal h-positive subspace W_+ of W which is $\frac{p}{2}$-dimensional and which has a distinguished basis produced by our procedure. Applying the same argument as above to the h-complement $W_+^\perp$ in W, one obtains an element $g \in G_0$ so that $W_0 := g(W)$ has the distinguished basis

$$(e_1 + ie_2, e_3 + ie_4, \ldots, e_{p-1} + ie_p, e_{p+1} + ie_{p+2}, \ldots e_{2n-3} + ie_{2n-2}).$$

Proposition 5.5. *If $\hat{W}$ is a b-isotropic n-plane in $\hat{V}$, then there exists an element $g \in G_0$ with $g(\hat{W}) = W_0 \oplus \mathbb{C}(e_{2n-1} + ie_{2n}) =: \hat{W}_0$.*

Proof. Let $g \in G_0$ be chosen as above with $g(W) = W_0$. It is then immediate that $g(\hat{W}) = W_0 \oplus \mathbb{C}\hat{w}$ where $\hat{w} = \pm e_{2n-1} + ie_{2n}$. We obtain the positive sign by, e.g., multiplying e_1 and e_2 by i, e_{2n-1} by -1 and e_j by $+1$ otherwise. Since this transformation is also in G_0, the desired result follows. $\qquad\square$

Theorem 5.6. *The Hermitian form h can be naturally extended to a nondegenerate Hermitian form $\hat{h}$ on $\hat{V}$ with signature $(p, q+1)$ (resp. $(p+1, q)$) if p is even (resp. odd) so that the unique open orbit D of the resulting real form $\hat{G}_0$ is the set of isotropic n-planes of signature $(\frac{p}{2}, \frac{q+1}{2})$ (resp. $(\frac{p+1}{2}, \frac{q}{2})$). Furthermore, the h-isometry group G_0 in $SO_{2n-1}(\mathbb{C})$ also acts transitively on D which is also its unique open orbit in Z.*

Proof. It is enough to consider the case where p is even and $\|e_{2n-1}\|_h^2 = -1$. Extending h to $\hat{h}$ on $\hat{V}$ with e_{2n} being orthogonal to V and $\|e_{2n}\|^2 = -1$, it follows that $\hat{h}$ is of signature $(p, q+1)$. Let $\hat{G}_0 = SO(p, q+1)$ be the real form of $\hat{G} = SO_{2n}(\mathbb{C})$ defined by $\hat{h}$. Arguing as above, we see that the unique open $\hat{G}_0$-orbit D in Z is the set of isotropic n-planes $\hat{W}$ with signature $(\frac{p}{2}, \frac{q+1}{2})$. A reformulation of Proposition 5.5 is that G_0 also acts transitively on D. $\qquad\square$

The following is a less technical formulation of this fact.

Corollary 5.7. *If Z is the complex flag manifold of isotropic n-planes in $\mathbb{C}^{2n}$ where both $\hat{G}$ and G act transitively, every real form G_0 of G has a unique open orbit D which is the unique open orbit of a canonically determined real form $\hat{G}_0 = \mathrm{Aut}(D)^0$ of $\hat{G} = \mathrm{Aut}(Z)^0$.*

References

[A] A. Andreotti, Théorèmes de dépendance algébrique sur les espaces complexes pseudo-concaves, *Bull. Soc. Math. France* **91** (1963) 1–38.

[B] J. Brun, Sur la simplification par les variétés homogènes, *Math. Ann.* 230, (1977) 175–182.

[FHW] G. Fels, A. Huckleberry, and J. A. Wolf, *Cycles Spaces of Flag Domains: A Complex Geometric Viewpoint*, Progress in Mathematics, Volume 245, Springer/Birkhäuser Boston, 2005.

[GRT] P. Griffiths, C. Robles and D. Toledo, *Quotients of non-classical flag domains are not algebraic* (arXiv1303.0252).

[Ha] R. Harvey, *Spinors and Calibrations*, Academic Press, 1990.

[H1] A. Huckleberry, Hyperbolicity of cycle spaces and automorphism groups of flag domains, *American Journal of Mathematics*, **136**, Nr. 2 (2013) 291–310.

[H2] A. Huckleberry, Remarks on homogeneous manifolds satisfying Levi-conditions, *Bollettino U.M.I.* (9) **III** (2010) 1–23.

[K] J. Kollar, *Neighborhoods of subvarieties in homogeneous spaces* (arXiv1308.5603).

[O1] A. Onishchik, Transitive compact transformation groups, *Math. Sb.* (N.S.) **60** (1963), 447–485 English Trans: AMS Trans. (2) 55 (1966) 5–58.

[O2] Arkadii L'vovich Onishchik (on his 70th birthday), *Russian Math. Surveys* **58.6** 1245–1253.

[W] J. A. Wolf, The action of a real semisimple Lie group on a complex manifold, I: Orbit structure and holomorphic arc components, *Bull. Amer. Math. Soc.* **75** (1969), 1121–1237.

Shintani Functions, Real Spherical Manifolds, and Symmetry Breaking Operators

Toshiyuki Kobayashi

Abstract For a pair of reductive groups $G \supset G'$, we prove a geometric criterion for the space $\mathrm{Sh}(\lambda, \nu)$ of Shintani functions to be finite-dimensional in the Archimedean case. This criterion leads us to a complete classification of the symmetric pairs (G, G') having finite-dimensional Shintani spaces. A geometric criterion for uniform boundedness of $\dim_{\mathbb{C}} \mathrm{Sh}(\lambda, \nu)$ is also obtained. Furthermore, we prove that symmetry breaking operators of the restriction of smooth admissible representations yield Shintani functions of moderate growth, of which the dimension is determined for $(G, G') = (O(n + 1, 1), O(n, 1))$.

Key words Branching law • Reductive group • Symmetry breaking • Real spherical variety • Shintani function

Mathematics Subject Classification (2010): Primary 22E46. Secondary 11F70, 14M27, 32N15, 53C35.

1 Introduction

The purpose of this article is to investigate Shintani functions for a pair of reductive groups $G \supset G'$ in the Archimedean case. Among others, we classify the reductive symmetric pairs (G, G') such that the Shintani spaces $\mathrm{Sh}(\lambda, \nu)$ are finite-dimensional for all $(\mathfrak{z}_G, \mathfrak{z}_{G'})$-infinitesimal character (λ, ν). Explicit dimension formulae for the Shintani spaces of moderate growth are determined for the pair $(G, G') = (O(n + 1, 1), O(n, 1))$.

*The author was partially supported by Grant-in-Aid for Scientific Research (B)(22340026) and (A)(25247006).

T. Kobayashi (✉)
Kavli IPMU (WPI) and Graduate School of Mathematical Sciences,
The University of Tokyo, Meguro-ku, Tokyo, 153-8914, Japan
e-mail: toshi@ms.u-tokyo.ac.jp

© Springer International Publishing Switzerland 2014

G. Mason et al. (eds.), *Developments and Retrospectives in Lie Theory*,
Developments in Mathematics 37, DOI 10.1007/978-3-319-09934-7_5

Let G be a real reductive linear Lie group. We write $\mathfrak{g}$ for the Lie algebra of G, and $U(\mathfrak{g}_\mathbb{C})$ for the universal enveloping algebra of the complexified Lie algebra $\mathfrak{g}_\mathbb{C} := \mathfrak{g} \otimes_\mathbb{R} \mathbb{C}$.

For $X \in \mathfrak{g}$ and $f \in C^\infty(G)$, we set

$$(L_X f)(g) := \frac{d}{dt}\Big|_{t=0} f(\exp(-tX)g), \quad (R_X f)(g) := \frac{d}{dt}\Big|_{t=0} f(g\exp(tX)),$$

(1.1)

and extend these actions to those of $U(\mathfrak{g}_\mathbb{C})$.

We denote by $\mathfrak{z}_G$ the $\mathbb{C}$-algebra of G-invariant elements in $U(\mathfrak{g}_\mathbb{C})$. Let $\mathfrak{j}$ be a Cartan subalgebra of $\mathfrak{g}$. Then any $\lambda \in \mathfrak{j}_\mathbb{C}^\vee$ gives rise to a $\mathbb{C}$-algebra homomorphism $\chi_\lambda : \mathfrak{z}_G \to \mathbb{C}$ via the Harish–Chandra isomorphism $\mathfrak{z}_G \xrightarrow{\sim} S(\mathfrak{j}_\mathbb{C})^{W(\mathfrak{j}_\mathbb{C})}$, where $W(\mathfrak{j}_\mathbb{C})$ is some finite group (see Sect. 3.3).

Suppose that G' is an algebraic reductive subgroup. Analogous notation will be applied to G'. For instance, $\mathrm{Hom}_{\mathbb{C}\text{-alg}}(\mathfrak{z}_{G'}, \mathbb{C}) \simeq (\mathfrak{j}_\mathbb{C}')^\vee / W(\mathfrak{j}_\mathbb{C}')$, $\chi_\nu \leftrightarrow \nu$, where $\mathfrak{j}'$ is a Cartan subalgebra of the Lie algebra $\mathfrak{g}'$ of G'.

We take a maximal compact subgroup K of G such that $K' := K \cap G'$ is a maximal compact subgroup. Following Murase–Sugano [19], we state:

Definition 1.1 (Shintani Function). We say $f \in C^\infty(G)$ is a *Shintani function* of $(\mathfrak{z}_G, \mathfrak{z}_{G'})$-infinitesimal characters (λ, ν) if f satisfies the following three properties:

(1) $f(k'gk) = f(g)$ for any $k' \in K', k \in K$.
(2) $R_u f = \chi_\lambda(u)f$ for any $u \in \mathfrak{z}_G$.
(3) $L_v f = \chi_\nu(v)f$ for any $v \in \mathfrak{z}_{G'}$.

We denote by $\mathrm{Sh}(\lambda, \nu)$ the space of Shintani functions of type (λ, ν).

For $G = G'$ and $\lambda = -\nu$, Shintani functions are nothing but Harish–Chandra's zonal spherical functions.

In this article, we provide the following three different realizations of the Shintani space $\mathrm{Sh}(\lambda, \nu)$:

- Matrix coefficients of symmetry breaking operators. (See Proposition 7.1.)
- $(K \times K')$-invariant functions on $(G \times G')/\mathrm{diag}\, G'$. (See Lemma 5.5.)
- G'-invariant functions on the Riemannian symmetric space $(G \times G')/(K \times K')$. (See Lemma 8.6.)

The first realization constructs Shintani functions having moderate growth (Definition 3.3) from the restriction of admissible smooth representations of G with respect to the subgroup G', whereas the second realization relates $\mathrm{Sh}(\lambda, \nu)$ with the theory of real spherical homogeneous spaces which was studied in [11–14]. Via the third realization, we can apply powerful methods (e.g., [9]) of harmonic analysis on Riemannian symmetric spaces for the study of Shintani functions.

By using these ideas, we give a characterization of the pair (G, G') for which the Shintani space $\text{Sh}(\lambda, \nu)$ is finite-dimensional for all (λ, ν):

Theorem 1.2 (see Theorem 4.1). *The following four conditions on a pair of real reductive algebraic groups $G \supset G'$ are equivalent:*

(i) (Shintani function) $\text{Sh}(\lambda, \nu)$ *is finite-dimensional for any pair (λ, ν) of $(\mathfrak{z}_G, \mathfrak{z}_{G'})$-infinitesimal characters.*

(ii) (Symmetry breaking) $\text{Hom}_{G'}(\pi^\infty, \tau^\infty)$ *is finite-dimensional for any pair $(\pi^\infty, \tau^\infty)$ of admissible smooth representations of G and G' (see Sect. 3.2).*

(iii) (Invariant bilinear form) *There exist at most finitely many linearly independent G'-invariant bilinear forms on $\pi^\infty \otimes \tau^\infty$ for any pair $(\pi^\infty, \tau^\infty)$ of admissible smooth representations of G and G'.*

(iv) (Geometric property (PP)) *There exist minimal parabolic subgroups P and P' of G and G', respectively, such that PP' is open in G.*

The dimension of the Shintani space $\text{Sh}(\lambda, \nu)$ depends on λ and ν in general. We give a characterization of the uniform boundedness property:

Theorem 1.3. *The following four conditions on a pair of real reductive algebraic groups $G \supset G'$ are equivalent:*

(i) (Shintani function) *There exists a constant C such that*

$$\dim_{\mathbb{C}} \text{Sh}(\lambda, \nu) \leq C$$

for any pair (λ, ν) of $(\mathfrak{z}_G, \mathfrak{z}_{G'})$-infinitesimal characters.

(ii) (Symmetry breaking) *There exists a constant C such that*

$$\dim_{\mathbb{C}} \text{Hom}_{G'}(\pi^\infty, \tau^\infty) \leq C$$

for any pair $(\pi^\infty, \tau^\infty)$ of admissible smooth representations of G and G'.

(iii) (Invariant bilinear form) *There exists a constant C such that*

$$\dim_{\mathbb{C}} \text{Hom}_{G'}(\pi^\infty \otimes \tau^\infty, \mathbb{C}) \leq C$$

for any pair $(\pi^\infty, \tau^\infty)$ of admissible smooth representations of G and G'.

(iv) (Geometric property (BB)) *There exist Borel subgroups B and B' of the complex Lie groups $G_{\mathbb{C}} \supset G'_{\mathbb{C}}$ with Lie algebras $\mathfrak{g}_{\mathbb{C}} \supset \mathfrak{g}'_{\mathbb{C}}$, respectively, such that BB' is open in $G_{\mathbb{C}}$.*

By using the geometric criterion (PP), we give a complete classification of the reductive symmetric pairs (G, G') for which one of (therefore any of) the equivalent conditions in Theorem 1.2 is fulfilled. See Theorem 2.3 for the classification. Among them, those satisfying the uniform boundedness property in Theorem 1.3 are listed in Theorem 2.4.

Example 1.4 (see Theorems 2.3 and 2.4).

(1) If (G, G') is

$$(GL(n + 1, \mathbb{C}), GL(n, \mathbb{C}) \times GL(1, \mathbb{C})) \qquad (n \geq 1),$$

$$(O(n + 1, \mathbb{C}), O(n, \mathbb{C})) \qquad\qquad\qquad (n \geq 1),$$

or any real form of them, then we have

$$\sup_{\lambda} \sup_{\nu} \dim_{\mathbb{C}} \mathrm{Sh}(\lambda, \nu) < \infty. \qquad (1.2)$$

(2) If (G, G') is

$$(Sp(n + 1, \mathbb{C}), Sp(n, \mathbb{C}) \times Sp(1, \mathbb{C})) \qquad (n \geq 2),$$

or its split real form, then $\mathrm{Sh}(\rho_{\mathfrak{g}}, \rho_{\mathfrak{g}'})$ is infinite-dimensional (see (3.4) for the notation). On the other hand, if (G, G') is a nonsplit real form, then $\mathrm{Sh}(\lambda, \nu)$ is finite-dimensional for all (λ, ν), but the dimension is not uniformly bounded, namely, (1.2) fails.

(3) If (G, G') is

$$(GL(n + 1, \mathbb{H}), GL(n, \mathbb{H}) \times GL(1, \mathbb{H})) \quad (n \geq 1),$$

then $\mathrm{Sh}(\lambda, \nu)$ is finite-dimensional for all (λ, ν), but (1.2) fails.

This article is organized as follows:

In Sect. 2, we give a complete list of the reductive symmetric pairs (G, G') such that the dimension of the Shintani space is finite/uniformly bounded.

After a brief review on basic results on continuous (infinite-dimensional) representations of real reductive Lie groups in Sect. 3, we enrich Theorem 1.2 by adding some more conditions that are equivalent to the finiteness of $\dim_{\mathbb{C}} \mathrm{Sh}(\lambda, \nu)$ in Theorem 4.1.

The upper estimate of $\dim_{\mathbb{C}} \mathrm{Sh}(\lambda, \nu)$ is proved in Sect. 5 by using the theory of *real spherical* homogeneous spaces which was established in [14].

In Sect. 7 we give a lower estimate of $\dim_{\mathbb{C}} \mathrm{Sh}(\lambda, \nu)$ by using the intertwining operators constructed in Sect. 6.

In Sect. 8 we apply the theory of harmonic analysis on Riemannian symmetric spaces, and investigate the relationship between symmetry breaking operators of the restriction of admissible smooth representations of G to G' and Shintani functions. Sect. 9 provides an example for $(G, G') = (O(n + 1, 1), O(n, 1))$ by using a recent work [16] with B. Speh on symmetry breaking operators.

2 Classification of (G, G') with $\dim_{\mathbb{C}} \mathrm{Sh}(\lambda, \nu) < \infty$

This section gives a complete classification of the reductive symmetric pairs (G, G') such that the dimension of the Shintani space $\mathrm{Sh}(\lambda, \nu)$ is finite/bounded for any $(\mathfrak{Z}_G, \mathfrak{Z}_{G'})$-infinitesimal characters (λ, ν). Owing to the criteria in Theorems 1.2 and 1.3, the classification is reduced to that of (real) spherical homogeneous spaces of the form $(G \times G')/\mathrm{diag}\, G'$, which was accomplished in [13].

Definition 2.1 (Symmetric Pair). Let G be a real reductive Lie group. We say (G, G') is a *reductive symmetric pair* if G' is an open subgroup of the fixed point subgroup G^σ of some involutive automorphism σ of G.

Example 2.2. (1) (Group case) Let G_1 be a Lie group. Then the pair

$$(G, G') = (G_1 \times G_1, \mathrm{diag}\, G_1)$$

forms a symmetric pair with the involution $\sigma \in \mathrm{Aut}(G)$ defined by $\sigma(x, y) = (y, x)$. Since the homogeneous space G/G' is isomorphic to the group manifold G_1 with $(G_1 \times G_1)$-action from the left and the right, the pair $(G_1 \times G_1, \mathrm{diag}\, G_1)$ is sometimes referred to as the group case.

(2) (Riemannian symmetric pair) Let K be a maximal compact subgroup of a real reductive linear Lie group G. Then the pair (G, K) is a symmetric pair because K is the fixed point subgroup of a Cartan involution θ of G. Since the homogeneous space G/K becomes a symmetric space with respect to the Levi-Civita connection of a G-invariant Riemannian metric on G/K, the pair (G, K) is sometimes referred to as a Riemannian symmetric pair.

The classification of reductive symmetric pairs was established by Berger [2] on the level of Lie algebras. Among them we list the pairs (G, G') such that the space of Shintani functions is finite-dimensional as follows:

Theorem 2.3. *Suppose* (G, G') *is a reductive symmetric pair. Then the following two conditions are equivalent:*

(i) $\mathrm{Sh}(\lambda, \nu)$ *is finite-dimensional for any* $(\mathfrak{Z}_G, \mathfrak{Z}_{G'})$*-infinitesimal characters* (λ, ν).
(ii) *The pair* $(\mathfrak{g}, \mathfrak{g}')$ *of the Lie algebras is isomorphic (up to outer automorphisms) to the direct sum of the following pairs:*

(A) Trivial case: $\mathfrak{g} = \mathfrak{g}'$.
(B) Abelian case: $\mathfrak{g} = \mathbb{R}$, $\mathfrak{g}' = \{0\}$.
(C) Compact case: $\mathfrak{g}$ *is the Lie algebra of a compact simple Lie group.*
D) Riemannian symmetric pair: $\mathfrak{g}'$ *is the Lie algebra of a maximal compact subgroup* K *of a non-compact simple Lie group* G.
(E) Split rank one case ($\mathrm{rank}_{\mathbb{R}}\, G = 1$):

> E1) $(\mathfrak{o}(p + q, 1), \mathfrak{o}(p) + \mathfrak{o}(q, 1))$ $(p + q \geq 2)$.
> E2) $(\mathfrak{su}(p + q, 1), \mathfrak{s}(\mathfrak{u}(p) + \mathfrak{u}(q, 1)))$ $(p + q \geq 1)$.
> E3) $(\mathfrak{sp}(p + q, 1), \mathfrak{sp}(p) + \mathfrak{sp}(q, 1))$ $(p + q \geq 1)$.
> E4) $(\mathfrak{f}_{4(-20)}, \mathfrak{o}(8, 1))$.

(F) Strong Gelfand pairs and their real forms:

    F1) $(\mathfrak{sl}(n+1,\mathbb{C}),\mathfrak{gl}(n,\mathbb{C}))$ $(n \geq 2)$.
    F2) $(\mathfrak{o}(n+1,\mathbb{C}),\mathfrak{o}(n,\mathbb{C}))$ $(n \geq 2)$.
    F3) $(\mathfrak{sl}(n+1,\mathbb{R}),\mathfrak{gl}(n,\mathbb{R}))$ $(n \geq 1)$.
    F4) $(\mathfrak{su}(p+1,q),\mathfrak{u}(p,q))$ $(p+q \geq 1)$.
    F5) $(\mathfrak{o}(p+1,q),\mathfrak{o}(p,q))$ $(p+q \geq 2)$.

(G) $(\mathfrak{g},\mathfrak{g}') = (\mathfrak{g}_1 + \mathfrak{g}_1, \mathrm{diag}\,\mathfrak{g}_1)$ Group case:

    G1) $\mathfrak{g}_1$ *is the Lie algebra of a compact simple Lie group.*
    G2) $(\mathfrak{o}(n,1)+\mathfrak{o}(n,1), \mathrm{diag}\,\mathfrak{o}(n,1))$ $(n \geq 2)$.

(H) Other cases:

    H1) $(\mathfrak{o}(2n,2),\mathfrak{u}(n,1))$ $(n \geq 1)$.
    H2) $(\mathfrak{su}^*(2n+2),\mathfrak{su}(2)+\mathfrak{su}^*(2n)+\mathbb{R})$ $(n \geq 1)$.
    H3) $(\mathfrak{o}^*(2n+2),\mathfrak{o}(2)+\mathfrak{o}^*(2n))$ $(n \geq 1)$.
    H4) $(\mathfrak{sp}(p+1,q),\mathfrak{sp}(p,q)+\mathfrak{sp}(1))$.
    H5) $(\mathfrak{e}_{6(-26)},\mathfrak{so}(9,1)+\mathbb{R})$.

We single out those pairs (G, G') having the uniform boundedness property as follows:

Theorem 2.4. *Suppose (G, G') is a reductive symmetric pair. Then the following conditions are equivalent:*

(i) *There exists a constant such that*

$$\dim_{\mathbb{C}} \mathrm{Sh}(\lambda, \nu) \leq C$$

    for any $(\mathfrak{Z}_G, \mathfrak{Z}_{G'})$-infinitesimal characters (λ, ν).
(ii) *The pair of the Lie algebras $(\mathfrak{g}, \mathfrak{g}')$ is isomorphic (up to outer automorphisms) to the direct sum of the pairs in (A), (B) and (F1)–(F5).*

Example 2.5. In connection with branching problems, some of the pairs appeared earlier in the literatures. For instance,

(1) (Strong Gelfand pairs [18]) (F1), (F2).
(2) (The Gross–Prasad conjecture [4]) (F2), (F5).
(3) (Finite-multiplicity for tensor products [11]) (G2).
(4) (Multiplicity-free restriction [1, 21]) (F1)–(F5).

Remark 2.6. The following pairs (G, G') are nonsymmetric pairs such that (G, G') satisfies the condition (i) of Theorem 2.4.

$$(G, G') = (SO(8, \mathbb{C}), \mathrm{Spin}(7, \mathbb{C})), \quad (SO(4, 4), \mathrm{Spin}(4, 3)).$$

In fact the Lie algebras $(\mathfrak{g}, \mathfrak{g}')$ are symmetric pairs, but the involution of $\mathfrak{g}$ does not lift to the group G.

Proof of Theorem 2.3. Direct from Theorem 1.2 and [13, Theorem 1.3].   □

Proof of Theorem 2.4. Direct from Theorem 1.3 and [13, Proposition 1.6].   □

3 Preliminary Results

We begin with a quick review of some basic results on (infinite-dimensional) continuous representations of real reductive Lie groups.

3.1 Continuous Representations and Frobenius Reciprocity

By a continuous representation π of a Lie group G on a topological vector space V we shall mean that $\pi : G \to GL_{\mathbb{C}}(V)$ is a group homomorphism such that the induced map $G \times V \to V$, $(g, v) \mapsto \pi(g)v$ is continuous. We say π is a (continuous) Hilbert [Banach, Fréchet, $\cdots$] representation if V is a Hilbert [Banach, Fréchet, $\cdots$] space. We note that a continuous Hilbert representation is not necessarily a unitary representation; a Hilbert representation π of G is said to be a unitary representation provided that all the operators $\pi(g)$ $(g \in G)$ are unitary.

Suppose π is a continuous representation of G on a Banach space V. A vector $v \in V$ is said to be *smooth* if the map $G \to V$, $g \mapsto \pi(g)v$ is of C^∞-class. Let V^∞ denote the space of smooth vectors of the representation (π, V). Then V^∞ carries a Fréchet topology with a family of seminorms $\|v\|_{i_1 \cdots i_k} := \|d\pi(X_{i_1}) \cdots d\pi(X_{i_k})v\|$, where $\{X_1, \cdots, X_n\}$ is a basis of $\mathfrak{g}$. Then V^∞ is a G-invariant subspace of V, and we obtain a continuous Fréchet representation (π^∞, V^∞) of G.

Suppose that G' is another Lie group. If π and τ are Hilbert representations of G and G' on the Hilbert spaces $\mathscr{H}_\pi$ and $\mathscr{H}_\tau$, respectively, then we can define a continuous Hilbert representation $\pi \boxtimes \tau$ of the direct product group on the Hilbert completion on $\mathscr{H}_\pi \widehat{\otimes} \mathscr{H}_\tau$ of the pre-Hilbert space $\mathscr{H}_\pi \otimes \mathscr{H}_\tau$.

Suppose further that G' is a subgroup of G. Then we may regard π as a representation of G' by the restriction. The resulting representation is denoted by $\pi|_{G'}$. The restriction of the outer tensor product $\pi \boxtimes \tau$ of $G \times G'$ to the subgroup $\text{diag } G' = \{(g', g') : g' \in G'\}$ is denoted by $\pi \otimes \tau$. By a *symmetry breaking operator* we mean a continuous G'-homomorphism from the representation space of π to that of τ. We write $\text{Hom}_{G'}(\pi|_{G'}, \tau)$ for the vector space of continuous G'-homomorphisms. Analogous notation is applied to smooth representations.

For the convenience of the reader, we review some basic properties of the restriction:

Lemma 3.1. *Suppose that π and τ are Hilbert representations of G and G' on Hilbert spaces $\mathscr{H}_\pi$ and $\mathscr{H}_\tau$, respectively.*

(1) *There is a canonical injective homomorphism:*

$$\text{Hom}_{G'}(\pi|_{G'}, \tau) \hookrightarrow \text{Hom}_{G'}(\pi^\infty|_{G'}, \tau^\infty), \qquad T \mapsto T|_{\mathscr{H}_\pi^\infty}. \tag{3.1}$$

(2) *Let $\tau^\vee$ be the contragredient representation of τ. Then we have a canonical isomorphism:*

$$\text{Hom}_{G'}(\pi|_{G'}, \tau) \simeq \text{Hom}_{G'}(\pi \otimes \tau^\vee, \mathbb{C}). \tag{3.2}$$

(3) *There is a canonical injective homomorphism if G and G' are real reductive:*

$$\text{Hom}_{G'}(\pi^\infty|_{G'}, \tau^\infty) \hookrightarrow \text{Hom}_{G'}(\pi^\infty \otimes (\tau^\vee)^\infty, \mathbb{C}).$$

Proof. (1) See [14, Lemma 5.1], for instance.

(2) We have a canonical isomorphism between the two vector spaces $\text{Hom}_\mathbb{C}(\mathscr{H}_\pi, \mathscr{H}_\tau)$ and $\text{Hom}_\mathbb{C}(\mathscr{H}_\pi \widehat{\otimes} \mathscr{H}_\tau^\vee, \mathbb{C})$, where $\text{Hom}_\mathbb{C}(\ ,\ )$ denotes the space of continuous linear maps. Taking G'-invariant elements, we get (3.2).

(3) See [1, Lemma A.0.8], for instance.   □

Proposition 3.2 (Frobenius Reciprocity). *Let H be a closed subgroup of a Lie group G. Suppose that π is a continuous representation of G on a topological vector space V. Then there is a canonical bijection*

$$\text{Hom}_H(\pi|_H, \mathbb{C}) \simeq \text{Hom}_G(\pi, C(G/H)), \qquad \lambda \mapsto T \tag{3.3}$$

defined by

$$T(v)(g) = \lambda(\pi(g^{-1})v) \qquad v \in V.$$

Furthermore, if π^∞ is a smooth representation, then we have

$$\text{Hom}_H(\pi^\infty|_H, \mathbb{C}) \simeq \text{Hom}_G(\pi^\infty, C^\infty(G/H)).$$

Proof. The linear map $T : V \to C(G/H)$ is continuous because $G \times V \to V$, $(g, v) \mapsto \pi(g^{-1})v$ is continuous. The last statement follows because $G \to V$, $g \mapsto \pi(g)^{-1}v$ is a C^∞-map.   □

3.2 Admissible Representations

In this subsection we review some basic terminologies for Harish–Chandra modules.

Let G be a real reductive linear Lie group, and K a maximal compact subgroup of G. Let $\mathscr{HC}$ denote the category of Harish–Chandra modules where the objects are $(\mathfrak{g}, K)$-modules of finite length, and the morphisms are $(\mathfrak{g}, K)$-homomorphisms.

Let π be a continuous representation of G on a Fréchet space V. Suppose that π is of finite length, namely, there are at most finitely many closed G-invariant subspaces in V. We say π is *admissible* if

$$\dim \text{Hom}_K(\tau, \pi|_K) < \infty$$

for any irreducible finite-dimensional representation τ of K. We denote by V_K the space of K-finite vectors. Then $V_K \subset V^\infty$ and the Lie algebra $\mathfrak{g}$ leaves V_K invariant. The resulting $(\mathfrak{g}, K)$-module on V_K is called the underlying $(\mathfrak{g}, K)$-module of π, and will be denoted by π_K.

An admissible representation (π, V) is said to be *spherical* if V contains a nonzero K-fixed vector, or equivalently, the underlying $(\mathfrak{g}, K)$-module V_K contains a nonzero K-fixed vector.

A vector $v \in V$ is said to be *cyclic* if the vector space $\mathbb{C}$-span$\{\pi(g)v : g \in G\}$ is dense in V. If W is a proper G-invariant closed subspace of V, then $v \mod W$ is a cyclic vector in the quotient representation on V/W. For a K-finite vector v, v is cyclic in π if and only if v is cyclic in the underlying $(\mathfrak{g}, K)$-module π_K in the sense that $U(\mathfrak{g}_\mathbb{C})v = V_K$.

3.3 Harish–Chandra Isomorphism

We review the standard normalization of the Harish–Chandra isomorphism of the $\mathbb{C}$-algebra $\mathfrak{Z}_G$, where we recall from Introduction that

$$\mathfrak{Z}_G = U(\mathfrak{g}_\mathbb{C})^G \equiv \{u \in U(\mathfrak{g}_\mathbb{C}) : \mathrm{Ad}(g)u = u \text{ for all } g \in G\}.$$

For a connected G, $\mathfrak{Z}_G$ is equal to the center $\mathfrak{Z}(\mathfrak{g}_\mathbb{C})$ of $U(\mathfrak{g}_\mathbb{C})$.

Let $\mathfrak{j}$ be a Cartan subalgebra of $\mathfrak{g}$, $\mathfrak{j}_\mathbb{C} = \mathfrak{j} \otimes_\mathbb{R} \mathbb{C}$, and $\mathfrak{j}_\mathbb{C}^\vee = \mathrm{Hom}_\mathbb{C}(\mathfrak{j}_\mathbb{C}, \mathbb{C})$. We set

$$W(\mathfrak{j}_\mathbb{C}) := N_{\tilde{G}}(\mathfrak{j}_\mathbb{C})/Z_{\tilde{G}}(\mathfrak{j}_\mathbb{C}),$$

where $\tilde{G}$ is the group generated by $\mathrm{Ad}(G)$ and the group $\mathrm{Int}(\mathfrak{g}_\mathbb{C})$ of inner automorphisms. For a connected G, $W(\mathfrak{j}_\mathbb{C})$ is the Weyl group for the root system $\Delta(\mathfrak{g}_\mathbb{C}, \mathfrak{j}_\mathbb{C})$.

Fix a positive system $\Delta^+(\mathfrak{g}_\mathbb{C}, \mathfrak{j}_\mathbb{C})$, and write $\mathfrak{n}_\mathbb{C}^+$ for the sum of the root spaces belonging to $\Delta^+(\mathfrak{g}_\mathbb{C}, \mathfrak{j}_\mathbb{C})$, and $\mathfrak{n}_\mathbb{C}^-$ for $\Delta^-(\mathfrak{g}_\mathbb{C}, \mathfrak{j}_\mathbb{C})$. We set

$$\rho_\mathfrak{g} := \frac{1}{2} \sum_{\alpha \in \Delta^+(\mathfrak{g}_\mathbb{C}, \mathfrak{j}_\mathbb{C})} \alpha \in \mathfrak{j}_\mathbb{C}^\vee. \tag{3.4}$$

Let $\gamma' : U(\mathfrak{g}_\mathbb{C}) \to U(\mathfrak{j}_\mathbb{C}) \simeq S(\mathfrak{j}_\mathbb{C})$ be the projection to the second factor of the decomposition $U(\mathfrak{g}_\mathbb{C}) = (\mathfrak{n}_\mathbb{C}^- U(\mathfrak{g}_\mathbb{C}) + U(\mathfrak{g}_\mathbb{C})\mathfrak{n}_\mathbb{C}^+) \oplus U(\mathfrak{j}_\mathbb{C})$. Then we have the Harish–Chandra isomorphism

$$\mathfrak{Z}_G = U(\mathfrak{g}_\mathbb{C})^G \xrightarrow{\sim}_{\gamma} S(\mathfrak{j}_\mathbb{C})^{W(\mathfrak{j}_\mathbb{C})}, \tag{3.5}$$

where $\gamma : U(\mathfrak{g}_\mathbb{C}) \to S(\mathfrak{j}_\mathbb{C})$ is defined by $\langle \gamma(u), \lambda \rangle = \langle \gamma'(u), \lambda - \rho_\mathfrak{g} \rangle$ for all $\lambda \in \mathfrak{j}_\mathbb{C}^\vee$.

Then any element $\lambda \in \mathfrak{j}_{\mathbb{C}}^{\vee}$ gives a $\mathbb{C}$-algebra homomorphism $\chi_{\lambda} : \mathfrak{Z}_G \to \mathbb{C}$ via the isomorphism (3.5), and $\chi_{\lambda} = \chi_{\lambda'}$ if and only if $\lambda' = w\lambda$ for some $w \in W(\mathfrak{j}_{\mathbb{C}})$. This correspondence yields a bijection:

$$\mathrm{Hom}_{\mathbb{C}\text{-alg}}(\mathfrak{Z}_G, \mathbb{C}) \simeq \mathfrak{j}_{\mathbb{C}}^{\vee}/W(\mathfrak{j}_{\mathbb{C}}), \quad \chi_{\lambda} \leftrightarrow \lambda. \tag{3.6}$$

In our normalization, the $\mathfrak{Z}_G$-infinitesimal character of the trivial representation $\mathbf{1}$ of G is given by $\rho_{\mathfrak{g}}$.

For $\lambda \in \mathfrak{j}_{\mathbb{C}}^{\vee}/W(\mathfrak{j}_{\mathbb{C}})$, we set

$$C^{\infty}(G; \chi_{\lambda}^{R}) := \{ f \in C^{\infty}(G) : R_u f = \chi_{\lambda}(u) f \text{ for any } u \in \mathfrak{Z}_G \},$$

$$C^{\infty}(G; \chi_{\lambda}^{L}) := \{ f \in C^{\infty}(G) : L_u f = \chi_{\lambda}(u) f \text{ for any } u \in \mathfrak{Z}_G \}.$$

Then we have $C^{\infty}(G; \chi_{\lambda}^{R}) = C^{\infty}(G; \chi_{-\lambda}^{L})$.

Let H be a closed subgroup of G. Since the action of $\mathfrak{Z}_G$ on $C^{\infty}(G)$ via R (and via L) commutes with the right H-action, R_u and L_u ($u \in \mathfrak{Z}_G$) induce differential operators on G/H. Thus, for $\lambda \in \mathfrak{j}_{\mathbb{C}}^{\vee}/W(\mathfrak{j}_{\mathbb{C}})$, we can define

$$C^{\infty}(G/H; \chi_{\lambda}^{R}) := \{ f \in C^{\infty}(G/H) : R_u f = \chi_{\lambda}(u) f \text{ for any } u \in \mathfrak{Z}_G \},$$

$$C^{\infty}(G/H; \chi_{\lambda}^{L}) := \{ f \in C^{\infty}(G/H) : L_u f = \chi_{\lambda}(u) f \text{ for any } u \in \mathfrak{Z}_G \}.$$

3.4 Shintani Functions of Moderate Growth

Without loss of generality, we may and do assume that a real reductive linear Lie group G is realized as a closed subgroup of $GL(n, \mathbb{R})$ such that G is stable under the transpose of matrix $g \mapsto {}^t g$ and $K = O(n) \cap G$. For $g \in G$ we define a map $\| \cdot \| : G \to \mathbb{R}$ by

$$\|g\| := \|g \oplus {}^t g^{-1}\|_{\mathrm{op}}$$

where $\| \cdot \|_{\mathrm{op}}$ is the operator norm of $M(2n, \mathbb{R})$. A continuous representation π of G on a Fréchet space V is said to be of *moderate growth* if for each continuous seminorm $| \cdot |$ on V there exist a continuous seminorm $| \cdot |'$ on V and a constant $d \in \mathbb{R}$ such that

$$|\pi(g)u| \leq \|g\|^{d} |u|' \quad \text{for } g \in G, u \in V.$$

For any admissible representation $(\pi, \mathscr{H})$ such that $\mathscr{H}$ is a Banach space, the smooth representation $(\pi^{\infty}, \mathscr{H}^{\infty})$ has moderate growth. We say $(\pi^{\infty}, \mathscr{H}^{\infty})$ is an *admissible smooth representation*. By the Casselman–Wallach globalization theory, there is a canonical equivalence of categories between the category $\mathscr{HC}$ of $(\mathfrak{g}, K)$-

modules of finite length and the category of admissible smooth representations of G [22, Chapter 11]. In particular, the Fréchet representation π^∞ is uniquely determined by its underlying $(\mathfrak{g}, K)$-module. We say π^∞ is the *smooth globalization* of $\pi_K \in \mathcal{HC}$.

For simplicity, by an *irreducible smooth representation* we shall mean an irreducible admissible smooth representation of G.

Definition 3.3. A smooth function f on G is said to have *moderate growth* if f satisfies the following three properties:

(1) f is right K-finite.
(2) f is $\mathfrak{Z}_G$-finite.
(3) There exists a constant $d \in \mathbb{R}$ (depending on f) such that if $u \in U(\mathfrak{g}_\mathbb{C})$, then there exists $C \equiv C(u)$ satisfying

$$|(R_u f)(x)| \leq C \|x\|^d \qquad (x \in G).$$

We denote by $C_{\text{mod}}^\infty(G)$ the space of all $f \in C^\infty(G)$ having moderate growth.

If (π, V) is an admissible representation of moderate growth, then the matrix coefficient $G \to \mathbb{C}$, $g \mapsto \langle \pi(g)v, u \rangle$ belongs to $C_{\text{mod}}^\infty(G)$ for any $v \in V_K$ and any linear functional u of the Fréchet space V.

We define the space of Shintani functions of moderate growth by

$$\text{Sh}_{\text{mod}}(\lambda, \nu) := \text{Sh}(\lambda, \nu) \cap C_{\text{mod}}^\infty(G). \tag{3.7}$$

4 Finite-Multiplicity Properties of Branching Laws

We are ready to make a precise statement of Theorem 1.2, and enrich it by adding some more equivalent conditions. The main results of this section is Theorem 4.1.

4.1 Finite-Multiplicity Properties of Branching Laws

Theorem 4.1. *The following twelve conditions on a pair of real reductive algebraic groups $G \supset G'$ are equivalent:*

(i) (PP) *There exist minimal parabolic subgroups P and P' of G and G', respectively, such that PP' is open in G.*

(ii) (Sh) $\dim_\mathbb{C} \text{Sh}(\lambda, \nu) < \infty$ *for any pair (λ, ν) of $(\mathfrak{Z}_G, \mathfrak{Z}_{G'})$-infinitesimal characters.*

(iii) (Sh$_{\text{mod}}$) $\dim_\mathbb{C} \text{Sh}_{\text{mod}}(\lambda, \nu) < \infty$ *for any pair (λ, ν) of $(\mathfrak{Z}_G, \mathfrak{Z}_{G'})$-infinitesimal characters.*

(iv) (Sh$_{\text{mod}}$)$_1$ $\dim_\mathbb{C} \text{Sh}_{\text{mod}}(\rho_\mathfrak{g}, \rho_{\mathfrak{g}'}) < \infty$.

(v) $(\infty \downarrow)$ $\dim_{\mathbb{C}} \operatorname{Hom}_{G'}(\pi^{\infty}|_{G'}, \tau^{\infty}) < \infty$ *for any pair* $(\pi^{\infty}, \tau^{\infty})$ *of admissible smooth representations of G and G'.*

(vi) $(\infty \quad \downarrow)_K$ $\dim_{\mathbb{C}} \operatorname{Hom}_{G'}(\pi^{\infty}|_{G'}, \tau^{\infty}) < \infty$ *for any pair* $(\pi^{\infty}, \tau^{\infty})$ *of admissible smooth representations of G and G' such that π^{∞} and $(\tau^{\infty})^{\vee}$ have cyclic spherical vectors.*

(vii) $(\mathscr{H} \downarrow)$ $\dim_{\mathbb{C}} \operatorname{Hom}_{G'}(\pi|_{G'}, \tau) < \infty$ *for any pair* (π, τ) *of admissible Hilbert representations of G and G'.*

(viii) $(\mathscr{H} \quad \downarrow)_K$ $\dim_{\mathbb{C}} \operatorname{Hom}_{G'}(\pi|_{G'}, \tau) < \infty$ *for any pair* (π, τ) *of admissible Hilbert representations of G and G' such that π and $\tau^{\vee}$ have cyclic spherical vectors.*

(ix) $(\infty \otimes)$ $\dim_{\mathbb{C}} \operatorname{Hom}_{G'}(\pi^{\infty} \otimes \tau^{\infty}, \mathbb{C}) < \infty$ *for any pair* $(\pi^{\infty}, \tau^{\infty})$ *of admissible smooth representations of G and G'.*

(x) $(\infty \otimes)_K$ $\dim_{\mathbb{C}} \operatorname{Hom}_{G'}(\pi^{\infty} \otimes \tau^{\infty}, \mathbb{C}) < \infty$ *for any pair* $(\pi^{\infty}, \tau^{\infty})$ *of admissible smooth representations of G and G' such that π^{∞} and τ^{∞} have cyclic spherical vectors.*

(xi) $(\mathscr{H} \otimes)$ $\dim_{\mathbb{C}} \operatorname{Hom}_{G'}(\pi \otimes \tau, \mathbb{C}) < \infty$ *for any pair* (π, τ) *of admissible Hilbert representations of G and G'.*

(xii) $(\mathscr{H} \otimes)_K$ $\dim_{\mathbb{C}} \operatorname{Hom}_{G'}(\pi \otimes \tau, \mathbb{C}) < \infty$ *for any pair* (π, τ) *of admissible Hilbert representations of G and G' such that π and τ have cyclic spherical vectors.*

4.2   Outline of the Proof of Theorem 4.1

The following implications are obvious:

$$\text{(ii) (Sh)} \Rightarrow \text{(iii) (Sh}_{\text{mod}}) \Rightarrow \text{(iv) (Sh}_{\text{mod}})_1.$$

By Lemma 3.1, we have the following inclusive relations and isomorphism.

$$\operatorname{Hom}_{G'}(\pi^{\infty} \otimes (\tau^{\vee})^{\infty}, \mathbb{C}) \supset \operatorname{Hom}_{G'}(\pi^{\infty}, \tau^{\infty}) \supset \operatorname{Hom}_{G'}(\pi, \tau)$$
$$\simeq \operatorname{Hom}_{G'}(\pi \otimes \tau^{\vee}, \mathbb{C}).$$

In turn, we have the obvious implications and equivalences as below.

$$
\begin{array}{ccccccc}
\text{(ix) } (\infty \otimes) & \Longrightarrow & \text{(v) } (\infty \downarrow) & \Longrightarrow & \text{(vii) } (\mathscr{H} \downarrow) & \Longleftrightarrow & \text{(xi) } (\mathscr{H} \otimes) \\
\Downarrow & & \Downarrow & & \Downarrow & & \Downarrow \\
\text{(x) } (\infty \otimes)_K & \Longrightarrow & \text{(vi) } (\infty \downarrow)_K & \Longrightarrow & \text{(viii) } (\mathscr{H} \downarrow)_K & \Longleftrightarrow & \text{(xii) } (\mathscr{H} \otimes)_K.
\end{array}
$$

The remaining nontrivial implications are

$$\text{(viii) } (\mathscr{H} \downarrow)_K \text{ or (iv) } (\text{Sh}_{\text{mod}})_1$$

$$\Downarrow$$

$$\text{(i) (PP)}$$

$$\Downarrow$$

$$\text{(ii) (Sh) and (ix) } (\infty\otimes).$$

We discuss the geometric property (PP) in Sect. 5.1. Then the implications

$$\text{(i) (PP)} \Rightarrow \text{(ii) (Sh) and (ix) } (\infty\otimes)$$

are given in Propositions 5.6 and 5.7, respectively.

The implication

$$\text{(viii) } (\mathscr{H} \downarrow)_K \Rightarrow \text{(i) (PP)}$$

is proved in Proposition 6.5, and the implication

$$\text{(iv) } (\text{Sh}_{\text{mod}})_1 \Rightarrow \text{(i) (PP)}$$

is proved in Corollary 7.3.

The relationship of $\text{Hom}_{G'}(\pi^\infty, \tau^\infty)$ (symmetry breaking operators) and $\text{Sh}(\lambda, \nu)$ (Shintani functions) will be discussed in Sects. 7 and 8.

4.3 Invariant Trilinear Forms

Suppose that π_i^∞ are admissible smooth representations of a Lie group G on Fréchet spaces $\mathscr{H}_i^\infty$ ($i = 1, 2, 3$). A continuous trilinear form

$$T : \mathscr{H}_1^\infty \times \mathscr{H}_2^\infty \times \mathscr{H}_3^\infty \to \mathbb{C}$$

is *invariant* if

$$T(\pi_1^\infty(g)u_1, \pi_2^\infty(g)u_2, \pi_3^\infty(g)u_3) = T(u_1, u_2, u_3)$$
$$\text{for all } g \in G \text{ and } u_i \in \mathscr{H}_i^\infty \ (i = 1, 2, 3).$$

Corollary 4.2. (1) *Suppose G is a real reductive Lie group. Then the following four conditions on G are equivalent:*

(i) *$(G \times G \times G)/\operatorname{diag} G$ is real spherical as a $(G \times G \times G)$-space.*

(ii) *(Shintani functions in the group case) The space $\text{Sh}((\lambda_1, \lambda_2), \lambda_3)$ of Shintani functions for $(G \times G, \operatorname{diag} G)$ is finite-dimensional for any triple of 3_G-infinitesimal characters λ_1, λ_2 and λ_3.*

(iii) (Symmetry breaking for the tensor product) *For any triple of admissible smooth representations* π_1^∞, π_2^∞, *and* π_3^∞ *of* G,

$$\dim_{\mathbb{C}} \operatorname{Hom}_G(\pi_1^\infty \otimes \pi_2^\infty, \pi_3^\infty) < \infty.$$

(iv) (Invariant trilinear form) *For any triple of admissible smooth representations* π_1^∞, π_2^∞ *and* π_3^∞ *of* G, *the space of invariant trilinear forms is finite-dimensional.*

(2) *Suppose that* G *is a simple Lie group. Then one of (therefore any of) the above four equivalent conditions is fulfilled if and only if either* G *is compact or* $\mathfrak{g}$ *is isomorphic to* $\mathfrak{o}(n, 1)$ $(n \geq 2)$.

Proof. The first and second statements are special cases of Theorems 4.1 and 2.3, respectively. $\square$

Remark 4.3. As in (vi) and (viii) of Theorem 4.1, the conditions (iii) and (iv) of Corollary 4.2 are equivalent to the analogous statements by replacing π_j^∞ ($j = 1, 2, 3$) with spherical ones.

Remark 4.4. The equivalence (i) $\Leftrightarrow$ (ii) was first formulated in [11] with a sketch of proof.

Example 4.5. For $G = O(n, 1)$, a meromorphic family of invariant trilinear forms for spherical principal series representations was constructed in [3].

Remark 4.6. It may happen that the restriction $\pi_{|G'}$ is discretely decomposable. We discussed in [10, 15] when $\operatorname{Hom}_{G'}(\tau, \pi_{|G'})$ is finite-dimensional for all irreducible representations τ of G'.

5 Real Spherical Manifolds and Shintani Functions

In this section we regard Shintani functions as smooth functions on the homogeneous space $(G \times G')/ \operatorname{diag} G'$, and apply the theory of real spherical homogeneous spaces [14]. In particular, we give a proof of the implication (i) (PP) $\Rightarrow$ (ii) (Sh) and (ix) ($\infty \otimes$) in Theorem 4.1 (see Proposition 5.6).

5.1 *Real Spherical Homogeneous Spaces and (PP)*

A complex manifold $X_{\mathbb{C}}$ with action of a complex reductive group $G_{\mathbb{C}}$ is called *spherical* if a Borel subgroup of $G_{\mathbb{C}}$ has an open orbit in $X_{\mathbb{C}}$. In the real setting, in search of a good framework for global analysis on homogeneous spaces which are broader than the usual (e.g. symmetric spaces), we proposed to call:

Definition 5.1 ([11]). Let G be a real reductive Lie group. We say a smooth manifold X with G-action is *real spherical* if a minimal parabolic subgroup P of G has an open orbit in X.

The significance of this geometric property is its application to the finite-multiplicity property in the regular representation of G on $C^\infty(X)$, which was proved by using the theory of hyperfunctions and regular singularities of a system of partial differential equations:

Proposition 5.2 ([14, Theorem A and Theorem 2.2]). *Suppose G is a real reductive linear Lie group, and H is a closed subgroup. If the homogeneous space G/H is real spherical, then the regular representation of G on the Fréchet space $C^\infty(G/H; \chi_\lambda^L)$ is admissible for any $\mathfrak{Z}_G$-infinitesimal character $\lambda \in \mathfrak{j}_{\mathbb C}^\vee / W(\mathfrak{j}_{\mathbb C})$. In particular,*

$$\mathrm{Hom}_G(\pi^\infty, C^\infty(G/H)) \text{ is finite-dimensional}$$

for any smooth admissible representation π^∞ of G.

Suppose that G' is an algebraic reductive subgroup of G. Let P' be a minimal parabolic subgroup of G'.

Definition 5.3 ([14]). We say the pair (G, G') satisfies (PP) if one of the following five equivalent conditions is satisfied.

(PP1) $(G \times G')/\operatorname{diag} G'$ is real spherical as a $(G \times G')$-space.
(PP2) G/P' is real spherical as a G-space.
(PP3) G/P is real spherical as a G'-space.
(PP4) G has an open orbit in $G/P \times G/P'$ via the diagonal action.
(PP5) There are finitely many G-orbits in $G/P \times G/P'$ via the diagonal action.

The above five equivalent conditions are determined only by the Lie algebras $\mathfrak{g}$ and $\mathfrak{g}'$. Therefore we also say that the pair $(\mathfrak{g}, \mathfrak{g}')$ of Lie algebras satisfies (PP).

Next we consider another property, to be denoted by (BB), which is stronger than (PP). Let $G_{\mathbb C}$ be a complex Lie group with Lie algebra $\mathfrak{g}_{\mathbb C} = \mathfrak{g} \otimes_{\mathbb R} \mathbb C$, and $G'_{\mathbb C}$ a subgroup of $G_{\mathbb C}$ with complexified Lie algebra $\mathfrak{g}'_{\mathbb C} = \mathfrak{g}' \otimes_{\mathbb R} \mathbb C$. Let B and B' be Borel subgroups of $G_{\mathbb C}$ and $G'_{\mathbb C}$, respectively.

Definition 5.4. We say the pair (G, G') (or the pair $(\mathfrak{g}, \mathfrak{g}')$) satisfies (BB) if one of the following five equivalent conditions is satisfied:

(BB1) $(G_{\mathbb C} \times G'_{\mathbb C})/\operatorname{diag} G'_{\mathbb C}$ is spherical as a $(G_{\mathbb C} \times G'_{\mathbb C})$-space.
(BB2) $G_{\mathbb C}/B'$ is spherical as a $G_{\mathbb C}$-space.
(BB3) $G_{\mathbb C}/B$ is real spherical as a $G'_{\mathbb C}$-space.
(BB4) $G_{\mathbb C}$ has an open orbit in $G_{\mathbb C}/B \times G_{\mathbb C}/B'$ via the diagonal action.
(BB5) There are finitely many $G_{\mathbb C}$-orbits in $G_{\mathbb C}/B \times G_{\mathbb C}/B'$ via the diagonal action.

The above five equivalent conditions are determined only by the complexified Lie algebras $\mathfrak{g}_{\mathbb C}$ and $\mathfrak{g}'_{\mathbb C}$. It follows from [14, Lemmas 4.2 and 5.3] that we have an implication that

$$(\text{BB}) \Rightarrow (\text{PP}).$$

5.2   Shintani Functions and Real Spherical Homogeneous Spaces

We return to Shintani functions for the pair $G \supset G'$. Let $(\lambda, \nu) \in j_{\mathbb{C}}^{\vee}/W(j_{\mathbb{C}}) \times (j_{\mathbb{C}}')^{\vee}/W(j_{\mathbb{C}}')$. We begin with an elementary and useful point of view:

Lemma 5.5. *The multiplication map*

$$\varphi : G \times G' \to G, \qquad (g, h) \mapsto gh^{-1}$$

induces the following linear isomorphism

$$\varphi^* : \mathrm{Sh}(\lambda, \nu) \xrightarrow{\sim} \mathrm{Hom}_{K \times K'}(\mathbf{1} \boxtimes \mathbf{1}, C^{\infty}((G \times G')/\mathrm{diag}\, G'; \chi_{\lambda,\nu}^{L})),$$

where $\mathbf{1}$ *denotes the trivial one-dimensional representation of the group* K *(or that of* K'*).*

Proof. The pull-back of functions

$$\varphi^* : C^{\infty}(G) \xrightarrow{\sim} C^{\infty}((G \times G')/\mathrm{diag}\, G')$$

satisfies

$$L_X L_Y(\varphi^* f) = \varphi^*(L_X R_Y f) \qquad \text{for all } X \in \mathfrak{g}, Y \in \mathfrak{g}' \text{ and } f \in C^{\infty}(G).$$

Hence φ^* maps $\mathrm{Sh}(\lambda, \nu)$ onto the space of $(K \times K')$-invariant functions of $C^{\infty}((G \times G')/\mathrm{diag}\, G'; \chi_{\lambda,\nu}^{L})$. $\qquad \square$

Proposition 5.6. *If* (G, G') *satisfies* (PP), *then* $\dim_{\mathbb{C}} \mathrm{Sh}(\lambda, \nu) < \infty$ *for any pair* (λ, ν) *of* $(\mathfrak{Z}_G, \mathfrak{Z}_{G'})$-*infinitesimal characters.*

Proof. Since (G, G') satisfies (PP1), the regular representation on the Fréchet space $C^{\infty}((G \times G')/\mathrm{diag}\, G'; \chi_{\lambda,\nu}^{L})$ is admissible as a representation of the direct product group $G \times G'$ by Proposition 5.2. Therefore, Proposition 5.6 follows from Lemma 5.5. $\qquad \square$

Proposition 5.7. *If* (G, G') *satisfies* (PP), *then* $\mathrm{Hom}_{G'}(\pi^{\infty} \otimes \tau^{\infty}, \mathbb{C})$ *is finite-dimensional for any pair* $(\pi^{\infty}, \tau^{\infty})$ *of admissible smooth representations of* G *and* G'.

Proof. Since $(G \times G')/\mathrm{diag}\, G'$ is real spherical,

$$\dim_{\mathbb{C}} \mathrm{Hom}_{G \times G'}(\pi^{\infty} \boxtimes \tau^{\infty}, C^{\infty}(G \times G'/\mathrm{diag}\, G')) < \infty$$

by Proposition 5.2. Therefore Proposition 5.7 follows from the Frobenius reciprocity (Proposition 3.2). $\qquad \square$

6 Construction of Intertwining Operators

In this section we give lower bounds of the dimension of the space of symmetry breaking operators for the restriction of admissible Hilbert representations.

6.1 A Generalization of the Poisson Integral Transform

We fix some general notation. Let H be a closed subgroup of G. Given a finite-dimensional representation τ of H on a vector space W_τ, we denote by $\mathscr{W}_\tau$ the G-equivariant vector bundle $G \times_H W_\tau$ over the homogeneous space G/H. Then we have a representation of G naturally on the space of sections

$$\mathscr{F}(G/H; \tau) \equiv \mathscr{F}(G/H; \mathscr{W}_\tau)$$
$$\simeq \{f \in \mathscr{F}(G) \otimes W : f(\cdot h) = \tau(h)^{-1} f(\cdot) \text{ for } h \in H\},$$

where $\mathscr{F} = \mathscr{A}, C^\infty, \mathscr{D}'$, or $\mathscr{B}$ denote the sheaves of analytic functions, smooth functions, distributions, or hyperfunctions, respectively.

Remark 6.1. We shall regard distributions as generalized functions à la Gelfand (or a special case of hyperfunctions à la Sato) rather than continuous linear forms on $C_c^\infty(G/H, \mathscr{W}_\tau)$.

We define a one-dimensional representation of H by

$$\chi_{G/H} : H \to \mathbb{R}^\times, \quad h \mapsto |\det(\mathrm{Ad}_\#(h) : \mathfrak{g}/\mathfrak{h} \to \mathfrak{g}/\mathfrak{h})|^{-1},$$

where $\mathrm{Ad}_\#(h)$ is the quotient representation of the adjoint representation $\mathrm{Ad}(h) \in GL_\mathbb{R}(\mathfrak{g})$. The bundle of volume densities of $X = G/H$ is given as a G-homogeneous line bundle $\Omega_X \simeq G \times_H \chi_{G/H}$. Then the dualizing bundle of $\mathscr{W}_\tau$ is given, as a homogeneous vector bundle, by

$$\mathscr{W}_\tau^* := (G \times_H W_\tau^\vee) \otimes \Omega_X \simeq G \times_H \tau^*,$$

where $(\tau^\vee, W_\tau^\vee)$ denotes the contragredient representation of (τ, W_τ), and τ^* is a complex representation of H given by

$$\tau^* := \tau^\vee \otimes \chi_{G/H}. \tag{6.1}$$

Suppose now that Q is a parabolic subgroup of a real reductive Lie group G, and $Q = LN$ a Levi decomposition. By an abuse of notation we write $\mathbb{C}_{2\rho}$ for $\chi_{G/Q}$. Then $\mathbb{C}_{2\rho}$ is trivial on the nilpotent subgroup N, and the restriction of $\mathbb{C}_{2\rho}$ to the Levi part L coincides with the one-dimensional representation defined by

$$L \to \mathbb{R}^\times \quad l \mapsto |\det(\mathrm{Ad}(l) : \mathfrak{n} \to \mathfrak{n})|.$$

In view of the isomorphism $K/(Q \cap K) \xrightarrow{\sim} G/Q$, $\mathscr{W}_\tau$ may be regarded as a K-equivariant vector bundle over $K/(Q \cap K)$. Then there exist a K-invariant Hermitian vector bundle structure on $\mathscr{W}_\tau$ and a K-invariant Radon measure on $K/(Q \cap K)$, and we can define a Hilbert representation of G on the Hilbert space $L^2(G/Q; \tau)$ of square integrable sections of $\mathscr{W}_\tau$. The underlying $(\mathfrak{g}, K)$-module of $\mathscr{F}(G/Q; \tau)$ does not depend on the choice of $\mathscr{F} = \mathscr{A}, C^\infty, \mathscr{D}', \mathscr{B}$, or L^2, and will be denoted by $E(G/Q; \tau)$.

We denote by $\hat{G}_f$ and $\hat{L}_f$ the sets of equivalence classes of finite-dimensional irreducible representations over $\mathbb{C}$ of the groups G and L, respectively. Then there is an injective map

$$\hat{G}_f \hookrightarrow \hat{L}_f, \qquad \sigma \mapsto \lambda(\sigma)$$

such that σ is the unique quotient of the $(\mathfrak{g}, K)$-module $E(G/Q; \lambda(\sigma))$. We note that $\lambda(\mathbf{1}) = \mathbb{C}_{2\rho}$.

Here is a Hilbert space analog of [14, Theorem 3.1] which was formulated in the category of $(\mathfrak{g}, K)$-modules (and was proved in the case where Q is a minimal parabolic subgroup of G).

Proposition 6.2. *Let Q be a parabolic subgroup of G, and H a closed subgroup of G. Suppose that there are m disjoint H-invariant open subsets in the real generalized flag variety G/Q. Then*

$$\dim \mathrm{Hom}_G(L^2(G/Q; \lambda(\sigma)), C(G/H; \tau)) \geq m \dim \mathrm{Hom}_H(\sigma|_H, \tau),$$

for any finite-dimensional representations σ and τ of G and H, respectively. In particular, we have

$$\dim \mathrm{Hom}_G(L^2(G/Q, \Omega_{G/Q}), C(G/H)) \geq m.$$

A key of the proof is the construction of integral intertwining operators formulated as follows:

Proposition 6.3. *Let τ and ζ be finite-dimensional representations of H and Q, respectively. We set $\zeta^* = \zeta^\vee \otimes \mathbb{C}_{2\rho}$. Let $(\mathscr{F}, \mathscr{F}')$ be one of the pairs*

$$(\mathscr{A}, \mathscr{B}), \ (C^\infty, \mathscr{D}'), \ (L^2, L^2), \ (\mathscr{D}', C^\infty), \ \text{or} \ (\mathscr{B}, \mathscr{A}).$$

Then there is a canonical injective map

$$\Phi : (\mathscr{F}'(G/Q; \zeta^*) \otimes \tau)^H \hookrightarrow \mathrm{Hom}_G(\mathscr{F}(G/Q; \zeta), C(G/H; \tau)).$$

Proof. The proof is essentially the same with that of [14, Lemma 3.2] which treated the case where $(\mathscr{F}, \mathscr{F}') = (\mathscr{A}, \mathscr{B})$ and where Q is a minimal parabolic subgroup of G. For the sake of completeness, we repeat the proof with appropriate modifications.

The natural G-invariant nondegenerate bilinear form

$$\langle \, , \, \rangle : \mathscr{F}(G/Q;\zeta) \times \mathscr{F}'(G/Q;\zeta^*) \to \mathbb{C}$$

induces an injective G-homomorphism

$$\Psi : \mathscr{F}'(G/Q;\zeta^*) \hookrightarrow \operatorname{Hom}_G(\mathscr{F}(G/Q;\zeta), C(G))$$

by

$$\Psi(\chi)(u)(g) := \langle \pi(g)^{-1}u, \chi \rangle \quad \text{for } \chi \in \mathscr{F}'(G/Q;\zeta^*) \text{ and } u \in \mathscr{F}(G/Q;\zeta),$$

where π is the regular representation of G on $\mathscr{F}(G/Q;\zeta)$.

Taking the tensor product with the finite-dimensional representation τ followed by collecting H-invariant elements, we get the linear map Φ in Proposition 6.3. □

Example 6.4 (Poisson Integral Transform). We apply Proposition 6.3 in the following setting:

$(\mathscr{F}, \mathscr{F}') = (\mathscr{B}, \mathscr{A})$,

$H = K$,

Q : a minimal parabolic subgroup of G,

τ : the trivial one-dimensional representation $\mathbf{1}$ of K,

ζ : a one-dimensional representation of Q such that $\zeta|_{Q \cap K}$ is trivial.

Then $\mathscr{A}(G/Q;\zeta^*)$ is identified with $\mathscr{A}(K/(Q \cap K))$ as a K-module, and the constant function $\mathbf{1}_K$ on $K/(Q \cap K)$ gives rise to an element of $(\mathscr{A}(G/Q;\zeta^*) \otimes \tau)^K$. Then $\mathscr{P}_\mu := \Phi(\mathbf{1}_K)$ in Proposition 6.3 coincides with the Poisson integral transform for the Riemannian symmetric space G/K [8, Chapter 2]:

$$\mathscr{P}_\mu : \mathscr{B}(G/Q;\zeta) \to C(G/K), \quad f \mapsto (\mathscr{P}_\mu f)(g) = \int_K f(gk)dk.$$

See Proposition 8.5 for the preceding results on the image of $\mathscr{P}_\mu$.

Proof of Proposition 6.2. The proof is parallel to that of [14, Theorem 3.17].

Let U_i ($i = 1, 2, \cdots, m$) be disjoint H-invariant open subsets in G/Q. We define

$$\chi_i(g) := \begin{cases} 1 & \text{if } g \in U_i, \\ 0 & \text{if } g \notin U_i. \end{cases}$$

Then $\chi_i \in L^2(G/Q) \simeq L^2(K/Q \cap K)$ ($i = 1, \cdots, m$), and they are H-invariant and linearly independent.

We take linearly independent elements $u_1, \ldots, u_n$ in $\mathrm{Hom}_H(\sigma|_H, \tau)$. Taking the dual of the surjective $(\mathfrak{g}, K)$-homomorphism $E(G/Q; \lambda(\sigma)) \twoheadrightarrow \sigma$, we have an injective $(\mathfrak{g}, K)$-homomorphism $\sigma^\vee \hookrightarrow E(G/Q; \lambda(\sigma)^*) \subset \mathcal{A}(G/Q; \lambda(\sigma)^*)$. Hence we may regard $u_j \in \mathrm{Hom}_H(\sigma|_H, \tau) \simeq (\sigma^\vee \otimes \tau)^H$ as H-invariant elements of $\mathcal{A}(G/Q; \lambda(\sigma)^*) \otimes \tau$. Then $\chi_i u_j \in (L^2(G/Q; \lambda(\sigma)^*) \otimes \tau)^H$ ($1 \le i \le m$, $1 \le j \le n$) are linearly independent.

Proposition 6.2 now follows from Proposition 6.3 with $(\mathcal{F}, \mathcal{F}') = (L^2, L^2)$. $\square$

Proposition 6.5. *Let Q and Q' be parabolic subgroups of G and G'. Suppose that there are m disjoint Q'-invariant open sets in G/Q. Then*

$$\dim \mathrm{Hom}_{G'}(L^2(G/Q, \Omega_{G/Q}), L^2(G'/Q')) \ge m.$$

Proof. We apply Proposition 6.2 to $(G \times G', \mathrm{diag}\, G', \mathbf{1}, \mathbf{1}, Q \times Q')$ for (G, H, σ, τ, Q). Then we have

$$\dim \mathrm{Hom}_{G \times G'}(\pi \boxtimes \tau, C(G \times G'/ \mathrm{diag}\, G')) \ge m,$$

where π is the Hilbert representation of G on $L^2(G/Q, \Omega_{G/Q})$ and τ is that of G' on $L^2(G'/Q', \Omega_{G'/Q'})$.

By Proposition 3.2, we have

$$\dim \mathrm{Hom}_{G'}((\pi \boxtimes \tau)|_{\mathrm{diag}\, G'}, \mathbb{C}) \ge m.$$

By Lemma 3.1 (2), we get the required lower bound. $\square$

6.2 Realization of Small Representations

We end this section with a refinement of [14, Theorem A (2)] which was formulated originally in the category of $(\mathfrak{g}, K)$-modules and was proved when Q is a minimal parabolic subgroup of G.

Definition 6.6. Let Q be a parabolic subgroup of a real reductive Lie group G. Let π be an irreducible admissible representation of G, and π_K the underlying $(\mathfrak{g}, K)$-module. We say π (or π_K) belongs to Q-*series* if π_K occurs as a subquotient of the induced $(\mathfrak{g}, K)$-module $E(G/Q; \tau)$ for some finite-dimensional representation τ of Q.

By Harish–Chandra's subquotient theorem [5], all irreducible admissible representations of G belong to P-series where P is a minimal parabolic subgroup of G. Loosely speaking, the larger a parabolic subgroup Q is, the "smaller" a representation belonging to Q-series becomes, as the following lemma indicates:

Lemma 6.7. *If π_K belongs to Q-series, then its Gelfand–Kirillov dimension, to be denoted by $\mathrm{DIM}(\pi_K)$, satisfies*

$$\mathrm{DIM}(\pi_K) \leq \dim G/Q.$$

The following result formulates that if a subgroup H is "small enough" then the space $(\pi^{-\infty})^H$ of H-invariant distribution vectors of π can be of infinite dimension even for a "small" admissible representations π:

Corollary 6.8. *Let H be an algebraic subgroup of G, and Q a parabolic subgroup of G. Assume that H does not have an open orbit in G/Q. Then for any algebraic finite-dimensional representation τ of H, there exists an irreducible admissible Hilbert representation π of G such that π satisfies the following two properties:*

- *π belongs to Q-series,*
- *$\dim \mathrm{Hom}_G(\pi, C(G/H; \tau)) = \infty$.*

In particular,

$$\dim \mathrm{Hom}_G(\pi^\infty, C^\infty(G/H; \tau)) = \dim \mathrm{Hom}_{\mathfrak{g},K}(\pi_K, \mathscr{A}(G/H; \tau)) = \infty.$$

Proof. There exist infinitely many disjoint H-invariant open sets in G/Q if H does not have an open orbit in G/Q (see [14, Lemma 3.5]). Hence Corollary 6.8 follows from Proposition 6.2 because there exist at most finitely many irreducible subquotients in the Hilbert representation of G on $L^2(G/Q, \Omega_{G/Q})$. $\qquad\square$

Corollary 6.9. *Let $G \supset G'$ be algebraic real reductive Lie groups and Q and Q' parabolic subgroups of G and G', respectively. Assume that Q' does not have an open orbit in G/Q. Then there exist irreducible admissible Hilbert representations π and τ of G and G', respectively, such that (π, τ) satisfies the following two properties:*

- *π belongs to Q-series, τ belongs to Q'-series.*
- *$\dim \mathrm{Hom}_{G'}(\pi|_{G'}, \tau) = \infty$.*

In particular, $\dim \mathrm{Hom}_{G'}(\pi^\infty|_{G'}, \tau^\infty) = \infty$.

Proof. Corollary 6.9 follows from Proposition 6.5. Since the argument is similar to the proof of Corollary 6.8, and we omit it. $\qquad\square$

7 Symmetry Breaking Operators and Construction of Shintani Functions

In this section we construct Shintani functions of moderate growth from symmetry breaking operators of the restriction of admissible smooth representations.

Proposition 7.1. *Let π^∞ be a spherical, admissible smooth representation of G, and τ^∞ that of G'. Suppose that π^∞ and τ^∞ have $\mathfrak{z}_G$ and $\mathfrak{z}_{G'}$-infinitesimal characters λ and $-\nu$, respectively.*

(1) *Let $\mathbf{1}_\pi$ and $\mathbf{1}_{\tau^\vee}$ be nonzero spherical vectors of π_K and $\tau^\vee_{K'}$, respectively. Then there is a natural linear map*

$$\mathrm{Hom}_{G'}(\pi^\infty, \tau^\infty) \to \mathrm{Sh}_{\mathrm{mod}}(\lambda, \nu), \qquad T \mapsto F \qquad (7.1)$$

defined by

$$F(g) := \langle T \circ \pi^\infty(g)\mathbf{1}_\pi, \mathbf{1}_{\tau^\vee}\rangle \quad \textit{for } g \in G.$$

(2) *Assume that the spherical vectors $\mathbf{1}_\pi$ and $\mathbf{1}_{\tau^\vee}$ are cyclic in π_K and $\tau^\vee_{K'}$, respectively. Then (7.1) is injective. In particular, if both π^∞ and τ^∞ are irreducible, (7.1) is injective.*

Remark 7.2. In the setting of Proposition 7.1, if we drop the assumption that $\mathbf{1}_\pi$ is cyclic, then the homomorphism (7.1) may not be injective. In fact, we shall see in Sect. 9 that there is a countable set of (λ, ν) for which the following three conditions are satisfied:

- $\dim_{\mathbb{C}} \mathrm{Hom}_{G'}(\pi^\infty, \tau^\infty) = 2$,
- $\dim_{\mathbb{C}} \mathrm{Sh}_{\mathrm{mod}}(\lambda, \nu) = 1$,
- $\mathbf{1}_{\tau^\vee}$ is cyclic in $\tau^\vee$.

Proof of Proposition 7.1. (1) Since $T \in \mathrm{Hom}_{G'}(\pi^\infty, \tau^\infty)$, the function $F \in C^\infty(G)$ satisfies

$$\begin{aligned}F(hg) &= \langle \tau^\infty(h) \circ \tau \circ \pi^\infty(g)\mathbf{1}_\pi, \mathbf{1}_{\tau^\vee}\rangle \\ &= \langle T \circ \pi^\infty(g)\mathbf{1}_\pi, (\tau^\vee)^\infty(h^{-1})\mathbf{1}_{\tau^\vee}\rangle \end{aligned} \qquad (7.2)$$

for all $h \in G'$ and $g \in G$. Therefore we have

$$\begin{aligned}F(k'gk) &= F(g) \quad \text{for } k' \in K' \text{ and } k \in K, \\ (L_Y F)(g) &= \langle T \circ \pi^\infty(g)\mathbf{1}_\pi, d\tau^\vee(Y)\mathbf{1}_{\tau^\vee}\rangle \quad \text{for } Y \in \mathfrak{g}' \subset U(\mathfrak{g}'_{\mathbb{C}}), \\ (R_X F)(g) &= \langle T \circ \pi^\infty(g)d\pi(X)\mathbf{1}_\pi, \mathbf{1}_{\tau^\vee}\rangle \quad \text{for } X \in \mathfrak{g} \subset U(\mathfrak{g}_{\mathbb{C}}).\end{aligned}$$

Since $u \in \mathfrak{z}_G$ acts on π^∞ as the scalar multiple of $\chi_\lambda(u)$, we have $d\pi^\infty(u)\mathbf{1}_\pi = \chi_\lambda(u)\mathbf{1}_\pi$, and therefore $R_u F = \chi_\lambda(u)F$. Likewise, for $v \in \mathfrak{z}_{G'}$, we have $d(\tau^\vee)^\infty(v)\mathbf{1}_{\tau^\vee} = \chi_\nu(v)\mathbf{1}_{\tau^\vee}$, and thus $L_v F = \chi_\nu(v)F$. Hence $F \in \mathrm{Sh}(\lambda, \nu)$.

Let V_π^∞ and W_τ^∞ be the representation spaces of π^∞ and τ^∞, respectively. First we find a continuous seminorm $|\cdot|_1$ on W_τ^∞ and a constant C_1 such that

$$|\langle w, \mathbf{1}_{\tau^\vee}\rangle| \le C_1|w|_1 \quad \text{for any } w \in W_\tau^\infty.$$

Second, since $T : V_\pi^\infty \to W_\tau^\infty$ is continuous, there exist a continuous seminorm $|\cdot|_2$ on V_π^∞ and a constant C_2 such that

$$|Tv|_1 \leq C_2 |v|_2 \quad \text{for any } v \in V_\pi^\infty.$$

Third, since π^∞ has moderate growth, there exist constants $C_3 > 0$, $d \in \mathbb{R}$ and a continuous seminorm $|\cdot|_3$ on V_π^∞ such that

$$|\pi^\infty(g)d\pi(u)\mathbf{1}_\pi|_2 \leq C_3 |d\pi^\infty(u)\mathbf{1}_\pi|_3 \|g\|^d$$

for any $g \in G$ and for any $u \in U(\mathfrak{g}_\mathbb{C})$.

Therefore $(R_u F)(g) = \langle T \circ \pi^\infty(g)d\pi(u)\mathbf{1}_\pi, \mathbf{1}_{\tau^\vee} \rangle$ satisfies the following inequality:

$$|(R_u F)(g)| \leq C_1 C_2 C_3 |d\pi^\infty(u)\mathbf{1}_\pi|_3 \|g\|^d \quad \text{for any } g \in G.$$

Hence $F \in C^\infty(G)$ has moderate growth.

(2) Suppose $F \equiv 0$. Since $\mathbf{1}_{\tau^\vee}$ is a cyclic vector, we have $T \circ \pi^\infty(g)\mathbf{1}_\pi = 0$ for any $g \in G$ by (7.2). Since $\mathbf{1}_\pi$ is a cyclic vector, we have $T = 0$. Therefore the map (7.1) is injective. □

Corollary 7.3. *Suppose that there are m disjoint P'-invariant open sets in G/P. Then*

$$\dim_\mathbb{C} \mathrm{Sh}_{\mathrm{mod}}(\rho_\mathfrak{g}, \rho_{\mathfrak{g}'}) \geq m. \tag{7.3}$$

In particular, if $\mathrm{Sh}_{\mathrm{mod}}(\rho_\mathfrak{g}, \rho_{\mathfrak{g}'})$ is finite-dimensional, then the pair (G, G') of reductive groups satisfies (PP).

Proof. We denote by π the Hilbert representation of G on $L^2(G/P, \Omega_{G/P})$, and by τ that of G' on $L^2(G'/P')$. By Proposition 6.5, we have

$$\dim_\mathbb{C} \mathrm{Hom}_{G'}(\pi|_{G'}, \tau) \geq m.$$

On the other hand, since both π and $\tau^\vee$ contain spherical cyclic vectors, we have

$$\dim_\mathbb{C} \mathrm{Sh}_{\mathrm{mod}}(\rho_\mathfrak{g}, -\rho_{\mathfrak{g}'}) \geq \dim_\mathbb{C} \mathrm{Hom}_{G'}(\pi^\infty|_{G'}, \tau^\infty)$$

from Proposition 7.1. Combining these inequalities with (3.1), we have obtained

$$\dim_\mathbb{C} \mathrm{Sh}_{\mathrm{mod}}(\rho_\mathfrak{g}, -\rho_{\mathfrak{g}'}) \geq m.$$

Since $-\rho_{\mathfrak{g}'}$ is conjugate to $\rho_{\mathfrak{g}'}$ by the longest element of the Weyl group $W(\mathfrak{j}'_\mathbb{C})$, we have proved (7.3).

Finally, if (G, G') does not satisfy (PP), then there exist infinitely many disjoint P'-invariant open sets in G/P, and therefore we get $\dim_\mathbb{C} \mathrm{Sh}_{\mathrm{mod}}(\rho_\mathfrak{g}, \rho_{\mathfrak{g}'}) = \infty$ from (7.3). □

8 Boundary Values of Shintani Functions

In this section we realize Shintani functions as joint eigenfunctions of invariant differential operators on the Riemannian symmetric space $X = (G \times G')/(K \times K')$, and then as hyperfunctions on the minimal boundary $Y = (G \times G')/(P \times P')$ of the compactification of X. The main results of this section are Theorems 8.1 and 8.2. We prove these theorems in Sects. 8.4 and 8.5, respectively, after giving a brief summary of the preceding results of harmonic analysis on Riemannian symmetric spaces in Sects. 8.2 and 8.3.

8.1 Symmetry Breaking of Principal Series Representations

Denote by θ the Cartan involution of the Lie algebra $\mathfrak{g}$ corresponding to the maximal compact subgroup K of G. We take a maximal abelian subspace $\mathfrak{a}$ in the vector space $\{X \in \mathfrak{g} : \theta X = -X\}$, and set

$$W(\mathfrak{a}) := N_K(\mathfrak{a})/Z_K(\mathfrak{a}).$$

We fix a positive system $\Sigma^+(\mathfrak{g}, \mathfrak{a})$ of the restricted root system $\Sigma(\mathfrak{g}, \mathfrak{a})$, and define a minimal parabolic subalgebra $\mathfrak{p}$ of $\mathfrak{g}$ by

$$\mathfrak{p} = \mathfrak{m} + \mathfrak{a} + \mathfrak{n} = \mathfrak{l} + \mathfrak{n},$$

where $\mathfrak{l} := Z_\mathfrak{g}(\mathfrak{a}) = \{X \in \mathfrak{a} : [H, X] = 0 \text{ for all } H \in \mathfrak{a}\}$, $\mathfrak{m} := \mathfrak{l} \cap \mathfrak{k}$, and $\mathfrak{n}$ is the sum of the root spaces for all $\alpha \in \Sigma^+(\mathfrak{g}, \mathfrak{a})$. Let $P = MAN$ be the minimal parabolic subgroup of G with Lie algebra $\mathfrak{p}$.

We take a Cartan subalgebra $\mathfrak{t}$ in $\mathfrak{m}$. Then $\mathfrak{j} := \mathfrak{t} + \mathfrak{a}$ is a maximally split Cartan subalgebra of $\mathfrak{g}$. We fix a positive system $\Delta^+(\mathfrak{m}_\mathbb{C}, \mathfrak{t}_\mathbb{C})$. Let $\rho_\mathfrak{n} \in \mathfrak{a}^\vee$ be half the sum of the elements in $\Sigma^+(\mathfrak{g}, \mathfrak{a})$ counted with multiplicities, and $\rho_\mathfrak{l} \in \mathfrak{t}_\mathbb{C}^\vee$ that of $\Delta^+(\mathfrak{m}_\mathbb{C}, \mathfrak{t}_\mathbb{C})$. The positive systems $\Sigma^+(\mathfrak{g}, \mathfrak{a})$ and $\Delta^+(\mathfrak{m}_\mathbb{C}, \mathfrak{t}_\mathbb{C})$ determine naturally a positive system $\Delta^+(\mathfrak{g}_\mathbb{C}, \mathfrak{j}_\mathbb{C})$. Then we have

$$\rho_\mathfrak{g} = \rho_\mathfrak{l} + \rho_\mathfrak{n} \in \mathfrak{j}_\mathbb{C}^\vee = \mathfrak{t}_\mathbb{C}^\vee + \mathfrak{a}_\mathbb{C}^\vee,$$

where we regard $\mathfrak{t}_\mathbb{C}^\vee$ and $\mathfrak{a}_\mathbb{C}^\vee$ as subspaces of $\mathfrak{j}_\mathbb{C}^\vee$ via the direct sum decomposition $\mathfrak{j} = \mathfrak{t} + \mathfrak{a}$. Then $\rho_\mathfrak{l} + \mathfrak{a}_\mathbb{C}^\vee = \rho_\mathfrak{g} + \mathfrak{a}_\mathbb{C}^\vee$ is an affine subspace of $\mathfrak{j}_\mathbb{C}^\vee$.

Analogous notation is applied to the reductive subgroup G'. In particular, $\mathfrak{j}' = \mathfrak{t}' + \mathfrak{a}'$ is a maximally split Cartan subalgebra of $\mathfrak{g}'$.

We recall $(\lambda, \nu) \in \mathfrak{j}_\mathbb{C}^\vee/W(\mathfrak{j}_\mathbb{C}) \times (\mathfrak{j}_\mathbb{C}')^\vee/W(\mathfrak{j}_\mathbb{C}') \simeq \mathrm{Hom}_{\mathbb{C}\text{-alg}}(\mathfrak{Z}_G \times \mathfrak{Z}_{G'}, \mathbb{C})$. Let us begin with a nonvanishing condition for $\mathrm{Sh}(\lambda, \nu)$.

Theorem 8.1. *If* $\mathrm{Sh}(\lambda, \nu) \neq \{0\}$, *then*

$$\lambda \in W(j_\mathbb{C})(\rho_{\mathfrak{l}} + \mathfrak{a}_\mathbb{C}^\vee) \quad \text{and} \quad \nu \in W(j'_\mathbb{C})(\rho_{\mathfrak{l}'} + (\mathfrak{a}'_\mathbb{C})^\vee). \tag{8.1}$$

We shall give a proof of Theorem 8.1 in Sect. 8.4.

Next we consider a construction of Shintani functions under the assumption (8.1). Suppose $\lambda \in j_\mathbb{C}^\vee$ satisfies $\lambda - \rho_{\mathfrak{l}} \in \mathfrak{a}_\mathbb{C}^\vee$. Then there exists $\lambda_+ \in \mathfrak{a}_\mathbb{C}^\vee$ such that λ_+ satisfies the following two conditions:

$$\lambda_+ - \rho_\mathrm{n} = w(\lambda - \rho_{\mathfrak{l}}) \quad \text{for some} \ \ w \in W(\mathfrak{a}). \tag{8.2}$$

$$\mathrm{Re}\langle \lambda_+ - \rho_\mathrm{n}, \alpha \rangle \geq 0 \quad \text{for any} \ \alpha \in \Sigma^+(\mathfrak{g}, \mathfrak{a}). \tag{8.3}$$

Similarly, suppose $\nu \in (j'_\mathbb{C})^\vee$ satisfies $\nu - \rho_{\mathfrak{l}'} \in (\mathfrak{a}'_\mathbb{C})^\vee$. Then there exists $\nu_- \in (\mathfrak{a}'_\mathbb{C})^\vee$ satisfying the following two conditions:

$$\nu_- - \rho_{\mathrm{n}'} = w'(-\nu + \rho_{\mathfrak{l}'}) \quad \text{for some} \ \ w' \in W(\mathfrak{a}').$$

$$\mathrm{Re}\langle \nu_- - \rho_{\mathrm{n}'}, \alpha \rangle \leq 0 \quad \text{for any} \ \alpha \in \Sigma^+(\mathfrak{g}, \mathfrak{a}'). \tag{8.4}$$

Theorem 8.2. *Suppose that* $\lambda \in j_\mathbb{C}^\vee$ *and* $\nu \in (j'_\mathbb{C})^\vee$ *satisfy* $\lambda + \rho_{\mathfrak{l}} \in \mathfrak{a}_\mathbb{C}^\vee$ *and* $\nu + \rho_{\mathfrak{l}'} \in (\mathfrak{a}'_\mathbb{C})^\vee$. *Let* λ_+ *and* ν_- *be defined as above.*

(1) *There is a natural injective linear map*

$$\mathrm{Hom}_{G'}(C^\infty(G/P; \lambda_+), C^\infty(G'/P'; \nu_-)) \hookrightarrow \mathrm{Sh}_\mathrm{mod}(\lambda, \nu). \tag{8.5}$$

(2) *If* G, G' *are classical groups, then* (8.5) *is a bijection:*

$$\mathrm{Hom}_{G'}(C^\infty(G/P; \lambda_+), C^\infty(G'/P'; \nu_-)) \xrightarrow{\sim} \mathrm{Sh}_\mathrm{mod}(\lambda, \nu). \tag{8.6}$$

We shall prove Theorem 8.2 in Sect. 8.5.

Remark 8.3. As the proof shows, the bijection (8.6) holds for generic (λ, ν) even when G or G' are exceptional groups.

8.2 Invariant Differential Operators

In this and next subsections we give a quick review of the preceding results of harmonic analysis on Riemannian symmetric spaces. We denote by $\mathbb{D}(G/K)$ the $\mathbb{C}$-algebra consisting of all G-invariant differential operators on the Riemannian symmetric space G/K. It is isomorphic to a polynomial ring of $(\dim_{\mathbb{R}} \mathfrak{a})$-generators. More precisely, let $\gamma' : U(\mathfrak{g}_{\mathbb{C}}) \to U(\mathfrak{a}_{\mathbb{C}}) = S(\mathfrak{a}_{\mathbb{C}})$ be the projection to the second factor of the decomposition $U(\mathfrak{g}_{\mathbb{C}}) = (\mathfrak{k}_{\mathbb{C}} U(\mathfrak{g}_{\mathbb{C}}) + U(\mathfrak{g}_{\mathbb{C}})\mathfrak{n}_{\mathbb{C}}) \oplus U(\mathfrak{a}_{\mathbb{C}})$. Then we have the Harish–Chandra isomorphism

$$\mathbb{D}(G/K) \xleftarrow[R]{\sim} U(\mathfrak{g}_{\mathbb{C}})^K / U(\mathfrak{g}_{\mathbb{C}})^K \cap U(\mathfrak{g}_{\mathbb{C}})\mathfrak{k}_{\mathbb{C}} \xrightarrow[\gamma]{\sim} S(\mathfrak{a}_{\mathbb{C}})^{W(\mathfrak{a})}, \qquad (8.7)$$

where $\gamma : U(\mathfrak{g}_{\mathbb{C}}) \to S(\mathfrak{a}_{\mathbb{C}})$ is defined by

$$\langle \gamma(u), \lambda \rangle = \langle \gamma'(u), \lambda - \rho_{\mathfrak{n}} \rangle \qquad \text{for all } \lambda \in \mathfrak{a}_{\mathbb{C}}^{\vee},$$

which is a generalization of (3.5), see [8, Chapter II]. Through (8.7), we have a bijection

$$\operatorname{Hom}_{\mathbb{C}\text{-alg}}(\mathbb{D}(G/K), \mathbb{C}) \simeq \mathfrak{a}_{\mathbb{C}}^{\vee}/W(\mathfrak{a}), \qquad \psi_{\mu} \leftrightarrow \mu, \qquad (8.8)$$

given by $\psi_{\mu}(R_v) = \langle \gamma(v), \mu \rangle = \langle \gamma'(v), \lambda - \rho_{\mathfrak{n}} \rangle$ for $v \in U(\mathfrak{g}_{\mathbb{C}})^K$.

Comparing the two bijections $\operatorname{Hom}_{\mathbb{C}\text{-alg}}(\mathfrak{Z}_G, \mathbb{C}) \simeq \mathfrak{j}_{\mathbb{C}}^{\vee}/W(\mathfrak{j}_{\mathbb{C}})$ (see (3.6)) and (8.8) via the $\mathbb{C}$-algebra homomorphism

$$\mathfrak{Z}_G \subset U(\mathfrak{g}_{\mathbb{C}})^K \xrightarrow{R} \mathbb{D}(G/K), \qquad (8.9)$$

we have

$$\psi_{\mu} \circ R = \chi_{\mu + \rho_{\mathfrak{n}}} \qquad \text{on } \mathfrak{Z}_G \text{ for all } \mu \in \mathfrak{a}_{\mathbb{C}}^{\vee}. \qquad (8.10)$$

By (8.10), we have

$$C^{\infty}(G/K; \mathcal{M}_{\mu}) \subset C^{\infty}(G/K; \chi_{\mu + \rho_{\mathfrak{l}}}^R). \qquad (8.11)$$

For a simple Lie group G, it is known [7] that the $\mathbb{C}$-algebra homomorphism (8.9) is surjective if and only if $(\mathfrak{g}, \mathfrak{k})$ is not one of the following pairs:

$$(\mathfrak{e}_{6(-14)}, \mathfrak{so}(10) + \mathbb{R}),$$
$$(\mathfrak{e}_{6(-26)}, \mathfrak{f}_{4(-52)}),$$
$$(\mathfrak{e}_{7(-25)}, \mathfrak{e}_{6(-78)} + \mathbb{R}),$$
$$(\mathfrak{e}_{8(-24)}, \mathfrak{e}_{7(-133)} + \mathfrak{su}(2)).$$

For $\mathscr{F} = \mathscr{A}, \mathscr{B}, C^{\infty}$, or $\mathscr{D}'$, we denote by $\mathscr{F}(G/K; \mathcal{M}_{\mu})$ the space of all $F \in \mathscr{F}(G/K)$ such that F satisfies the system of the following partial differential equations:

$$DF = \psi_\mu(D)F \quad \text{for all } D \in \mathbb{D}(G/K). \qquad (\mathcal{M}_\mu)$$

Since the Laplacian Δ on the Riemannian symmetric space G/K is an elliptic differential operator and belongs to $\mathbb{D}(G/K)$, we have

$$\mathscr{A}(G/K;\mathcal{M}_\mu) = \mathscr{B}(G/K;\mathcal{M}_\mu) = C^\infty(G/K;\mathcal{M}_\mu) = \mathscr{D}'(G/K;\mathcal{M}_\mu)$$

by the elliptic regularity theorem.

8.3 Poisson Transform and Boundary Maps

Given $\mu \in \mathfrak{a}_{\mathbb{C}}^\vee$, we lift and extend it to a one-dimensional representation of the minimal parabolic subgroup $P = MAN$ by

$$P \to \mathbb{C}^\times, \quad m \exp H n \mapsto e^{\langle \mu, H \rangle}$$

for $m \in M$, $H \in \mathfrak{a}$, and $n \in N$.

Remark 8.4. In the field of harmonic analysis on symmetric spaces people sometimes adopt the opposite signature of the (normalized) parabolic induction which is used in the representation theory of real reductive groups. Since our definition of parabolic induction does not involve the "ρ-shift" (*i.e.*, unnormalized parabolic induction where $\sqrt{-1}\mathfrak{a}^\vee + \rho_\mathfrak{n}$ is the unitary axis), the G-module $C^\infty(G/P;\mu)$ in our notation corresponds to $C^\infty(G/P;\mathscr{L}_{\rho_\mathfrak{n}-\mu})$ in [8, 20].

With this remark in mind, we summarize some known results that we need:

Proposition 8.5. (1) *The* $(\mathfrak{g}, K)$-*module* $E(G/P;\mu)$ *has* $\mathfrak{Z}_G$-*infinitesimal character* $\mu + \rho_\mathfrak{g} = \mu + \rho_\mathfrak{n} + \rho_\mathfrak{l} \in \mathfrak{j}_{\mathbb{C}}^\vee/W(\mathfrak{j}_{\mathbb{C}})$.
(2) *The* $(\mathfrak{g}, K)$-*module* $E(G/P;\mu)$ *is spherical for all* $\mu \in \mathfrak{a}_{\mathbb{C}}^\vee$. *Furthermore, the unique (up to scalar) nonzero spherical vector is cyclic if* μ *satisfies*

$$\mathrm{Re}\langle \mu - \rho_\mathfrak{n}, \alpha \rangle \geq 0 \quad \text{for any } \alpha \in \Sigma^+(\mathfrak{g}, \mathfrak{a}).$$

(3) *For all* $\mu \in \mathfrak{a}_{\mathbb{C}}^\vee$, *the Poisson transform* $\mathscr{P}_\mu$ *maps into the space of joint eigenfunctions of the* $\mathbb{C}$-*algebra* $\mathbb{D}(G/K)$:

$$\mathscr{P}_\mu : \mathscr{B}(G/P;\mu) \to \mathscr{A}(G/K;\mathcal{M}_{\rho_\mathfrak{n}-\mu}). \qquad (8.12)$$

(4) *The Poisson transform* (8.12) *is bijective if* μ *satisfies*

$$\mathrm{Re}\langle \mu - \rho_\mathfrak{n}, \alpha \rangle \leq 0 \quad \text{for any } \alpha \in \Sigma^+(\mathfrak{g}, \mathfrak{a}). \qquad (8.13)$$

(5) *The Poisson transform $\mathscr{P}_\mu$ induces a bijection*

$$\mathscr{P}_\mu : \mathscr{D}'(G/P; \mu) \to \mathscr{A}_{\mathrm{mod}}(G/K; \mathscr{M}_{\rho_n - \mu})$$

if (8.13) *is satisfied.*

Proof. The first statement is elementary. See Kostant [17] for (2), and Helgason [6] for (3). The fourth statement was proved in Kashiwara et.al. [9] by using the theory of regular singularity of a system of partial differential equations. For the proof of the fifth statement, see Oshima–Sekiguchi [20] or Wallach [22, Theorem 11.9.4]. We note that for $f \in \mathscr{A}(G/K; \mathscr{M}_{\rho_n - \mu})$, f has moderate growth (Definition 3.3) if and only if f has at most exponential growth in the sense that there exist constants $d \in \mathbb{R}$ and $C > 0$ such that $|f(x)| \le C \|x\|^d$ for all $x \in G$.   □

8.4 Proof of Theorem 8.1

In Lemma 5.5, we realized the Shintani space $\mathrm{Sh}(\lambda, \nu)$ in $C^\infty((G \times G')/\mathrm{diag}\, G')$. We give another realization of $\mathrm{Sh}(\lambda, \nu)$:

Lemma 8.6. *The multiplication map*

$$\psi : G \times G' \to G, \quad (g, g') \mapsto (g')^{-1} g$$

induces the following bijection:

$$\psi^* : \mathrm{Sh}(\lambda, \nu) \xrightarrow{\sim} C^\infty((G \times G')/(K \times K'); \chi_{\lambda,\nu}^R)^{\mathrm{diag}\, G'}. \tag{8.14}$$

Proof. We set $C^\infty(K'\backslash G/K) := \{f \in C^\infty(G) : f(k'gk) = f(g) \quad \text{for all } k' \in K' \text{ and } k \in K\}$. The pull-back ψ^* of functions induces a bijective linear map

$$
\begin{array}{ccc}
C^\infty(G) & \xrightarrow{\sim} & C^\infty(G \times G')^{\mathrm{diag}\, G'} \\
\cup & & \cup \\
C^\infty(K'\backslash G/K) & \xrightarrow[\psi^*]{\sim} & C^\infty((G \times G')/(K \times K'))^{\mathrm{diag}\, G'}.
\end{array}
$$

On the other hand, for $X \in \mathfrak{g}$ and $Y \in \mathfrak{g}'$, we have

$$R_X R_Y(\psi^* f) = \psi^*(R_X L_Y f).$$

Thus Lemma 8.6 is proved.   □

Proof of Theorem 8.1. Suppose $\mathrm{Sh}(\lambda, \nu) \ne \{0\}$. Then, by Lemma 8.6, we have

$$V_{\lambda,\nu} := C^\infty((G \times G')/(K \times K'); \chi_{\lambda,\nu}^R) \ne \{0\}.$$

Since $R(\mathfrak{Z}_{G \times G'})$ is an ideal in $\mathbb{D}((G \times G')/(K \times K'))$ of finite-codimension, we can take the boundary values of $V_{\lambda, \nu}$ inductively to the hyperfunction-valued principal series representations of $G \times G'$ as in [14, Section 2]. To be more precise, there exist $\mu_1, \cdots, \mu_N \in \mathfrak{a}_{\mathbb{C}}^{\vee} \times (\mathfrak{a}_{\mathbb{C}}')^{\vee}$ and $(G \times G')$-invariant subspaces

$$\{0\} = V(0) \subset V(1) \subset \cdots \subset V(N) = V_{\lambda, \nu}$$

such that the quotient space $V(j)/V(j-1)$ is isomorphic to a subrepresentation of the spherical principal series representation $\mathscr{B}((G \times G')/(P \times P'); \mu_j)$ as $(G \times G')$-modules.

Comparing the $\mathfrak{Z}_{G \times G'}$-infinitesimal characters of $V_{\lambda, \nu}$ and $\mathscr{B}((G \times G')/(P \times P'); \mu_j)$, we get Theorem 8.1.   □

8.5 Proof of Theorem 8.2

Proof of Theorem 8.2. (1) Since λ_+ satisfies (8.3), the $(\mathfrak{g}, K)$-module $E(G/P; \lambda_+)$ contains a cyclic spherical vector by Proposition 8.5. Similarly, the $(\mathfrak{g}', K')$-module

$$E(G'/P'; \nu_-)_{K'}^{\vee} \simeq E(G'/P'; \nu_-^*)_{K'}$$

has a cyclic vector because $\nu_-^* = -\nu_- + 2\rho_{\mathfrak{n}'}$ satisfies

$$\mathrm{Re}\langle \nu_-^* - \rho_{\mathfrak{n}'}, \alpha \rangle \geq 0 \quad \text{for any } \alpha \in \Sigma^+(\mathfrak{g}', \mathfrak{a}')$$

by (8.4). Hence the first statement follows from Proposition 7.1.

(2) In view of the definition of moderate growth (Definition 3.3), we see that the bijection ψ^* in (8.14) induces the following bijection:

$$\mathrm{Sh}_{\mathrm{mod}}(\lambda, \nu) \xrightarrow{\sim} C_{\mathrm{mod}}^{\infty}((G \times G')/(K \times K'); \chi_{\lambda, \nu}^R)^{\mathrm{diag}\, G'}. \tag{8.15}$$

Since the $\mathbb{C}$-algebra homomorphism $R : \mathfrak{Z}_{G \times G'} \to \mathbb{D}(G \times G'/K \times K')$ is surjective for classical groups G and G', the isomorphism (8.15) implies the following bijection

$$\mathrm{Sh}_{\mathrm{mod}}(\lambda, \nu) \simeq C_{\mathrm{mod}}^{\infty}((G \times G')/(K \times K'); \mathscr{M}_{(\lambda + \rho_{\mathfrak{l}}, \nu + \rho_{\mathfrak{l}'})})^{\mathrm{diag}\, G'}$$

by (8.11). In turn, combining with the Poisson transform, we have obtained the following natural isomorphism by Proposition 8.5 (5):

$$\mathrm{Sh}_{\mathrm{mod}}(\lambda, \nu) \simeq \mathscr{D}'((G \times G')/(P \times P'); \lambda_+^* \boxtimes \nu_-)^{\mathrm{diag}\, G'}.$$

By [16, Proposition 3.2], we proved the following natural bijection:

$$\mathrm{Hom}_{G'}(C^\infty(G/P;\lambda_+), C^\infty(G'/P';\nu_-)) \simeq \mathscr{D}'((G\times G')/(P\times P');\lambda_+^*\boxtimes\nu_-)^{\mathrm{diag}\, G'}.$$

Hence we have completed the proof of Theorem 8.2.   □

9 Shintani Functions for $(O(n+1,1), O(n,1))$

It has been an open problem to find $\dim_{\mathbb{C}} \mathrm{Sh}(\lambda,\nu)$ in the Archimedean case [19, Remark 5.6]. In this section, by using a classification of symmetry breaking operators between spherical principal series representations of the pair $(G, G') = (O(n+1,1), O(n,1))$ in a recent work [16] with B. Speh, we determine the dimension of $\mathrm{Sh}(\lambda,\nu)$ in this case.

We denote by $[x]$ the greatest integer that does not exceed x. For the pair $(G, G') = (O(n+1,1), O(n,1))$, $(\mathfrak{z}_G, \mathfrak{z}_{G'})$-infinitesimal characters (λ,ν) are parametrized by

$$\mathfrak{j}_{\mathbb{C}}^\vee/W(\mathfrak{j}_{\mathbb{C}}) \times (\mathfrak{j}_{\mathbb{C}}')^\vee/W(\mathfrak{j}_{\mathbb{C}}') \simeq \mathbb{C}^{[\frac{n+2}{2}]}/W_{[\frac{n+2}{2}]} \times \mathbb{C}^{[\frac{n+1}{2}]}/W_{[\frac{n+1}{2}]}$$

in the standard coordinates, where $W_k := \mathfrak{S}_k \ltimes (\mathbb{Z}/2\mathbb{Z})^k$.

Theorem 9.1. *Let* $(G, G') = (O(n+1,1), O(n,1))$.

(1) *The following three conditions on* (λ, ν) *are equivalent:*

  (i) $\mathrm{Sh}(\lambda,\nu) \neq \{0\}$.
  (ii) $\mathrm{Sh}_{\mathrm{mod}}(\lambda,\nu) \neq \{0\}$.
  (iii) *In the standard coordinates*

$$\lambda = w(\frac{n}{2}+t, \frac{n}{2}-1, \frac{n}{2}-2, \cdots, \frac{n}{2}-[\frac{n}{2}]), \tag{9.1}$$

$$\nu = w'(\frac{n-1}{2}+s, \frac{n-1}{2}-1, \cdots, \frac{n-1}{2}-[\frac{n-1}{2}]),$$

  for some $t, s \in \mathbb{C}$, $w \in W_{[\frac{n+2}{2}]}$, *and* $w' \in W_{[\frac{n+1}{2}]}$.

(2) *If* (λ, ν) *satisfies* (iii) *in* (1), *then*

$$\dim_{\mathbb{C}} \mathrm{Sh}_{\mathrm{mod}}(\lambda,\nu) = 1.$$

Proof. It is sufficient to prove the implication (i) $\Rightarrow$ (iii) and (2).

For $\lambda \in \mathfrak{j}_{\mathbb{C}}^\vee/W(\mathfrak{j}_{\mathbb{C}}) \simeq \mathbb{C}^{[\frac{n+1}{2}]}/W_{[\frac{n+1}{2}]}$, λ belongs to $\rho_{\mathfrak{l}} + \mathfrak{a}_{\mathbb{C}}^\vee$ mod $W(\mathfrak{j}_{\mathbb{C}})$ if and only if λ is of the form (9.1) for some $t \in \mathbb{C}$ and $w \in W_{[\frac{n+2}{2}]}$. Similarly for $\nu \in (\mathfrak{j}_{\mathbb{C}}')^\vee/W(\mathfrak{j}_{\mathbb{C}}')$. Hence the implication (i) $\Rightarrow$ (iii) holds as a special case of Theorem 8.1.

Next, suppose that (λ, ν) satisfies (iii). Without loss of generality, we may and do assume $\operatorname{Re} t \geq \frac{n}{2}$ and $\operatorname{Re} s \leq \frac{n-1}{2}$. In this case the unique element $\lambda_+ \in \mathfrak{a}_{\mathbb{C}}$ satisfying (8.2) and (8.3) is equal to t if we identify $\mathfrak{a}_{\mathbb{C}}^{\vee}$ with $\mathbb{C}$ via the standard basis $\{e_1\}$ of $\mathfrak{a}^{\vee}$ such that $\Sigma(\mathfrak{g}, \mathfrak{a}) = \{\pm e_1\}$. Similarly, $\nu_- = s$ via $(\mathfrak{a}_{\mathbb{C}}')^{\vee} \simeq \mathbb{C}$.

We define a discrete subset of $\mathfrak{a}_{\mathbb{C}}^{\vee} \oplus (\mathfrak{a}_{\mathbb{C}}')^{\vee} \simeq \mathbb{C}^2$ by

$$L_{\text{even}} := \{(a, b) \in \mathbb{Z}^2 : a \leq b \leq 0, \ a \equiv b \mod 2\}.$$

According to [16, Theorem 1.1], we have

$$\dim_{\mathbb{C}} \operatorname{Hom}_{G'}(C^{\infty}(G/P; a), C^{\infty}(G'/P'; b)) \simeq \begin{cases} 1 & \text{if } (a, b) \in \mathbb{C}^2 \setminus L_{\text{even}}, \\ 2 & \text{if } (a, b) \in L_{\text{even}}. \end{cases}$$

Since $(\lambda_+, \nu_-) = (t, s) \notin L_{\text{even}}$, we conclude that

$$\dim_{\mathbb{C}} \operatorname{Sh}_{\text{mod}}(\lambda, \nu) = \dim_{\mathbb{C}} \operatorname{Hom}_{G'}(C^{\infty}(G/P; \lambda_+), C^{\infty}(G'/P'; \nu_-)) = 1$$

by Theorem 8.2 (2). Thus Theorem 9.1 is proved. □

10 Concluding Remarks

We raise the following two related questions:

Problem 10.1. Find a condition on a pair of real reductive linear Lie groups $G \supset G'$ such that the following properties (A) and (B) are satisfied.

(A) All Shintani functions have moderate growth (Definition 3.3), namely, $\operatorname{Sh}_{\text{mod}}(\lambda, \nu) = \operatorname{Sh}(\lambda, \nu)$ for all $(\mathfrak{z}_G, \mathfrak{z}_{G'})$-infinitesimal characters (λ, ν).
(B) The natural injective homomorphism

$$\operatorname{Hom}_{G'}(\pi^{\infty}|_{G'}, \tau^{\infty}) \hookrightarrow \operatorname{Hom}_{\mathfrak{g}', K'}(\pi_K, \tau_{K'}) \tag{10.1}$$

is bijective for any admissible smooth representations π^{∞} and τ^{∞} of G and G', respectively.

Remark 10.2. (1) If $G' = \{e\}$, then neither (A) nor (B) holds.
(2) If $G = G'$, then (A) holds by the theory of asymptotic behaviors of Harish–Chandra's zonal spherical functions and (B) holds by the Casselman–Wallach theory of the Fréchet globalization [22, Chapter 11].
(3) If $G' = K$, then both (A) and (B) hold.
(4) It is plausible that if (G, G') satisfies the geometric condition (PP) (Definition 5.3), then both (A) and (B) hold.

By using the argument in Sects. 7 and 8 on the construction of Shintani functions from symmetry breaking operators, we have the following:

Proposition 10.3. *For a pair of real reductive classical Lie groups $G \supset G'$, (B) implies (A).*

Proof. Let λ_+ and ν_- be as in Theorem 8.2. We denote by π^∞ the admissible smooth representation of G on $C^\infty(G/P; \lambda_+)$ and τ^∞ the admissible smooth representation of G' on $C^\infty(G'/P'; \nu_-)$. Then by Theorem 8.2 (2), we have the following linear isomorphism:

$$\mathrm{Hom}_{G'}(\pi^\infty|_{G'}, \tau^\infty) \xrightarrow{\sim} \mathrm{Sh}_{\mathrm{mod}}(\lambda, \nu).$$

Similarly to the proof of Theorem 8.2 (2), we have the natural bijection

$$\mathrm{Hom}_{G'}(\pi^\omega|_{G'}, \tau^\omega) \simeq \mathrm{Sh}(\lambda, \nu),$$

where π^ω is a continuous representation of G on the space of real analytic vectors of π^∞, and τ^ω that of τ^∞.

In view of the canonical injective homomorphisms

$$\mathrm{Hom}_{G'}(\pi^\infty|_{G'}, \tau^\infty) \hookrightarrow \mathrm{Hom}_{G'}(\pi^\omega|_{G'}, \tau^\omega) \hookrightarrow \mathrm{Hom}_{\mathfrak{g}',K'}(\pi_K, \tau_K),$$

we see that if (B) holds, then the inclusion

$$\mathrm{Sh}_{\mathrm{mod}}(\lambda, \nu) \hookrightarrow \mathrm{Sh}(\lambda, \nu)$$

is bijective. Hence the implication (B) $\Rightarrow$ (A) is proved.   □

Acknowledgements Parts of the results and the idea of the proof were delivered in various occasions including Summer School on Number Theory in Nagano (Japan) in 1995 organized by F. Sato, Distinguished Sackler Lectures organized by J. Bernstein at Tel Aviv University (Israel) in May 2007, the conference in honor of E. B. Vinberg's 70th birthday in Bielefeld (Germany) in July 2007 organized by H. Abels, V. Chernousov, G. Margulis, D. Poguntke, and K. Tent, "Lie Groups, Lie Algebras and their Representations" in November 2011 in Berkeley (USA) organized by J. Wolf, "Group Actions with Applications in Geometry and Analysis" at Reims University (France) in June, 2013 organized by J. Faraut, J. Hilgert, B. Ørsted, M. Pevzner, and B. Speh, and "Representations of Reductive Groups" at University of Utah (USA) in July, 2013 organized by J. Adams, P. Trapa, and D. Vogan. The author is grateful for their warm hospitality.

References

1. A. Aizenbud, D. Gourevitch, Multiplicity one theorem for $(GL_{n+1}(\mathbb{R}), GL_n(\mathbb{R}))$, Selecta Math. **15** (2009), 271–294.
2. M. Berger, Les espaces symétriques non compacts, Ann. Sci. École Norm. Sup. **74** (1957), 85–177.

3. J.-L. Clerc, T. Kobayashi, B. Ørsted, and M. Pevzner, *Generalized Bernstein–Reznikov integrals*, Math. Ann. **349** (2011), 395–431.

4. B. Gross, D. Prasad, On the decomposition of a representations of SO_n when restricted to SO_{n-1}, Canad. J. Math. **44** (1992), 974–1002.

5. Harish–Chanrda, Representations of semisimple Lie groups II, Trans. Amer. Math. Soc. **76** (1954), 26–65.

6. S. Helgason, A duality for symmetric spaces with applications to group representations, II, Adv. Math. **22** (1976), 187–219.

7. S. Helgason, Some results on invariant differential operators on symmetric spaces, Amer. J. Math., **114** (1992), 789–811.

8. S. Helgason, Groups and geometric analysis. Integral geometry, invariant differential operators, and spherical functions, Mathematical Surveys and Monographs **83**, American Mathematical Society, Providence, RI, 2000.

9. M. Kashiwara, A. Kowata, K. Minemura, K. Okamoto, T. Oshima and M. Tanaka, Eigenfunctions of invariant differential operators on a symmetric space, Ann. of Math. **107** (1978), 1–39.

10. T. Kobayashi, Discrete decomposability of the restriction of $A_q(\lambda)$ with respect to reductive subgroups and its applications, Invent. Math. **117** (1994), 181–205; Part II, Ann. of Math. (2) **147** (1998), 709–729; Part III, Invent. Math. **131** (1998), 229–256.

11. T. Kobayashi, Introduction to harmonic analysis on real spherical homogeneous spaces, Proceedings of the 3rd Summer School on Number Theory, Homogeneous Spaces and Automorphic Forms, in Nagano (F. Sato, ed.), 1995, 22–41 (in Japanese).

12. T. Kobayashi, F-method for symmetry breaking operators, Differential Geom. Appl. **33** (2014), 272–289, Special issue in honour of M. Eastwood, doi:10.1016/j.difgeo.2013.10.003. Published online 20 November 2013, (available at arXiv:1303.3545).

13. T. Kobayashi, T. Matsuki, Classification of finite-multiplicity symmetric pairs, Transform. Groups **19** (2014), 457–493, special issue in honour of Professor Dynkin for his 90th birthday, doi:10.1007/s00031-014-9265-x (available at arXiv:1312.4246).

14. T. Kobayashi, T. Oshima, Finite multiplicity theorems for induction and restriction, Adv. Math. **248** (2013), 921–944, (available at arXiv:1108.3477).

15. T. Kobayashi, Y. Oshima, Classification of symmetric pairs with discretely decomposable restrictions of $(\mathfrak{g}, K)$-modules, Crelles Journal, published online 13 July, 2013. 19 pp. doi: 10.1515/crelle-2013-0045.

16. T. Kobayashi, B. Speh, Symmetry breaking for representations of rank one orthogonal groups, 131 pages, to appear in Mem. Amer. Math. Soc. arXiv:1310.3213; (announcement: Intertwining operators and the restriction of representations of rank one orthogonal groups, C. R. Acad. Sci. Paris, Ser. I, **352** (2014), 89–94).

17. B. Kostant, In the existence and irreducibility of certain series of representations, Bull. Amer. Math. Soc., **75** (1969), 627–642.

18. M. Krämer, Multiplicity free subgroups of compact connected Lie groups, Arch. Math. (Basel) **27** (1976), 28–36.

19. A. Murase, T. Sugano, Shintani functions and automorphic L-functions for $GL(n)$. Tohoku Math. J. (2) **48** (1996), no. 2, 165–202.

20. T. Oshima, J. Sekiguchi, Eigenspaces of invariant differential operators on an affine symmetric space, Invent. Math. **57** (1980), 1–81.

21. B. Sun and C.-B. Zhu, Multiplicity one theorems: the Archimedean case, Ann. of Math. **175** (2012), 23–44.

22. N. R. Wallach, Real reductive groups. I, II, Pure and Applied Mathematics, **132** Academic Press, Inc., Boston, MA, 1988. xx+412 pp.

Harmonic Spinors on Reductive Homogeneous Spaces

Salah Mehdi and Roger Zierau

Abstract An integral intertwining operator is given from certain principal series representations into spaces of harmonic spinors for Kostant's cubic Dirac operator. This provides an integral representation for harmonic spinors on a large family of reductive homogeneous spaces.

Key words Cubic Dirac operator • Harmonic spinors • Reductive homogeneous spaces • Fundamental series representations • Principal series representations

Mathematics Subject Classification (2010): Primary 22E46. Secondary 22F30.

1 Introduction

A realization of the discrete series representations of a semisimple Lie group as an L^2-space of harmonic spinors was given in [11] and [1]. More precisely, suppose G is a noncompact connected semisimple real Lie group with finite center and K a maximal compact subgroup of G. Write S for the spin representation of K (after passing to a cover if necessary). For a finite-dimensional K-representation E, the tensor product $S \otimes E$ determines a homogeneous vector bundle $\mathscr{S} \otimes \mathscr{E}$ over G/K and a geometric Dirac operator (defined in terms of an invariant connection) acting on smooth sections:

$$\mathscr{D}_{G/K}(\mathscr{E}) : C^\infty(G/K, \mathscr{S} \otimes \mathscr{E}) \to C^\infty(G/K, \mathscr{S} \otimes \mathscr{E}).$$

S. Mehdi
Université de Lorraine - Metz, Institut Elie Cartan de Lorraine, Metz cedex,
UMR 7502 - CNRS, France
e-mail: salah.mehdi@univ-lorraine.fr

R. Zierau (✉)
Oklahoma State University, Stillwater OK 74078, USA
e-mail: zierau@math.okstate.edu

© Springer International Publishing Switzerland 2014
G. Mason et al. (eds.), *Developments and Retrospectives in Lie Theory*,
Developments in Mathematics 37, DOI 10.1007/978-3-319-09934-7_6

If G has a nonempty discrete series, then the kernel of $\mathscr{D}_{G/K}(\mathscr{E})$ on L^2-sections is an irreducible unitary representation in the discrete series of G, and every discrete series representation of G occurs this way for some K-representation E. (See [7] for a thorough discussion.) A similar construction of tempered representations is given by "partially harmonic" spinors in [15].

It is therefore natural to study the kernels of Dirac operators on other homogeneous spaces. In [8] and [9], we addressed this problem in a more general context where G is a connected real reductive Lie group and K is replaced by connected closed reductive subgroups H for which the complex ranks of H and G are equal, but H is not necessarily compact. The Dirac operator $\mathscr{D}_{G/K}(\mathscr{E})$ is replaced by Kostant's cubic Dirac operator [6]:

$$\mathscr{D}_{G/H}(\mathscr{E}) : C^\infty(G/H, \mathscr{S} \otimes \mathscr{E}) \to C^\infty(G/H, \mathscr{S} \otimes \mathscr{E}).$$

This operator is the sum of a first-order term and a zero-order term, which comes from a degree three element in the Clifford algebra of the orthogonal complement of the complexified Lie algebra of H. (The zero-order term vanishes when H is any symmetric subgroup.) Integral formulas for harmonic spinors are given in [8] and [9]. The L^2-theory is begun in [2].

In the present article we consider a larger class of homogeneous spaces by removing the equal rank condition. Suppose that $E = E_\mu$ is a finite-dimensional representation of H with highest weight μ (with respect to some positive system). Under certain conditions on H and μ we prove the following theorem. This is Theorem 29 of Sect. 4.

Theorem A. *There is a parabolic subgroup P in G, a representation W of P and a nonzero G-equivariant map*

$$\mathscr{P} : C^\infty(G/P, \mathscr{W}) \to C^\infty(G/H, \mathscr{S} \otimes \mathscr{E}_\mu)$$

with $\mathscr{D}_{G/H}(\mathscr{E}_\mu) \circ \mathscr{P} = 0$, where $C^\infty(G/P, \mathscr{W})$ denotes the space of smooth vectors of the principal series representation $\mathrm{Ind}_P^G(W)$. In particular, the kernel of $\mathscr{D}_{G/H}(\mathscr{E}_\mu)$ contains a smooth representation of G.

The intertwining operator $\mathscr{P}$ is an integral operator; the formula for $\mathscr{P}$ is analogous to the classical Poisson integral formula giving harmonic functions on the disk. The condition on H referred to above is stated in Section 4.3 as Assumption 24. It is that H is not too small; it guarantees that certain Dirac cohomology spaces are nonzero. The conditions on μ are very mild regularity conditions.

The representation of $P = MAN$ on W is formed from a fundamental series of M, a character of A and the trivial action of N. As an important ingredient the fundamental series is realized as a space of harmonic spinors on $M/M \cap H$. This is the content of the following proposition (which occurs as Proposition 19 in Sect. 3).

Proposition B. *Every fundamental series representation occurs in the kernel of a Dirac operator $\mathscr{D}_{G/H}(\mathscr{E})$.*

The paper is organized as follows. In Sect. 2 we fix the notation and give some well-known facts about the spin representations. We also describe a reciprocity for geometric and algebraic Dirac operators. This is an important technique for relating spaces of harmonic spinors and Dirac cohomology. In Sect. 3, we realize certain cohomologically induced representations as submodules of kernels of geometric Dirac operators and prove Proposition B. Finally, Sect. 4 is devoted to the construction of the parabolic subgroup P and the proof of Theorem A. A proof of the reciprocity between geometric and algebraic harmonic spinors is provided in an appendix.

2 Preliminaries

2.1 The Groups and Homogeneous Spaces

Let G be a connected real reductive Lie group. We will denote the complexification of Lie(G), the Lie algebra of G, by $\mathfrak{g}$ (and similarly for other Lie groups). By a real reductive group we mean that $\mathfrak{g}$ is reductive, i.e., $\mathfrak{g} = [\mathfrak{g}, \mathfrak{g}] + \mathfrak{z}$, where $\mathfrak{z}$ denotes the center of $\mathfrak{g}$. Fix a G-invariant nondegenerate bilinear symmetric form $\langle \, , \, \rangle$ on $\mathfrak{g}$. If K/Z is a maximal compact subgroup of G/Z, where Z is the center of G, then K is the fixed point group of a Cartan involution θ of G. Write $\mathfrak{g} = \mathfrak{k} + \mathfrak{s}$ for the corresponding Cartan decomposition of $\mathfrak{g}$, where $\mathfrak{k}$ is the Lie algebra of K and $\mathfrak{s} = \mathfrak{k}^{\perp}$. We take H to be a closed subgroup of G, with complexified Lie algebra denoted by $\mathfrak{h}$, satisfying the following conditions:

$$H \text{ is connected and reductive,}$$

$$H \text{ is } \theta\text{-stable,} \tag{1}$$

$$\langle \, , \, \rangle \text{ remains nondegenerate when restricted to } \mathfrak{h}.$$

In this situation, there is a decomposition

$$\mathfrak{g} = \mathfrak{h} + \mathfrak{q}, \text{ where } \mathfrak{q} = \mathfrak{h}^{\perp}.$$

The restriction of $\langle \, , \, \rangle$ to $\mathfrak{q}$ remains nondegenerate and $[\mathfrak{h}, \mathfrak{q}] \subset \mathfrak{q}$.

2.2 Spin Representation

The construction of the spin representation is briefly reviewed here; we follow the discussion of [3, Ch. 9]. Let $\mathscr{C}l(\mathfrak{q})$ be the Clifford algebra of the complexification of $\mathfrak{q}$, i.e., the quotient of the tensor algebra of $\mathfrak{q}$ by the ideal generated by the elements $X \otimes Y + Y \otimes X - \langle X, Y \rangle$ with $X, Y \in \mathfrak{q}$. Let

$$\mathfrak{so}(\mathfrak{q}) = \{ T \in \mathrm{End}(\mathfrak{q}) : \langle T(X), Y \rangle + \langle X, T(Y) \rangle = 0, \ \forall \ X, Y \in \mathfrak{q} \}.$$

Then the endomorphisms $R_{X,Y} : W \mapsto \langle Y, W \rangle X - \langle X, W \rangle Y$ span $\mathfrak{so}(\mathfrak{q})$. The linear extension of the map $R_{X,Y} \mapsto \frac{1}{2}(XY - YX)$ is an injective Lie algebra homomorphism of $\mathfrak{so}(\mathfrak{q})$ into $\mathscr{C}l_2(\mathfrak{q})$, the subalgebra of degree-two elements in the Clifford algebra. Let $\mathfrak{q}^+$ be a maximally isotropic subspace of $\mathfrak{q}$ and write $S_\mathfrak{q}$ for the exterior algebra $\wedge \mathfrak{q}^+$ of $\mathfrak{q}^+$. The spin representation $(s_\mathfrak{q}, S_\mathfrak{q})$ of $\mathfrak{h}$ is defined as the composition map

$$\mathfrak{h} \overset{\mathrm{ad}}{\rightarrow} \mathfrak{so}(\mathfrak{q}) \hookrightarrow \mathscr{C}l_2(\mathfrak{q}) \subset \mathscr{C}l(\mathfrak{q}) \overset{\gamma_\mathfrak{q}}{\rightarrow} \mathrm{End}(S_\mathfrak{q})$$

where $\gamma_\mathfrak{q}$ denotes the Clifford multiplication. Although the construction is independent of maximal isotropic subspace $\mathfrak{q}^+$, the explicit description of a particularly useful $\mathfrak{q}^+$ will be given in Sect. 4.3 below.

There is an hermitian inner product $\langle \, , \, \rangle_{S_\mathfrak{q}}$ on $S_\mathfrak{q}$ for which $\gamma_\mathfrak{q}(X)$, $X \in \mathfrak{q} \subset \mathscr{C}l(\mathfrak{q})$, is skew-hermitian [14, Lemma 9.2.3]:

$$\langle \gamma_\mathfrak{q}(X)u, v \rangle_{S_\mathfrak{q}} = -\langle u, \gamma_\mathfrak{q}(X)v \rangle_{S_\mathfrak{q}}, \quad \forall \ X \in \mathfrak{q}, \ \forall u, v \in S_\mathfrak{q}. \tag{2}$$

2.3 Geometric Dirac Operators

Let E be a finite-dimensional representation of $\mathfrak{h}$ such that the tensor product $S_\mathfrak{q} \otimes E$ lifts to a representation of the group H. There is an associated smooth homogeneous vector bundle over G/H, which we denote by $\mathscr{S}_\mathfrak{q} \otimes \mathscr{E}$, whose space of smooth sections is

$$C^\infty(G/H, \mathscr{S}_\mathfrak{q} \otimes \mathscr{E}) \simeq \left\{ C^\infty(G) \otimes (S_\mathfrak{q} \otimes E) \right\}^H$$

$$\simeq \{ f : G \rightarrow S_\mathfrak{q} \otimes E \mid f \text{ is smooth and } f(gh) = h^{-1} \cdot f(g), h \in H \}.$$

We remark that G acts by left translations on each of these function spaces.

Let $\{X_j\}$ be a fixed basis of $\mathfrak{q}$ satisfying

$$\langle X_j, X_k \rangle = \delta_{jk}. \tag{3}$$

Denote the universal enveloping algebra of $\mathfrak{g}$ by $\mathscr{U}(\mathfrak{g})$. Let $\mathbf{c_q}$ be the degree three element in $\mathscr{C}l(\mathfrak{q})$ defined as the image under the Chevalley isomorphism of the 3-form

$$(X, Y, Z) \mapsto \langle X, [Y, Z]\rangle$$

on $\mathfrak{q}$. The element $\sum X_j \otimes (\gamma(X_j) \otimes 1) - 1 \otimes (\gamma(\mathbf{c_q}) \otimes 1)$ in $\mathscr{U}(\mathfrak{g}) \otimes \mathrm{End}(S_{\mathfrak{q}} \otimes E)$ is H-invariant, so defines a G-invariant differential operator

$$\mathscr{D}_{G/H}(\mathscr{E}) : C^{\infty}(G/H, \mathscr{S}_{\mathfrak{q}} \otimes \mathscr{E}) \to C^{\infty}(G/H, \mathscr{S}_{\mathfrak{q}} \otimes \mathscr{E})$$

acting on $C^{\infty}(G/H, \mathscr{S}_{\mathfrak{q}} \otimes \mathscr{E})$. We refer to $\mathscr{D}_{G/H}(\mathscr{E})$ as the geometric (cubic) Dirac operator; it is given by the following formula:

$$\mathscr{D}_{G/H}(\mathscr{E}) = \sum R(X_j) \otimes \gamma(X_j) \otimes 1 - 1 \otimes \gamma(\mathbf{c_q}) \otimes 1. \tag{4}$$

Note that $\mathscr{D}_{G/H}(\mathscr{E})$ is independent of the basis $\{X_j\}$ satisfying (3) (since each of the two terms is, by itself, independent of the basis). We will often write $\mathscr{D}_{G/H}$ for $\mathscr{D}_{G/H}(\mathscr{E})$.

2.4 Dirac Cohomology

Associated with a $\mathfrak{g}$-module (π, V), there is an *algebraic cubic Dirac operator* D_V: $V \otimes S_{\mathfrak{q}} \longrightarrow V \otimes S_{\mathfrak{q}}$ defined by

$$D_V = \sum_j \pi(X_j) \otimes \gamma(X_j) - 1 \otimes \gamma(\mathbf{c_q}). \tag{5}$$

The following formula[1] for the square of D_V is due to Kostant [6, Theorem 2.16]:

$$2D_V^2 = \Omega_{\mathfrak{g}} \otimes 1 - \Omega_{\Delta\mathfrak{h}} + \|\rho(\mathfrak{g})\|^2 - \|\rho(\mathfrak{h})\|^2, \tag{6}$$

where $\Omega_{\mathfrak{g}}$ is the Casimir element for $\mathfrak{g}$ acting on V and $\Omega_{\Delta\mathfrak{h}}$ is the Casimir element of $\mathfrak{h}$ acting in $V \otimes S_{\mathfrak{q}}$. In the case when $\mathfrak{h} = \mathfrak{k}$, the cubic term $\mathbf{c_q}$ vanishes and formula (6) is due to Parthasarathy (see [11]).

The (cubic) Dirac cohomology of the $\mathfrak{g}$-module V is the $\mathfrak{h}$-module defined as the quotient

$$H^{(\mathfrak{h},\mathfrak{g})}(V) = \ker(D_V)/\ker(D_V) \cap \mathrm{Im}(D_V).$$

---

[1] The factor of 2 in this formula does not appear in [6]. This is because we are taking $xy + yx = \langle x, y\rangle$ in the definition of the Clifford algebra, while $xy + yx = 2\langle x, y\rangle$ is used in [6].

The Dirac cohomology will also be denoted by $H_D(V)$ when the pair $(\mathfrak{h}, \mathfrak{g})$ is understood.

Finally, we note that in the case when V is a unitarizable $(\mathfrak{g}, K)$-module, i.e., V has a nondegenerate $(\mathfrak{g}, K)$-invariant positive definite hermitian form $\langle \, , \, \rangle_V$, then one gets a nondegenerate hermitian form on $V \otimes S_q$ defined by

$$\langle \, , \, \rangle_{V \otimes S_q} = \langle \, , \, \rangle_V \otimes \langle \, , \, \rangle_{S_q} \qquad (7)$$

with respect to which D_V is selfadjoint [14, Lemma 9.3.3]. In the case where $\mathfrak{h} = \mathfrak{k}$, since the hermitian form on $S_{\mathfrak{s}}$ is positive definite, the form $\langle \, , \, \rangle_{V \otimes S_{\mathfrak{s}}}$ is also positive definite and it follows that

$$H_D(V) = \ker(D_V).$$

When $\mathfrak{h} \subset \mathfrak{k}$, the same conclusion holds for $H_D^{(\mathfrak{h}, \mathfrak{k})}(V)$ for a finite-dimensional representation V.

2.5 Algebraic Dirac Operators vs. Geometric Dirac Operators

Let V be a smooth admissible representation of G, V_K the space of K-finite vectors in V, and V_K^* the K-finite dual of V_K. Let E be a finite-dimensional representation of $\mathfrak{k}$ such that the tensor product $S_{\mathfrak{s}} \otimes E$ with the spin representation of $\mathfrak{k}$ lifts to a representation of the group K. Then $S_{\mathfrak{s}} \otimes E$ induces a homogeneous bundle $\mathscr{S}_{\mathfrak{s}} \otimes \mathscr{E} \longrightarrow G/K$. There is a vector space isomorphism

$$\mathrm{Hom}_G(V, C^\infty(G/K, S_{\mathfrak{s}} \otimes \mathscr{E})) \simeq \mathrm{Hom}_K(E^*, S_{\mathfrak{s}} \otimes V_K^*)$$

given by $T \mapsto T_1$, with $T_1(e^*)(v) = \langle e^*, T(v)(1) \rangle$, where $e^* \in E^*, v \in V$ and $1 \in G$ is the identity element. In addition one has $\left(D_{V_K^*} T_1(e^*)\right)(v) = \langle e^*, \left(\mathscr{D}_{G/K} T(v)\right)(1) \rangle$, for all $e^* \in E^*$ and $v \in V$. One may conclude the following.

Proposition 8. $\mathrm{Hom}_G(V, \ker(\mathscr{D}_{G/K}(\mathscr{E}))) \simeq \mathrm{Hom}_K(E^*, \ker(D_{V_K^*}))$.

See the appendix for details.

There is also an isomorphism

$$\mathrm{Hom}_K(E^*, V_K^* \otimes S_{\mathfrak{s}}) \simeq \mathrm{Hom}_K(V_K \otimes S_{\mathfrak{s}}, E)$$

given by $B \mapsto b$, with $\langle s, B(e^*)(v) \rangle = \langle e^*, b(v \otimes s) \rangle$. We also have the identity

$$\langle s, \left(D_{V_K^*} B(e^*)\right)(v) \rangle = \langle e^*, b\left(D_{V_K}(v \otimes s)\right) \rangle.$$

The pairing on the left-hand side is a nondegenerate pairing of the selfdual representation $S_\mathfrak{s}$ with itself. From this we may conclude that

$$\mathrm{Hom}_K(E^*, \ker(D_{V_K^*})) \simeq \mathrm{Hom}_K((V_K \otimes S)/\mathrm{Im}(D_{V_K}), E).$$

Now assume that V_K is unitarizable. Then D_{V_K} is selfadjoint and

$$(V_K \otimes S)/\mathrm{Im}(D_{V_K}) \simeq \mathrm{Im}(D_{V_K})^\perp \simeq \ker(D_{V_K}).$$

We may conclude that $\ker(D_{V_K^*}) \simeq \big(\ker(D_{V_K})\big)^*$, as K-modules. Therefore, by Sect. 2.4,

$$H_D(V_K^*) \simeq \big(H_D(V_K)\big)^*. \tag{9}$$

The above discussion applies to the Dirac operator on the homogeneous space K/H, when $H \subset K$, resulting in the statement that

$$\mathrm{Hom}_K(E, \ker(\mathscr{D}_{K/H}(\mathscr{F}))) \simeq \mathrm{Hom}_H(F, H_D^{(\mathfrak{h},\mathfrak{k})}(E)). \tag{10}$$

3 The Fundamental Series

An important special case of our main result occurs when H is compact. In this section we see that in this case (under certain mild conditions) the kernel of $\mathscr{D}_{G/H}$ is nonzero. In fact the kernel contains certain fundamental series representations. This is analogous to the well-known fact that $\ker(\mathscr{D}_{G/K})$ contains a discrete series representation [1, 11] when $\mathrm{rank}_C(\mathfrak{g}) = \mathrm{rank}_C(\mathfrak{k})$.

3.1 Cartan Subalgebras and Roots

We assume that $\mathfrak{h}$ is as in Sect. 2.1 and that $\mathfrak{h} \subset \mathfrak{k}$. Let $\mathfrak{t}_\mathfrak{h}$ be a Cartan subalgebra of $\mathfrak{h}$. Extend to a Cartan subalgebra $\mathfrak{t} = \mathfrak{t}_\mathfrak{h} + \mathfrak{t}_\mathfrak{q}$ of $\mathfrak{k}$ by choosing $\mathfrak{t}_\mathfrak{q} \subset \mathfrak{k} \cap \mathfrak{q}$. Now extend to a Cartan subalgebra $\mathfrak{t} + \mathfrak{a}$ of $\mathfrak{g}$ by choosing $\mathfrak{a}$ abelian in $\mathfrak{s}$.

Let $\Delta^+ \subset \Delta(\mathfrak{t} + \mathfrak{a}, \mathfrak{g})$ be defined by a lexicographic order with $\mathfrak{t}_\mathfrak{h}$ first, then $\mathfrak{t}_\mathfrak{q}$, then $\mathfrak{a}$. Such a positive system has the property that

$$\Delta^+(\mathfrak{h}) := \{\alpha|_{\mathfrak{t}_\mathfrak{h}} : \mathfrak{g}^{(\alpha)} \subset \mathfrak{h}, \alpha \in \Delta^+ \text{ and } \alpha|_{\mathfrak{t}_\mathfrak{h}} \neq 0\}$$

is a positive system of roots in $\mathfrak{h}$. Here we are denoting the α-root space in $\mathfrak{g}$ by $\mathfrak{g}^{(\alpha)}$.

Note that $\mathfrak{t} + \mathfrak{a}$ is a fundamental Cartan subalgebra of $\mathfrak{g}$, i.e., is maximally compact. The positive systems described above may also be described as follows.

There is $\Lambda_0 \in (\mathfrak{t} + \mathfrak{a})^*$ with $\Lambda_0|_\mathfrak{a} = 0$ so that $\Delta^+ = \{\alpha : \langle \Lambda_0, \alpha \rangle > 0\}$. The Borel subalgebras that arise from such a positive system are the θ-stable Borel subalgebras containing $\mathfrak{t} + \mathfrak{a}$. This gives a positive system of $\mathfrak{t}$-roots in $\mathfrak{k}$: $\Delta^+(\mathfrak{k}) = \{\beta \in \Delta(\mathfrak{t}, \mathfrak{k}) : \langle \Lambda_0, \beta \rangle > 0\}$.

Suppose that $\mu \in \mathfrak{t}_\mathfrak{h}^*$ and that μ is dominant for an *arbitrary* positive system. Then by choosing μ as the first basis vector defining a lexicographic order as above, we arrive at positive systems Δ^+ and $\Delta^+(\mathfrak{h})$ with the property that μ, extended to be 0 on $\mathfrak{t}_\mathfrak{q} + \mathfrak{a}$, is dominant for both Δ^+ and $\Delta^+(\mathfrak{h})$. In Sect. 3.3, where we begin with a finite-dimensional representation E of $\mathfrak{h}$, we may therefore assume that the highest weight μ of E is Δ^+-dominant.

We make the following assumption on H. This is the assumption on H not being too small mentioned in the introduction; it will be necessary for certain Dirac cohomology spaces to be nonzero. See Sect. 3.3.

Assumption 11. *There is no root $\alpha \in \Delta(\mathfrak{g})$ so that $\alpha|_{\mathfrak{t}_\mathfrak{h}} = 0$.*

Note that this assumption automatically holds when either $H = K$ or $\mathfrak{h}$ and $\mathfrak{g}$ have equal rank.

Under this assumption we may construct the spin representation $S_\mathfrak{q}$ by choosing a maximal isotropic subspace $\mathfrak{q}^+$ of $\mathfrak{q}$ as follows. Choose any maximal isotropic subspace $(\mathfrak{t}_\mathfrak{q} + \mathfrak{a})^+$ of $\mathfrak{t}_\mathfrak{q} + \mathfrak{a}$, then set

$$\mathfrak{q}^+ = (\mathfrak{t}_\mathfrak{q} + \mathfrak{a})^+ + \sum_\gamma \mathfrak{q}^{(\gamma)},$$

where the sum is over all $\mathfrak{t}_\mathfrak{h}$-weights γ in $\mathfrak{q}$. The assumption tells us that no such γ is zero, so $\mathfrak{q}^+$ is indeed maximally isotropic.

We will use the notation of $\rho(\mathfrak{g})$ for one half the sum of the roots in Δ^+. Similarly $\rho(\mathfrak{h})$ denotes one half the sum of the roots in $\Delta^+(\mathfrak{h})$. We also use the notation $\rho(\mathfrak{q})$ for one half the sum of the $\mathfrak{t}_\mathfrak{h}$-weights in $\mathfrak{q}^+$, and similarly for $\rho(\mathfrak{k} \cap \mathfrak{q})$.

3.2   The Fundamental Series and Its Dirac Cohomology

The fundamental series representations are cohomologically induced representations. They arise as follows. Let $\mathfrak{b}$ be a θ-stable Borel subalgebra in $\mathfrak{g}$ which contains the fundamental Cartan subalgebra $\mathfrak{t} + \mathfrak{a}$. The positive system associated with $\mathfrak{b}$ is described in the previous subsection. Write $\mathfrak{b} = \mathfrak{t} + \mathfrak{a} + \mathfrak{u}$ for the Levi decomposition of $\mathfrak{b}$. Then a fundamental series representation is a cohomologically induced representation $A_\mathfrak{b}(\lambda)$, for $\lambda \in (\mathfrak{t} + \mathfrak{a})^*$, with $\lambda|_\mathfrak{a} = 0$ and λ a Δ^+-dominant and analytically integral weight, having the following properties.

(a) The infinitesimal character is $\lambda + \rho(\mathfrak{g})$.
(b) $\lambda|_\mathfrak{t} + 2\rho(\mathfrak{s} \cap \mathfrak{u})$ is the highest weight of a lowest K-type with respect to $\Delta^+(\mathfrak{k})$, where $\rho(\mathfrak{s} \cap \mathfrak{u})$ denotes half the sum of the $\mathfrak{t}$-weights in $\mathfrak{s} \cap \mathfrak{u}$.

(c) Each K-type has highest weight of the form $\lambda_{|_t} + 2\rho(\mathfrak{s} \cap \mathfrak{u}) + \sum_{\gamma \in \Delta(\mathfrak{s} \cap \mathfrak{u})} n_\gamma \gamma$, where n_γ are non negative integers.

It is known from Vogan and Zuckerman [13, Theorem 2.5] that these properties uniquely determine $A_\mathfrak{b}(\lambda)$. The fundamental series representations are irreducible and unitarizable [12], [14, Ch. 6].

The computation of the Dirac cohomology, with respect to $\mathfrak{k} \subset \mathfrak{g}$, is straightforward using Kostant's formula (6) for the square of $D_{A_\mathfrak{b}(\lambda)} : A_\mathfrak{b}(\lambda) \otimes S_\mathfrak{s} \to A_\mathfrak{b}(\lambda) \otimes S_\mathfrak{s}$ and the properties (a)–(c) above. Although this is contained in Theorem 5.2 of [5], we give the short proof here. First, by the unitarizability of $A_\mathfrak{b}(\lambda)$, the Dirac cohomology is $\ker(D_{A_\mathfrak{b}(\lambda)}) = \ker(D^2_{A_\mathfrak{b}(\lambda)})$. Let F_τ be the finite-dimensional highest weight representation of $\mathfrak{k}$ with highest weight τ with respect to $\Delta^+(\mathfrak{k})$. By Kostant's formula (6), F_τ occurs in $\ker(D_{A_\mathfrak{b}(\lambda)})$ if F_τ occurs in $A_\mathfrak{b}(\lambda) \otimes S_\mathfrak{s}$ and $\|\lambda + \rho(\mathfrak{g})\| = \|\tau + \rho(\mathfrak{k})\|$.

The t-weights of $S_\mathfrak{s}$ are all weights of the form $\frac{1}{2}(\pm \gamma_1 \pm \cdots \pm \gamma_k)$ with $\gamma_i \in \Delta(\mathfrak{s} \cap \mathfrak{u})$. Each weight occurs with multiplicity 2^d, where d is the greatest integer in $\dim(\mathfrak{a})/2$. We use the notation $\langle A \rangle = \sum_{\alpha \in A} \alpha$, for any set A of weights. With this notation the t-weights in $S_\mathfrak{s}$ are

$$\Delta(S_\mathfrak{s}) = \{\langle A \rangle - \rho(\mathfrak{s} \cap \mathfrak{u}) : A \subset \Delta(\mathfrak{s} \cap \mathfrak{u})\}$$
$$= \{\rho(\mathfrak{s} \cap \mathfrak{u}) - \langle A \rangle : A \subset \Delta(\mathfrak{s} \cap \mathfrak{u})\}.$$

Each component of $A_\mathfrak{b}(\lambda) \otimes S_\mathfrak{s}$ has highest weight of the form

$$\tau = \lambda_{|_t} + 2\rho(\mathfrak{s} \cap \mathfrak{u}) + \sum_{\gamma \in \Delta(\mathfrak{s} \cap \mathfrak{u})} n_\gamma \gamma + (\langle A \rangle - \rho(\mathfrak{s} \cap \mathfrak{u}))$$

$$= \lambda_{|_t} + \rho(\mathfrak{s} \cap \mathfrak{u}) + \sum_{\gamma \in \Delta(\mathfrak{s} \cap \mathfrak{u})} m_\gamma \gamma, \quad \text{for some nonnegative integers } m_\gamma.$$

Now

$$\|\tau + \rho(\mathfrak{k})\|^2 = \|\lambda + \rho(\mathfrak{g}) + \sum_{\gamma \in \Delta(\mathfrak{s} \cap \mathfrak{u})} m_\gamma \gamma\|^2$$

$$= \|\lambda + \rho(\mathfrak{g})\|^2 + \|\sum_{\gamma \in \Delta(\mathfrak{s} \cap \mathfrak{u})} m_\gamma \gamma\|^2 + \sum_{\gamma \in \Delta(\mathfrak{s} \cap \mathfrak{u})} m_\gamma \langle \lambda + \rho(\mathfrak{g}), \gamma \rangle.$$

For this to equal $\|\lambda + \rho(\mathfrak{g})\|^2$ one must have $m_\gamma = 0$ (because $\langle \lambda + \rho(\mathfrak{g}), \gamma \rangle > 0$ and $m_\gamma \geq 0$). Therefore, all $n_\gamma = 0$ and $\langle A \rangle = 0$. So $\tau = \lambda_{|_t} + \rho(\mathfrak{s} \cap \mathfrak{u})$ and the multiplicity is 2^d, where d is the greatest integer in $\dim(\mathfrak{a})/2$. This proves the following statement.

Proposition 12. *Let $A_\mathfrak{b}(\lambda)$ be the cohomologically induced representation described in (a)–(c) above. Then*

$$H^{(\mathfrak{k},\mathfrak{g})}(A_\mathfrak{b}(\lambda)) = 2^d \, F_{\lambda|_\mathfrak{t} + \rho(\mathfrak{s} \cap \mathfrak{u})}. \tag{13}$$

Now let $\mu \in \mathfrak{t}^*$ be $\Delta^+(\mathfrak{k})$-dominant and integral. Let E_μ be the irreducible finite-dimensional representation of $\mathfrak{k}$ with highest weight μ. Define $\lambda_\mu \in (\mathfrak{t} + \mathfrak{a})^*$ by $\lambda_\mu|_\mathfrak{t} = \mu - \rho(\mathfrak{s} \cap \mathfrak{u})$ and $\lambda_\mu|_\mathfrak{a} = 0$. Then, as described at the end of Sect. 3.1, the positive system Δ^+ can be chosen so that λ_μ is Δ^+-dominant. With such a choice of Δ^+ we have the following corollary, which is a consequence of Proposition 8 and Eq. (9).

Corollary 14. *Suppose μ is $\Delta^+(\mathfrak{k})$-dominant and $S_\mathfrak{s} \otimes E_\mu$ lifts to a representation of K. Then the kernel of $\mathscr{D}_{G/K} : C^\infty(G/K, \mathscr{S} \otimes \mathscr{E}_\mu) \to C^\infty(G/K, \mathscr{S} \otimes \mathscr{E}_\mu)$ contains a smooth G-representation infinitesimally equivalent to $A_\mathfrak{b}(\lambda_\mu)$.*

Proof. By Proposition 12 gives $\mathrm{Hom}_K\big(E_\mu, H^{(\mathfrak{g},\mathfrak{k})}(A_\mathfrak{b}(\lambda_\mu))\big) \neq 0$, since $\lambda_\mu + \rho$ is Δ^+-dominant. Now Proposition 8 (along with (9)) gives

$$\mathrm{Hom}_{(\mathfrak{g},K)}\big(A_\mathfrak{b}(\lambda_\mu), \ker(\mathscr{D}_{G/K}(\mathscr{E}_\mu))\big) \neq 0.$$

$\square$

3.3 Induction in Stages

Now assume that $H \subset K$. Let F_μ be the irreducible finite-dimensional representation of H having highest weight $\mu \in \mathfrak{t}_\mathfrak{h}^*$. As described at the end of Sect. 3.1, we may assume that the extension of μ (by 0 on $\mathfrak{t}_\mathfrak{q} + \mathfrak{a}$) is Δ^+-dominant, and therefore its restriction to $\mathfrak{t}$ is $\Delta^+(\mathfrak{k})$-dominant. Let E_ξ be the irreducible finite-dimensional representation of $\mathfrak{k}$ of highest weight $\xi \in \mathfrak{t}^*$.

In [10] it is shown that $F_\mu \subset H_D^{(\mathfrak{h},\mathfrak{k})}(E_\xi)$ if and only if $\mu = w(\xi + \rho(\mathfrak{k})) - \rho(\mathfrak{h})$ for some $w \in W(\mathfrak{k})$ having the property that $w(\xi + \rho(\mathfrak{k}))|_{\mathfrak{t}_\mathfrak{q}} = 0$.
Define

$$\xi|_{\mathfrak{t}_\mathfrak{h}} = \mu - (\rho(\mathfrak{k}) - \rho(\mathfrak{h}))|_{\mathfrak{t}_\mathfrak{h}} \tag{15}$$

$$\xi|_{\mathfrak{t}_\mathfrak{q}} = -\rho(\mathfrak{k})|_{\mathfrak{t}_\mathfrak{q}}.$$

Then $\xi + \rho(\mathfrak{k})$ is $\Delta^+(\mathfrak{k})$-dominant. By taking $w = e$ above, we have $F_\mu \subset H_D^{(\mathfrak{h},\mathfrak{k})}(E_\xi)$. We may conclude from (10) that $E_\xi \subset \ker\big(\mathscr{D}_{K/H}(\mathscr{F}_\mu)\big)$.

Our goal now is to realize a fundamental series representation in the kernel of $\mathscr{D}_{G/H}(\mathscr{F}_\mu)$. The induction in stages argument begins with the identification

$$C^\infty(G/H, S_\mathfrak{q} \otimes \mathscr{F}_\mu) \simeq C^\infty(G/K, S_\mathfrak{s} \otimes C^\infty(K/H, S_{\mathfrak{k} \cap \mathfrak{q}} \otimes \mathscr{F}_\mu)), \quad (16)$$

$$f \longleftrightarrow F_f,$$

with $(F_f(g))(k) = (k \otimes 1) \cdot f(gk)$. In [9] it is shown that

$$\left(F_{\mathscr{D}_{G/H}f}\right)(g) = (\mathscr{D}_{G/K} F_f)(g) + \mathscr{D}_{K/H}\left(F_f(g)\right). \quad (17)$$

With ξ defined as in (15), we have seen that E_ξ may be realized as a subspace of $\ker\left(\mathscr{D}_{K/H}(\mathscr{F}_\mu)\right) \subset C^\infty(K/H, S_{\mathfrak{k} \cap \mathfrak{q}} \otimes \mathscr{F}_\mu)$. Therefore, when restricted to $C^\infty(G/K, \mathscr{E}_\xi)$, under the identification of (16),

$$\mathscr{D}_{G/K}(F_f) = F_{\mathscr{D}_{G/H}(f)}. \quad (18)$$

Proposition 19. *The kernel of $\mathscr{D}_{G/H}(\mathscr{F}_\mu)$ contains a smooth representation infinitesimally equivalent to a cohomologically induced representation $A_\mathfrak{b}(\lambda_\mu)$ with*

$$\lambda_\mu|_{\mathfrak{t}_\mathfrak{h}} = \mu - \rho(\mathfrak{q}), \quad \lambda|_\mathfrak{a} = 0 \quad \text{and} \quad \lambda|_{\mathfrak{t}_\mathfrak{q}} = -\rho(\mathfrak{k})|_{\mathfrak{t}_\mathfrak{q}}.$$

Proof. Let ξ be as in (15) and realize $E_\xi \subset \ker(\mathscr{D}_{K/H}(\mathscr{F}_\mu))$. So we may consider

$$C^\infty(G/K, S_\mathfrak{s} \otimes \mathscr{E}_\xi) \subset C^\infty(G/K, S_\mathfrak{s} \otimes C^\infty(K/H, S_{\mathfrak{k} \cap \mathfrak{q}} \otimes \mathscr{F}_\mu))$$

$$\simeq C^\infty(G/H, S_\mathfrak{q} \otimes \mathscr{F}_\mu).$$

By (18)

$$\ker\left(\mathscr{D}_{G/K}(\mathscr{E}_\xi)\right) \subset \ker\left(\mathscr{D}_{G/H}(\mathscr{F}_\mu)\right).$$

Applying (14),

$$A_\mathfrak{b}(\lambda_\mu) \subset \ker(\mathscr{D}_{G/K}(\mathscr{E}_\xi)) \subset \ker(\mathscr{D}_{G/H}(\mathscr{F}_\mu)),$$

since $\lambda_\mu|_\mathfrak{t} = \xi|_\mathfrak{t} - \rho(\mathfrak{s})$. $\qquad\square$

Note that the λ_μ appearing in the proposition is such that $\lambda_\mu + \rho(\mathfrak{g})$ is dominant, but need not be regular. Therefore, $A_\mathfrak{b}(\lambda_\mu)$ may equal 0. The further condition that λ_μ be dominant will ensure that $A_\mathfrak{b}(\lambda_\mu)$ is nonzero.

The following lemma will be important in Sect. 4.4. Note that the map $eval_e : C^\infty(G/H, S_\mathfrak{q} \otimes \mathscr{F}_\mu) \to S_\mathfrak{q} \otimes F_\mu$ given by $eval_e(f) = f(e)$ (with e the identity element of G) is an H-homomorphism. Assuming

(i) $\langle \mu + \rho(\mathfrak{h}) - \langle B \rangle, \beta \rangle \geq 0$, for all $\beta \in \Delta^+(\mathfrak{h})$ and $B \subset \Delta(\mathfrak{q}^+)$;

(ii) $\langle \mu + \rho(\mathfrak{h}) - 2\rho(\mathfrak{k} \cap \mathfrak{q}), \beta \rangle > 0$, for all $\beta \in \Delta^+(\mathfrak{h})$,

Steinberg's formula for the decomposition of the tensor product of two finite-dimensional representations of H tells us that $S_\mathfrak{q} \otimes F_\mu$ contains the irreducible representation of H having highest weight $\mu + \rho(\mathfrak{q}) - 2\rho(\mathfrak{k} \cap \mathfrak{q})$. Let V_0 be the corresponding isotypic subspace of $S_\mathfrak{q} \otimes F_\mu$.

Lemma 20. *Realizing $A_\mathfrak{b}(\lambda_\mu) \subset \ker(\mathscr{D}_{G/H}(\mathscr{F}_\mu))$, we have $eval_e(A_\mathfrak{b}(\lambda_\mu)) \subset V_0$.*

Proof. We first give the proof for the case $H = K$. In this case $\mathfrak{q} = \mathfrak{s}$ and $\mathfrak{k} \cap \mathfrak{q} = 0$, so $\lambda_\mu = \mu - \rho(\mathfrak{s})$. The possible K-types in $A_\mathfrak{b}(\lambda_\mu)$ have the form $\lambda_\mu + 2\rho(\mathfrak{s}) + \sum n_\gamma \gamma = \mu + \rho(\mathfrak{s}) + \sum n_\gamma \gamma$, with $\gamma \in \Delta^+(\mathfrak{s}), n_\gamma \geq 0$. On the other hand, the K-components of $S_\mathfrak{q} \otimes F_\mu = S_\mathfrak{s} \otimes F_\mu$ all have highest weights of the form $\mu + \rho(\mathfrak{s}) - \langle B \rangle, B \subset \Delta^+(\mathfrak{s})$. Since the image of $eval_e$ must consist of K-types common to $A_\mathfrak{b}(\lambda_\mu)$ and $S_\mathfrak{q} \otimes F_\mu$, the only possibility is that all $n_\gamma = 0$ and $\langle B \rangle = 0$. Therefore, $eval_e(A_\mathfrak{b}(\lambda_\mu))$ is contained in the isotypic subspace with highest weight $\mu + \rho(\mathfrak{s}) = \mu + \rho(\mathfrak{q}) - 2\rho(\mathfrak{k} \cap \mathfrak{q})$, i.e., $eval_e(A_\mathfrak{b}(\lambda_\mu)) \subset V_0$.

Now consider arbitrary $H \subset K$. Since $f(e) = (F_f(e))(e)$, we first consider $F_f \mapsto F_f(e)$. By the $H = K$ case, $F_f(e)$ is in the isotypic subspace of $S_\mathfrak{s} \otimes E_\xi$ of type $E_{\xi+\rho(\mathfrak{s})}$ (with ξ as in (15)). Now, evaluation at e gives an H-homomorphism $E_{\xi+\rho(\mathfrak{s})} \to S_\mathfrak{q} \otimes F_\mu$. Again we compare the H-types. The highest weights of H-components of $E_{\xi+\rho(\mathfrak{s})}$ are of the form $\xi + \rho(\mathfrak{s}) - \langle A \rangle, A \subset \Delta^+(\mathfrak{h})$. But $\xi + \rho(\mathfrak{s}) - \langle A \rangle = \mu - \rho(\mathfrak{k} \cap \mathfrak{q}) + \rho(\mathfrak{s}) - \langle A \rangle = \mu + \rho(\mathfrak{q}) - 2\rho(\mathfrak{k} \cap \mathfrak{q}) - \langle A \rangle$. The H-types in $S_\mathfrak{q} \otimes F_\mu$ are of the form $\mu + \rho(\mathfrak{q}) - \langle B \rangle, B \subset \Delta(\mathfrak{k} \cap \mathfrak{q}^+)$. The only way for us to have $\mu + \rho(\mathfrak{q}) - 2\rho(\mathfrak{k} \cap \mathfrak{q}) - \langle A \rangle = \mu + \rho(\mathfrak{q}) - \langle B \rangle$ is for $\langle B \rangle = 2\rho(\mathfrak{k} \cap \mathfrak{q}) + \langle A \rangle$. As $B \subset \Delta(\mathfrak{k} \cap \mathfrak{q}^+)$, this means that $B = \Delta(\mathfrak{k} \cap \mathfrak{q}^+)$ and $\langle A \rangle = 0$. We conclude that the image of $eval_e|_{A_\mathfrak{b}(\lambda_\mu)}$ is contained in V_0. $\qquad\square$

4 The Main Theorem

Let H be an arbitrary subgroup of G satisfying (1). We associate to H a parabolic subgroup P in G. The main result is the construction of an integral intertwining map $\mathscr{P} : C^\infty(G/P, \mathscr{W}) \to \ker(\mathscr{D}_{G/H})$. The precise statement is contained in Theorem 29.

4.1 Roots and Positive Systems

We need to make some choices of Cartan subalgebras and positive root systems, which will be used in the construction of our intertwining operator. These choices are compatible with those made in Sect. 3.1, where the special case of $\mathfrak{h} \subset \mathfrak{k}$ was considered.

Consider the complexified Lie algebra $\mathfrak{h}$ of the reductive group H and the decomposition $\mathfrak{g} = \mathfrak{h} + \mathfrak{q}$. Choose a maximal abelian subspace $\mathfrak{a}_\mathfrak{h}$ of $\mathfrak{h} \cap \mathfrak{s}$. Define $\mathfrak{l} = \mathfrak{z}_\mathfrak{g}(\mathfrak{a}_\mathfrak{h})$, the centralizer of $\mathfrak{a}_\mathfrak{h}$ in $\mathfrak{g}$. A Cartan subalgebra of $\mathfrak{g}$ is chosen as follows.

- Let $\mathfrak{t}_\mathfrak{h}$ be a Cartan subalgebra of $\mathfrak{h} \cap \mathfrak{k} \cap \mathfrak{l}$. Note that $\mathfrak{a}_\mathfrak{h} + \mathfrak{t}_\mathfrak{h}$ is a Cartan subalgebra of $\mathfrak{h}$, since $\mathfrak{a}_\mathfrak{h}$ is maximal abelian in $\mathfrak{h} \cap \mathfrak{s}$.
- Extend $\mathfrak{t}_\mathfrak{h}$ to a Cartan subalgebra $\mathfrak{t} = \mathfrak{t}_\mathfrak{h} + \mathfrak{t}_\mathfrak{q}$ of $\mathfrak{k} \cap \mathfrak{l}$ with $\mathfrak{t}_\mathfrak{q} \subset \mathfrak{q} \cap \mathfrak{k} \cap \mathfrak{l}$. Note that $\mathfrak{t}$ is not necessarily a Cartan subalgebra of $\mathfrak{k}$.
- Finally, choose $\mathfrak{a}_\mathfrak{q} \subset \mathfrak{q} \cap \mathfrak{s} \cap \mathfrak{l}$ so that $\mathfrak{a}_\mathfrak{h} + \mathfrak{t} + \mathfrak{a}_\mathfrak{q}$ is a Cartan subalgebra of $\mathfrak{l}$. Write $\mathfrak{a} := \mathfrak{a}_\mathfrak{h} + \mathfrak{a}_\mathfrak{q}$. Since $\mathrm{rank}_C(\mathfrak{g}) = \mathrm{rank}_C(\mathfrak{l})$, we see that $\mathfrak{a} + \mathfrak{t}$ is also a Cartan subalgebra of $\mathfrak{g}$.

Remarks. (1) When $\mathfrak{h} \subset \mathfrak{k}$, $\mathfrak{a}_\mathfrak{h} = \{0\}$. Therefore $\mathfrak{l} = \mathfrak{g}$ and $\mathfrak{a} + \mathfrak{t}$ is a fundamental Cartan subalgebra of both $\mathfrak{g}$ and $\mathfrak{l}$, as in Sect. 3.1.

(2) When $\mathrm{rank}_C(\mathfrak{h}) = \mathrm{rank}_C(\mathfrak{g})$ (as in [9]) $\mathfrak{t} = \mathfrak{t}_\mathfrak{h}$ and $\mathfrak{a} = \mathfrak{a}_\mathfrak{h}$.

Let $\Delta := \Delta(\mathfrak{a} + \mathfrak{t}, \mathfrak{g})$ be the set of $\mathfrak{a} + \mathfrak{t}$-roots in $\mathfrak{g}$. For any $\alpha \in \Delta$ we will write $\mathfrak{g}^{(\alpha)}$ for the corresponding root space.

Let Δ^+ be any positive system of roots in $\Delta := \Delta(\mathfrak{a} + \mathfrak{t}, \mathfrak{g})$ given by a lexicographic order with a basis of $\mathfrak{a}_\mathfrak{h}$ first, then (in order) bases of $\mathfrak{t}_\mathfrak{h}$, $\mathfrak{t}_\mathfrak{q}$ and $\mathfrak{a}_\mathfrak{q}$.

A positive system of $(\mathfrak{a}_\mathfrak{h} + \mathfrak{t}_\mathfrak{h})$-roots in $\mathfrak{h}$ is chosen using the lexicographic order with the same basis of $\mathfrak{a}_\mathfrak{h}$ as above, followed by the basis of $\mathfrak{t}_\mathfrak{h}$. Call this positive system $\Delta^+(\mathfrak{h})$.

4.2 The Parabolic Subgroup

Having fixed a positive system of roots Δ^+ in $\mathfrak{g}$ we may define a parabolic subalgebra of $\mathfrak{g}$ as follows. Set

$$\Sigma^+ := \{\alpha \in \Delta^+ : \alpha|_{\mathfrak{a}_\mathfrak{h}} \neq 0\}.$$

Then

$$\mathfrak{p} := \mathfrak{l} + \mathfrak{n}, \text{ where } \mathfrak{n} := \sum_{\alpha \in \Sigma^+} \mathfrak{g}^{(\alpha)},$$

is a parabolic subalgebra of $\mathfrak{g}$. Thanks to the choice of Δ^+, $\mathfrak{p}$ is the complexification of a (real) subalgebra of $Lie(G)$. We define P to be the *connected* subgroup of G corresponding to this real parabolic Lie algebra.

It will be convenient for us to write $\mathfrak{l} = \mathfrak{m} + \mathfrak{a}_\mathfrak{h}$ with

$$\mathfrak{m} = \sum_{\alpha \in \Delta, \alpha|_{\mathfrak{a}_\mathfrak{h}} = 0} \mathfrak{g}^{(\alpha)} + (\mathfrak{a}_\mathfrak{q} + \mathfrak{t}).$$

Therefore,

$$\mathfrak{p} = \mathfrak{m} + \mathfrak{a}_\mathfrak{h} + \mathfrak{n}. \tag{21}$$

However, this is *not* (typically) the Langlands decomposition of $\mathfrak{p}$. Note that $\mathfrak{a}_\mathfrak{q} \subset \mathfrak{l} \cap \mathfrak{s}$, but it can happen that some (but not all) of $\mathfrak{a}_\mathfrak{q}$ lies in the center of $\mathfrak{l}$. The decomposition (21) gives a corresponding decomposition of P, which we write as $P = MA_\mathfrak{h}N$.

Lemma 22. $\mathfrak{p} \cap \mathfrak{h}$ *is a minimal parabolic subalgebra of* $\mathfrak{h}$. *In particular,* $\mathfrak{m} \cap \mathfrak{h} \subset \mathfrak{k}$ *and* $\mathfrak{l} \cap \mathfrak{h} = \mathfrak{l} \cap \mathfrak{h} \cap \mathfrak{k} + \mathfrak{a}_\mathfrak{h}$.

Proof. The Borel subalgebra of $\mathfrak{h}$ defined by $\mathfrak{a}_\mathfrak{h} + \mathfrak{t}_\mathfrak{h}$ and $\Delta^+(\mathfrak{h})$ is contained in $(\mathfrak{p} \cap \mathfrak{h})$, so $\mathfrak{p} \cap \mathfrak{h}$ is a parabolic subalgebra. Since $\mathfrak{a}_\mathfrak{h}$ is maximal abelian in $\mathfrak{h} \cap \mathfrak{s}$, $\mathfrak{l} \cap \mathfrak{h} = \mathfrak{z}_\mathfrak{h}(\mathfrak{a}_\mathfrak{h}) = \mathfrak{l} \cap \mathfrak{h} \cap \mathfrak{k} + \mathfrak{a}_\mathfrak{h}$. Therefore, $\mathfrak{p} \cap \mathfrak{h}$ is minimal.   □

We set $\bar{\mathfrak{n}} = \sum_{\alpha \in \Sigma^+} \mathfrak{g}^{(-\alpha)}$, so $\mathfrak{l} + \bar{\mathfrak{n}}$ is the parabolic subalgebra opposite to $\mathfrak{p}$.

Lemma 23. *The following hold.*

(a) $\mathfrak{l} = \mathfrak{l} \cap \mathfrak{h} + \mathfrak{l} \cap \mathfrak{q}$.
(b) $\mathfrak{m} = \mathfrak{m} \cap \mathfrak{h} + \mathfrak{m} \cap \mathfrak{q}$.
(c) $\mathfrak{q} = \mathfrak{q} \cap \mathfrak{l} + \mathfrak{q} \cap \mathfrak{n} + \mathfrak{q} \cap \bar{\mathfrak{n}}$.
(d) $\mathfrak{m} \cap \mathfrak{q} = \mathfrak{m} \cap \mathfrak{q} \cap \mathfrak{k} + \mathfrak{m} \cap \mathfrak{s}$.

Proof. (a) This is clear since $\mathfrak{h}$ and $\mathfrak{q}$ are $\mathfrak{a}_\mathfrak{h}$-stable and $\mathfrak{l}$ is the 0-weight space of $\mathfrak{a}_\mathfrak{h}$.
(b) This follows from (a) since

$$\mathfrak{m} \subset \mathfrak{l} = \mathfrak{l} \cap \mathfrak{h} + \mathfrak{l} \cap \mathfrak{q}$$

$$= \mathfrak{m} \cap \mathfrak{h} + \mathfrak{a}_\mathfrak{q} + \mathfrak{m} \cap \mathfrak{q}.$$

(c) Since $\mathfrak{a}_\mathfrak{h}$ acts on $\mathfrak{q}$, $\mathfrak{q}$ is a sum of $\mathfrak{a}_\mathfrak{h}$-weight spaces. If $X \in \mathfrak{q}$ is a weight vector, then the weight is $\alpha|_{\mathfrak{a}_\mathfrak{h}}$, for some $\alpha \in \Delta \cup \{0\}$. Therefore, $X \in \mathfrak{l}, \mathfrak{n}$ or $\bar{\mathfrak{n}}$.
(d) Since $\mathfrak{a}_\mathfrak{h}$ is θ-stable, so is $\mathfrak{m}$. Therefore, $\mathfrak{m} \cap \mathfrak{q}$ is also θ-stable. It follows that $\mathfrak{m} \cap \mathfrak{q} \cap \mathfrak{k} + \mathfrak{m} \cap \mathfrak{q} \cap \mathfrak{s}$. By Lemma 22, $\mathfrak{m} \cap \mathfrak{h} \cap \mathfrak{s} = 0$. Therefore, $\mathfrak{m} \cap \mathfrak{q} \cap \mathfrak{s} = \mathfrak{m} \cap \mathfrak{s}$ and (d) follows.   □

We need to fix a positive system in $\Delta(\mathfrak{t} + \mathfrak{a}_\mathfrak{q}, \mathfrak{m})$. Since we will be applying Sect. 3 to $H \cap M \subset M$ in place of $H \subset G$, we will need a choice of $\Delta^+(\mathfrak{m})$ as in Sect. 3.1. We therefore define $\Delta^+(\mathfrak{m})$ using the lexicographic order with the same bases of $\mathfrak{t}_\mathfrak{h}, \mathfrak{t}_\mathfrak{q}$ and $\mathfrak{a}_\mathfrak{q}$ (in that order) that were used in the lexicographic order defining Δ^+ earlier. Observe that $\mathfrak{t} + \mathfrak{a}_\mathfrak{q}$ is a fundamental Cartan subalgebra of $\mathfrak{m}$. Using the basis of $\mathfrak{t}_\mathfrak{h}$ gives a lexicographic order that in turn gives a positive system $\Delta^+(\mathfrak{m} \cap \mathfrak{h})$.

We make the following assumption on H, which is analogous to and consistent with Assumption 11.

Assumption 24. *There is no root* $\beta \in \Delta(\mathfrak{t}_\mathfrak{h} + \mathfrak{t}_\mathfrak{q} + \mathfrak{a}_\mathfrak{q}, \mathfrak{m})$ *so that* $\beta|_{\mathfrak{t}_\mathfrak{h}} = 0$.

4.3 The Representation $S_q \otimes E_\mu$

Let $\mu \in (\mathfrak{a}_\mathfrak{h} + \mathfrak{t}_\mathfrak{h})^*$ be $\Delta^+(\mathfrak{h})$-dominant and integral. Let E_μ be the irreducible finite-dimensional $\mathfrak{h}$-representation with highest weight μ. We consider the tensor product $S_q \otimes E_\mu$, a representation of $\mathfrak{h}$.

The construction of the spin representation in Sect. 2.2 requires a choice of maximal isotropic subspace of $\mathfrak{q}$. This is done as follows. Choose some maximal isotropic subspace $(\mathfrak{a}_q + \mathfrak{t}_q)^+$ of $\mathfrak{a}_q + \mathfrak{t}_q$ and set

$$\mathfrak{q}^+ = (\mathfrak{a}_q + \mathfrak{t}_q)^+ + \sum_{\beta \in \Delta^+(\mathfrak{m} \cap \mathfrak{q})} \mathfrak{m}^{(\gamma)} + \mathfrak{q} \cap \mathfrak{n}.$$

Then $\mathfrak{q}^+$ is maximally isotropic in $\mathfrak{q}$ by Lemma 23(c) and the fact that $\mathfrak{l} \cap \mathfrak{q} = \mathfrak{m} \cap \mathfrak{q}$.

Note that the $(\mathfrak{a}_\mathfrak{h} + \mathfrak{t}_\mathfrak{h})$-weights in $\mathfrak{q}^+$ are the weights in $\mathfrak{q}$ that are positive with respect to the lexicographic order for the same bases of $\mathfrak{a}_\mathfrak{h}$ and $\mathfrak{t}_\mathfrak{h}$ used in the definition of Δ^+ above. It follows that

$$\rho(\mathfrak{m} \cap \mathfrak{q}) := \rho(\mathfrak{m})|_{\mathfrak{t}_\mathfrak{h}} - \rho(\mathfrak{m} \cap \mathfrak{h})$$

is $1/2$ the sum of the $\mathfrak{t}_\mathfrak{h}$-weights in $\mathfrak{m} \cap \mathfrak{q}^+$.

The weight

$$\rho(\mathfrak{q}) := \frac{1}{2} \sum_{\gamma \in \Delta(\mathfrak{q}^+)} \gamma,$$

with each weight occurring as many times as its multiplicity in $\mathfrak{q}^+$, is $\Delta^+(\mathfrak{h})$-dominant. The set of weights of S_q is

$$\Delta(S_q) = \{\rho(\mathfrak{q}) - \langle A \rangle : A \subset \Delta(\mathfrak{q}^+)\}$$
$$= \{\langle A \rangle - \rho(\mathfrak{q}) : A \subset \Delta(\mathfrak{q}^+)\}.$$

The subgroup $H \cap M \subset M$ satisfies the conditions of (1). Therefore, there is a spin representation $S_{\mathfrak{m} \cap \mathfrak{q}}$. Using the maximal isotropic subspace $\mathfrak{m} \cap \mathfrak{q}^+$ of $\mathfrak{m} \cap \mathfrak{q}$ one easily sees that $S_{q \cap m}$ is naturally contained in S_q as $\mathfrak{h} \cap \mathfrak{m}$-representation. The set of weights of $S_{\mathfrak{m} \cap \mathfrak{q}}$ with respect to the Cartan subalgebra $\mathfrak{t}_\mathfrak{h}$ of $\mathfrak{m} \cap \mathfrak{h}$ is

$$\Delta(S_{\mathfrak{m} \cap \mathfrak{q}}) = \{\rho(\mathfrak{m} \cap \mathfrak{q}) - \langle B \rangle : B \subset \Delta(\mathfrak{m} \cap \mathfrak{q}^+)\}.$$

Lemma 25. *The following hold.*

(a) $S_{q \cap m} \subset (S_q)^{\mathfrak{h} \cap \mathfrak{n}}$, *the* $\mathfrak{n} \cap \mathfrak{h}$-*invariants in* S_q.
(b) *As a subspace of* S_q, $\mathfrak{a}_\mathfrak{h}$ *acts on* $S_{q \cap m}$ *by* $\rho(\mathfrak{q})|_{\mathfrak{a}_\mathfrak{h}}$.
(c) $\rho(\mathfrak{q})|_{\mathfrak{t}_\mathfrak{h}} = \rho(\mathfrak{m} \cap \mathfrak{q})$.

Proof. The proof of the first statement is as in [8, Lemma 3.8]. The second follows since the weights of $S_{\mathfrak{m}\cap\mathfrak{q}} \subset S_{\mathfrak{q}}$ are of the form $\rho(\mathfrak{q}) - \langle B \rangle$ with $B \subset \Delta(\mathfrak{m} \cap \mathfrak{q}^{+})$. But $\langle B \rangle|_{\mathfrak{a}_{\mathfrak{h}}} = 0$, since the weights in $\mathfrak{m}$ vanish on $\mathfrak{a}_{\mathfrak{h}}$. The last statement follows from $\rho(\mathfrak{q})|_{\mathfrak{t}_{\mathfrak{h}}} - \rho(\mathfrak{m} \cap \mathfrak{q}) = \rho(\mathfrak{n} \cap \mathfrak{q})|_{\mathfrak{t}_{\mathfrak{h}}} = 0$, which follows from the fact that $\Delta(\mathfrak{n} \cap \mathfrak{q})$ is stable under $-\theta$.   □

Now consider E_{μ} (with $\Delta^{+}(\mathfrak{h})$-dominant integral $\mu \in (\mathfrak{a}_{\mathfrak{h}} + \mathfrak{t}_{\mathfrak{h}})^{*}$). Set $F_{\mu|_{\mathfrak{t}_{\mathfrak{h}}}} := (E_{\mu})^{\mathfrak{n}\cap\mathfrak{h}}$, an irreducible representation of $\mathfrak{m} \cap \mathfrak{h}$ of highest weight $\mu|_{\mathfrak{t}_{\mathfrak{h}}}$.

For the remainder of this article we make the following assumptions on μ.

(i) $\langle \mu + \rho(\mathfrak{m} \cap \mathfrak{h}) - \langle B \rangle , \beta \rangle \geq 0$, for all $\beta \in \Delta^{+}(\mathfrak{m} \cap \mathfrak{h})$ and $B \subset \Delta^{+}(\mathfrak{m} \cap \mathfrak{q})$.

(ii) $\langle \mu + \rho(\mathfrak{m} \cap \mathfrak{h}) - 2\rho(\mathfrak{m} \cap \mathfrak{k} \cap \mathfrak{q}) , \beta \rangle > 0$, for all $\beta \in \Delta^{+}(\mathfrak{m} \cap \mathfrak{h})$.

$$(26)$$

By Steinberg's formula for the decomposition of a tensor product of finite-dimensional representations, we see that $S_{\mathfrak{m}\cap\mathfrak{q}} \otimes F_{\mu|_{\mathfrak{t}_{\mathfrak{h}}}}$ contains the irreducible highest weight representation of $\mathfrak{m} \cap \mathfrak{h}$ having highest weight $\mu + \rho(\mathfrak{m} \cap \mathfrak{q}) - 2\rho(\mathfrak{m} \cap \mathfrak{k} \cap \mathfrak{q})$. Let V_{0} be the isotypic subspace of type $F_{\mu+\rho(\mathfrak{m}\cap\mathfrak{q})-2\rho(\mathfrak{m}\cap\mathfrak{k}\cap\mathfrak{q})}$. Note that the assumption of (11) and the definition of V_{0} are consistent with Sect. 3.3.

Observe that

$$V_{0} \subset S_{\mathfrak{m}\cap\mathfrak{q}} \otimes F_{\mu|_{\mathfrak{t}_{\mathfrak{h}}}} \subset \left(S_{\mathfrak{q}} \otimes E_{\mu}\right)^{\mathfrak{n}\cap\mathfrak{h}},$$

by Lemma 25. It also follows from Lemma 25 parts (a) and (c), that $\mathfrak{a}_{\mathfrak{h}}$ acts on V_{0} (as a subspace of $S_{\mathfrak{q}} \otimes E_{\mu}$) by the weight $(\mu + \rho(\mathfrak{q}))|_{\mathfrak{a}_{\mathfrak{h}}}$.

4.4 Harmonic Spinors

Let P be the parabolic subgroup of G associated to H as in Sect. 4.2. Fix $\mu \in (\mathfrak{a}_{\mathfrak{h}} + \mathfrak{t}_{\mathfrak{h}})^{*}$ and assume that $S_{\mathfrak{q}} \otimes E_{\mu}$ lifts to a representation of the group H. Therefore, we have a smooth homogeneous vector bundle $S_{\mathfrak{q}} \otimes \mathscr{E}_{\mu} \to G/H$ and a cubic Dirac operator

$$\mathscr{D}_{G/H}(\mathscr{E}_{\mu}) : C^{\infty}(G/H, S_{\mathfrak{q}} \otimes \mathscr{E}_{\mu}) \to C^{\infty}(G/H, S_{\mathfrak{q}} \otimes \mathscr{E}_{\mu}).$$

Our goal is to construct an intertwining operator from a representation induced from P to $\ker\left(\mathscr{D}_{G/H}(\mathscr{E}_{\mu})\right)$. Let W be a representation of P. Write the induced representation as

$$C^{\infty}(G/P, \mathscr{W}) = \{\varphi : G \to W : \varphi \text{ is smooth and } \varphi(gp) = p^{-1}\varphi(g), p \in P, g \in G\}.$$

The action of $g \in G$ is by left translation on functions: $(g \cdot f)(g_1) = f(g^{-1}g_1)$. The following is essentially Lemma 4.2 of [9], it is easily proved using a standard change of variables formula (e.g., [4, Lem. 5.19]).

Lemma 27. *Let W be some representation of P. If*

$$t \in \mathrm{Hom}_{P \cap H}(W \otimes C_{-2\rho(\mathfrak{h})|_{\mathfrak{a}_\mathfrak{h}}}, S_\mathfrak{q} \otimes E_\mu)$$

is nonzero, then

$$(\mathscr{P}_t \varphi)(g) = \int_{H \cap K} \ell \cdot t(\varphi(g\ell)) \, d\ell$$

is a nonzero G-intertwining map

$$\mathscr{P}_t : C^\infty(G/P, \mathscr{W}) \to C^\infty(G/H, S_\mathfrak{q} \otimes \mathscr{E}_\mu).$$

Given the bundle $S_\mathfrak{q} \otimes \mathscr{E}_\mu \to G/H$ we now make our choice of P-representation W and homomorphism t.

Let $F_{\mu|_{\mathfrak{t}_\mathfrak{h}}}$ be the irreducible representation of $H \cap M$ of highest weight $\mu|_{\mathfrak{t}_\mathfrak{h}}$. Suppose that μ satisfies the assumptions of (26). Then, by Sect. 3.3 applied to $H \cap M \subset M$, $\ker(\mathscr{D}_{M/H \cap M}(\mathscr{F}_\mu))$ contains a representation W infinitesimally equivalent to $A_{\mathfrak{b} \cap \mathfrak{m}}(\lambda_\mu)$ with

$$\lambda_\mu|_{\mathfrak{t}_\mathfrak{h}} = \mu|_{\mathfrak{t}_\mathfrak{h}} - \rho(\mathfrak{m} \cap \mathfrak{q}), \quad \lambda_\mu|_{\mathfrak{a}_\mathfrak{q}} = 0 \quad \text{and} \quad \lambda_\mu|_{\mathfrak{t}_\mathfrak{q}} = -\rho(\mathfrak{m} \cap \mathfrak{k}).$$

By Lemma 20, evaluation at the identity is nonzero on W and has an image in V_0. Give W the trivial N-action and define $\mathfrak{a}_\mathfrak{h}$ to act by $(\mu + \rho(\mathfrak{q}) + 2\rho(\mathfrak{h}))|_{\mathfrak{a}_\mathfrak{h}}$. Take t to be evaluation at the identity: $t(w \otimes 1) = w(e)$.

Lemma 28. $t \in \mathrm{Hom}_{P \cap H}(W \otimes C_{-2\rho(\mathfrak{h})|_{\mathfrak{a}_\mathfrak{h}}}, S_\mathfrak{q} \otimes E_\mu)$.

Proof. Evaluation is an $M \cap H$-homomorphism. The action of $\mathfrak{a}_\mathfrak{h}$ on $W \otimes C_{-2\rho(\mathfrak{h})}$ is by $(\mu + \rho(\mathfrak{q}))|_{\mathfrak{a}_\mathfrak{h}}$. The action on the image of t is by $(\mu + \rho(\mathfrak{q}))|_{\mathfrak{a}_\mathfrak{h}}$, as pointed out at the end of Sect. 4.3. The action of $N \cap H$ on both $W \otimes C_{-2\rho(\mathfrak{h})}$ and the image of t is trivial. $\square$

When these conditions are satisfied and t is chosen as above, our main theorem, stated as Theorem A in the introduction, is the following.

Theorem 29. *The intertwining map $\mathscr{P}_t$ has image in the kernel of $\mathscr{D}_{G/H}(\mathscr{E}_\mu)$. In particular, $\ker(\mathscr{D}_{G/H}(\mathscr{E}_\mu)) \neq 0$.*

Proof. The first observation is that

$$(\mathscr{D}_{G/H}(\mathscr{P}_t \varphi))(g) = \int_{H \cap K} \left(\sum_i a_i R(X_i) \otimes \gamma(X_i) \right) \ell \cdot t(\varphi(\cdot \ell))|_g \, d\ell$$

$$\int_{H \cap K} \ell \cdot \left(\sum_i a_i R(X_i) \otimes \gamma(X_i) \right) t(\varphi(\cdot))|_{g\ell} \, d\ell.$$

As t is evaluation at e, it suffices to show that

$$\sum_i \left(a_i \, R(X_i) \otimes \gamma(X_i)\right) \varphi(\cdot)(e) = 0 \tag{30}$$

for $\varphi \in C^\infty(G/P, \mathscr{W})$.

We now choose the basis X_i in a suitable way. Let $\{E_j\}$ be a basis of $\mathfrak{q} \cap \mathfrak{n}$ and $\{\overline{E}_j\}$ a basis of $\mathfrak{q} \cap \bar{\mathfrak{n}}$ such that

$$\langle E_j , \overline{E}_k \rangle \quad = \delta_{jk}$$
$$\langle E_j , E_k \rangle = \langle \overline{E}_j , \overline{E}_k \rangle = 0,$$

and let $\{Z_j\}$ be a basis of $\mathfrak{m} \cap \mathfrak{q}$ so that

$$\langle Z_j , Z_k \rangle = \delta_{jk}.$$

Setting

$$Y_j^+ = \frac{1}{\sqrt{2}}(E_j + \overline{E}_j) \text{ and } Y_j^- = \frac{1}{\sqrt{2}}(E_j - \overline{E}_j),$$

we get an orthogonal basis $\{Z_j\} \cup \{Y_j^\pm\}$ of $\mathfrak{q}$ as required in (3). The (geometric) Dirac operator $\mathscr{D}_{G/H}(\mathscr{E}_\mu)$ may be written as follows (equation (4.7) in [9]):

$$
\begin{aligned}
\mathscr{D}_{G/H}(\mathscr{E}_\mu) = {} & \sum_i \left(R(Z_i) \otimes \gamma(Z_i) \otimes 1 - \gamma(\mathbf{c}_{\mathfrak{m} \cap \mathfrak{q}}) \right) \\
& + \sum_j \left(R(E_j) \otimes \epsilon(\overline{E}_j) \otimes 1 + R(\overline{E}_j) \otimes \iota(E_j) \otimes 1 \right) \\
& - 1 \otimes \Big(\sum_j \langle Z_i , Z_i \rangle \langle Z_i , [E_j, \overline{E}_k] \rangle \gamma(Z_i) \epsilon(\overline{E}_j) \iota(E_k) \\
& \quad + \sum \langle E_j , [E_k, \overline{E}_\ell] \rangle \epsilon(\overline{E}_j) \epsilon(\overline{E}_k) \iota(E_\ell) \\
& \quad + \sum \langle E_j , [\overline{E}_k, \overline{E}_\ell] \rangle \epsilon(\overline{E}_j) \iota(E_k) \iota(E_\ell) \Big) \otimes 1,
\end{aligned} \tag{31}
$$

where ι (resp. ϵ) stands for the interior (resp. exterior) product (resp. multiplication) and $\mathbf{c}_{\mathfrak{m} \cap \mathfrak{q}}$ is the cubic term for $H \cap M \subset M$.

Now insert (31) into (30). The first term vanishes as follows. First note that

$$
\begin{aligned}
\left(R(Z_i) \varphi(\cdot) \right)(e)|_g &= \frac{d}{ds} \varphi(g \exp(s Z_i))(e)|_{s=0} \\
&= \frac{d}{ds} \left(\exp(s Z_i) \right)^{-1} \varphi(g))(e)|_{s=0}, \text{ by } M\text{-equivariance of } \varphi,
\end{aligned}
$$

$$= \frac{d}{ds}\varphi(g)(\exp(sZ_i))|_{s=0}, \text{ by the definition of } M \text{ action on } W,$$

$$= R(Z_i)\big(\varphi(g)(\cdot)\big)|_e.$$

Now,

$$\big(\textstyle\sum_i R(Z_i) \otimes \gamma(Z_i) - \gamma(\mathfrak{c}_{\mathfrak{m}\cap\mathfrak{q}})\big)\varphi(\cdot)(e)|_g$$

$$= \sum_i a_i\big(R(Z_i)\varphi(g)\big)|_e - \gamma(\mathfrak{c}_{\mathfrak{m}\cap\mathfrak{q}})\varphi(g)(e)$$

$$= \big(\mathscr{D}_{M/M\cap H}\varphi(g)\big)(e)$$

$$= 0, \quad \text{since } \varphi(g) \in W \subset \ker(\mathscr{D}_{M/M\cap H}).$$

By the right $P = MA_\mathfrak{h}N$-equivariance defining $C^\infty(G/P, \mathscr{W})$, $R(E_j)\varphi = 0$, so the next terms in (31) vanish. Since the image of t is contained in $S_{\mathfrak{m}\cap\mathfrak{q}} \otimes F_{\mu|_{\mathfrak{t}_\mathfrak{h}}}$ and each E_j is orthogonal to $\mathfrak{m} \cap \mathfrak{q}$ (so that $\iota(E_j)S_{\mathfrak{m}\cap\mathfrak{q}} = 0$), the remaining terms are 0. $\qquad\square$

Appendix: Geometric vs. Algebraic Dirac Operators

For the convenience of the reader we provide here a proof of Proposition 8. Recall that V denotes a smooth admissible representation of G, V_K the space of K-finite vectors in V, V_K^* the K-finite dual of V_K and $S_\mathfrak{s}$ is the spin representation for $\mathfrak{k}$. Let E be a finite-dimensional representation of $\mathfrak{k}$ such that the tensor product $E \otimes S_\mathfrak{s}$ lifts to a representation of the group K, and denote by E^* the dual of E. The K-representation $E \otimes S_\mathfrak{s}$ induces a homogeneous bundle $\mathscr{S}_\mathfrak{s} \otimes \mathscr{E} \longrightarrow G/K$ over G/K whose space of smooth sections, on which G acts by left translations, is denoted by $C^\infty(G/K, \mathscr{S}_\mathfrak{s} \otimes \mathscr{E})$. The map

$$\Psi : \mathrm{Hom}_G(V, C^\infty(G/K, \mathscr{S}_\mathfrak{s} \otimes \mathscr{E})) \longrightarrow \mathrm{Hom}_K(E^*, S_\mathfrak{s} \otimes V_K^*)$$

defined by $\Psi(T)(e^*)(v) = 1 \otimes e^*T(v)(1)$, is an isomorphism, where 1 denotes the identity G.

Next, as in Sect. 2, consider the (cubic) Dirac operators

$$\mathscr{D}_{G/K}(\mathscr{E}) : C^\infty(G/K, \mathscr{S}_\mathfrak{s} \otimes \mathscr{E}) \longrightarrow C^\infty(G/K, \mathscr{S}_\mathfrak{s} \otimes \mathscr{E}) \text{ and}$$

$$D_{V_K^*} : S_\mathfrak{s} \otimes V_K^* \longrightarrow S_\mathfrak{s} \otimes V_K^*,$$

and define the maps

$$\mathscr{D}_* : \mathrm{Hom}_G(V, C^\infty(G/K, \mathscr{S}_\mathfrak{s} \otimes \mathscr{E})) \longrightarrow \mathrm{Hom}_G(V, C^\infty(G/K, \mathscr{S}_\mathfrak{s} \otimes \mathscr{E}))$$

and

$$D_* : \mathrm{Hom}_K(E^\star, S_{\mathfrak{s}} \otimes V_K^*) \longrightarrow \mathrm{Hom}_K(E^\star, S_{\mathfrak{s}} \otimes V_K^*)$$

by

$$(D_*(T))(v) = \mathscr{D}_{G/K}(\mathscr{E})(T(v)),$$
$$(D_*(A))(e^*) = D_{V_K^*}(A(e^*)).$$

We claim that the following diagram is commutative:

$$
\begin{array}{ccc}
\mathrm{Hom}_G(V, C^\infty(G/K, \mathscr{S}_{\mathfrak{s}} \otimes \mathscr{E})) & \xrightarrow{\ \Psi\ } & \mathrm{Hom}_K(E^\star, S_{\mathfrak{s}} \otimes V_K^*) \\
\Big\downarrow{\mathscr{D}_*} & & \Big\downarrow{D_*} \\
\mathrm{Hom}_G(V, C^\infty(G/K, \mathscr{S}_{\mathfrak{s}} \otimes \mathscr{E})) & \xrightarrow{\ \Psi\ } & \mathrm{Hom}_K(E^\star, S_{\mathfrak{s}} \otimes V_K^*)
\end{array}
$$

Indeed one has

$$\Psi(D_*(T))(e^*)(v) = (1 \otimes e^*)\big((D_*(T))(v)(1)\big) = (1 \otimes e^*)\big(\mathscr{D}_{G/K}(T(v))(1)\big)$$

and

$$\mathscr{D}_{G/K}(T(v))(1)$$
$$= \sum_i \frac{d}{dt}\Big|_{t=0}(\gamma(X_i) \otimes 1)(T(v)(\exp(tX_i)(1)) - (\gamma(\mathbf{c}_{\mathfrak{s}}) \otimes 1)(T(v)(1))$$
$$= \sum_i \frac{d}{dt}\Big|_{t=0}(\gamma(X_i) \otimes 1)(T(\exp(-tX_i)v)(1)) - (\gamma(\mathbf{c}_{\mathfrak{s}}) \otimes 1)(T(v)(1))$$
$$= -\sum_i (\gamma(X_i) \otimes 1)(T(X_i v)(1)) - (1 \otimes \gamma(\mathbf{c}_{\mathfrak{s}}))(T(v)(1))$$

which means that

$$\Psi(D_*(T))(e^*)(v) = -\sum_i (\gamma(X_i) \otimes e^*)(T(X_i v)(1)) - (\gamma(\mathbf{c}_{\mathfrak{s}}) \otimes e^*)(T(v)(1)).$$

On the other hand, one has

$$\Big((D_*(\Psi(T))(e^*)\Big)(v) = \Big(D_{V_K^*}(\Psi(T)(e^*))\Big)(v)$$
$$= -\sum_i \big((\gamma(X_i) \otimes X_i)(\Psi(T)(e^*))(v) - (\gamma(\mathbf{c}_{\mathfrak{s}}) \otimes 1)(\Psi(T)(e^*)(v))$$

$$= -\sum_i (\gamma(X_i) \otimes 1)(\Psi(T)(e^\star)(X_i v)) - (\gamma(\mathbf{c}_5) \otimes 1)(\Psi(T)(e^\star)(v))$$

$$= -\sum_i (\gamma(X_i) \otimes e^\star)(T(X_i v)(1)) - (\gamma(\mathbf{c}_5) \otimes e^\star)(T(v)(1)).$$

We deduce the following isomorphism relating algebraic and geometric harmonic spinors:

$$\Psi : \mathrm{Hom}_G(V, \ker(\mathscr{D}_{G/K}(\mathscr{E}))) \simeq \mathrm{Hom}_K(E^\star, \ker(D_{V_K^\star})),$$

therefore proving Proposition 8.

References

1. M. Atiyah and W. Schmid, *A geometric construction of the discrete series for semisimple Lie groups*, Inv. Math. **42** (1977), 1–62.
2. A. Huckleberry, I. Penkov, and G. Zuckerman (Eds.). *Lie groups: structure, actions, and representations*. Process in Mathematics, Birkhäuser, **306** (2013), 41–57.
3. R. Goodman and N. R. Wallach, *Representations and Invariants of the Classical Groups*. Encyclopedia of Mathematics and its Applications, **68**, Cambridge University Press, 1998.
4. S. Helgason, *Groups and Geometric Analysis*. Academic Press, 1984.
5. J.-S. Huang, Y.-F. Kang, and P. Pandžić, *Dirac cohomology of some Harish–Chandra modules*, Trans. Groups **14** (2009), 163–173.
6. B. Kostant, *A cubic Dirac operator and the emergence of Euler number multiplets of representations for equal rank subgroups*, Duke Math. J. **100** (1999), no. 3, 447–501.
7. S. Mehdi, *Parthasarathy Dirac operators and discrete series representations*, Connected at infinity II: A Selection of Mathematics by Indians. Edited by R. Bhatia, C. S. Rajan and A. I. Singh. Text and Readings in Mathematics **67**. Hindustan Book Agency, New Delhi, 2013.
8. S. Mehdi and R. Zierau, *Harmonic spinors on semisimple symmetric spaces*. J. Funct. Anal. **198** (2003), 536–557.
9. S. Mehdi and R. Zierau, *Principal series representations and harmonic spinors*. Adv. Math. **1999** (2006), 1–28.
10. S. Mehdi and R. Zierau, *The Dirac cohomology of a finite dimensional representation*. To appear in the Proceedings of the AMS. Proc. Amer. Math. Soc. **142** (2014), 1507–1512.
11. R. Parthasarathy, *Dirac operator and the discrete series*. Ann. of Math. **96** (1972), 1–30.
12. D. A. Vogan, *Unitarizability of certain series of representations*. Ann. of Math. **120** (1984), 141–187.
13. D. A. Vogan and G. J. Zuckerman, *Unitary representations with non-zero cohomology*. Comp. Math., **53** (1984), 51–90.
14. N. R. Wallach, *Real Reductive Groups I. Pure and Applied Mathematics* **132**, Academic press, 1988.
15. J. A. Wolf, *Partially harmonic spinors and representations of reductive Lie groups*. J. Functional Analysis **15** (1974), 117–154.

Twisted Harish–Chandra Sheaves and Whittaker Modules: The Nondegenerate Case

Dragan Miličić and Wolfgang Soergel

Abstract In this paper we develop a geometric approach to the study of the category of Whittaker modules. As an application, we reprove a well-known result of B. Kostant on the structure of the category of nondegenerate Whittaker modules.

Key words Whittaker modules • Localization of representations

Mathematics Subject Classification (2010): Primary 22E47. Secondary 14F10.

Introduction

Let $\mathfrak{g}$ be a complex semisimple Lie algebra, $\mathscr{U}(\mathfrak{g})$ its enveloping algebra and $\mathscr{Z}(\mathfrak{g})$ the center of $\mathscr{U}(\mathfrak{g})$. Let $\mathfrak{b}$ be a fixed Borel subalgebra of $\mathfrak{g}$ and $\mathfrak{n} = [\mathfrak{b}, \mathfrak{b}]$ its nilpotent radical. A Whittaker module is a finitely generated $\mathscr{U}(\mathfrak{g})$-module which is also $\mathscr{U}(\mathfrak{n})$-finite and $\mathscr{Z}(\mathfrak{g})$-finite. The category of Whittaker modules contains as a full subcategory the category of highest weight modules, and at the other extreme, the category of nondegenerate Whittaker modules (for the precise definition see Sect. 4). In his paper [5], Kostant shows that the category of nondegenerate Whittaker modules has an extremely simple structure. The main goal of this paper is to explain Kostant's result using geometric methods—we reprove it in Sect. 5.

Our idea was to use the localization theory of Beilinson and Bernstein [1] to transfer the study of Whittaker modules to the study of a particular category of $\mathscr{D}$-modules on the flag variety X of $\mathfrak{g}$. As explained in the first four sections of the paper, our methods actually work for arbitrary Whittaker modules. The localizations of Whittaker modules are holonomic, which immediately implies that

D. Miličić (✉)
Department of Mathematics, University of Utah, Salt Lake City, UT 84112, USA
e-mail: milicic@math.utah.edu

W. Soergel
Universität Freiburg, Mathematisches Institut, Eckerstraße 1, D-79104 Freiburg, Germany
e-mail: wolfgang.soergel@math.uni-freiburg.de

© Springer International Publishing Switzerland 2014
G. Mason et al. (eds.), *Developments and Retrospectives in Lie Theory*,
Developments in Mathematics 37, DOI 10.1007/978-3-319-09934-7_7

Whittaker modules are of finite length—this was proven before by McDowell [6]. He also proved that any irreducible Whittaker module is a quotient of a "standard" Whittaker module—these are a generalization of Verma modules. This leads to the natural problem of determining multiplicities of irreducible constituents of standard Whittaker modules.

Our project was started at Mathematical Sciences Research Institute in Berkeley, CA, in 1987–1988, during a special year in representation theory. The first draft of this paper and some of the early results on the multiplicity questions were obtained there. In particular, we realized at that time how simple is the geometric explanation of Kostant's result.

At that time the success of the geometric approach to prove the Kazhdan–Lusztig conjectures for Verma modules was based on the fact that the localizations of highest weight modules are holonomic modules with regular singularities; this made the standard techniques used in the study of composition series questions (like the decomposition theorem) applicable. We realized immediately that the localizations of Whittaker modules have irregular singularities. Therefore, at that time, we were unable to pursue the geometric analysis of the multiplicity problem any further. Still, assuming that the decomposition theorem holds for arbitrary irreducible holonomic modules, we were able to get a number of interesting conjectural statements about the structure of the category of Whittaker modules. The most important of these statements was later proved, by completely different algebraic methods, in [10].

Recently, Mochizuki proved the decomposition theorem in full generality and made our old geometric approach rigorous [11]. Still, we decided to publish this paper in its more-or-less original form to stress the simplicity of Kostant's result, deferring the general case to a future publication.

We were informed by Joseph Bernstein that he was aware that Kostant's result follows easily from localization theory.

1   Twisted Harish–Chandra Sheaves

Let K be a connected algebraic group with Lie algebra $\mathfrak{k}$ and ϕ a morphism of K into the group of inner automorphisms $\mathrm{Int}(\mathfrak{g})$ of $\mathfrak{g}$ such that its differential induces an injection of $\mathfrak{k}$ into $\mathfrak{g}$. Hence we can identify $\mathfrak{k}$ with a subalgebra of $\mathfrak{g}$. We say that $(\mathfrak{g}, K)$ is a *Harish–Chandra pair* if K acts by finitely many orbits on the flag variety X of $\mathfrak{g}$.

Fix a Harish–Chandra pair $(\mathfrak{g}, K)$ in the following.

Let $\eta : \mathfrak{k} \longrightarrow \mathbb{C}$ be a morphism of Lie algebras, i.e., a linear form on $\mathfrak{k}$ which vanishes on $[\mathfrak{k}, \mathfrak{k}]$.

An η-twisted Harish–Chandra module is a triple (π, ν, V) where:

(i) (π, V) is a finitely generated $\mathscr{U}(\mathfrak{g})$-module;
(ii) (ν, V) is an algebraic K-module;

(iii) the differential of the K-action on V induces a $\mathscr{U}(\mathfrak{k})$-module structure on V such that

$$\pi(\xi) = \nu(\xi) + \eta(\xi)$$

for any $\xi \in \mathfrak{k}$.

We denote by $\mathscr{M}_{fg}(\mathfrak{g}, K, \eta)$ the category of all η-twisted Harish–Chandra modules.

Let $\mathfrak{h}$ be the (abstract) Cartan algebra of $\mathfrak{g}$ [8, §2]. Let Σ be the root system in $\mathfrak{h}^*$ and W the corresponding Weyl group. Let $\lambda \in \mathfrak{h}^*$ and $\theta \in W \cdot \lambda$. By a theorem of Harish–Chandra, θ determines a maximal ideal J_θ in $\mathscr{Z}(\mathfrak{g})$. Let $\mathscr{U}_\theta$ be the quotient of $\mathscr{U}(\mathfrak{g})$ by the ideal generated by J_θ. Then we denote by $\mathscr{M}_{fg}(\mathscr{U}_\theta, K, \eta)$ the full subcategory of $\mathscr{M}_{fg}(\mathfrak{g}, K, \eta)$ consisting of modules which are actually $\mathscr{U}_\theta$-modules, i.e., they are annihilated by J_θ.

In [1], Beilinson and Bernstein construct, for each $\lambda \in \mathfrak{h}^*$, a twisted sheaf of differential operators $\mathscr{D}_\lambda$ on the flag variety X of $\mathfrak{g}$. For any $\lambda \in \theta$, the global sections $\Gamma(X, \mathscr{D}_\lambda)$ of $\mathscr{D}_\lambda$ are equal to $\mathscr{U}_\theta$.

As above, one can define the category $\mathscr{M}_{coh}(\mathscr{D}_\lambda, K, \eta)$ of coherent $\mathscr{D}_\lambda$-modules which also admit an algebraic action of K. Differentiation of the K-action gives an action of the Lie algebra $\mathfrak{k}$, we assume that it satisfies a compatibility condition analogous to (iii) (compare [4, Appendix B], [9, Section 4]). We call the objects of $\mathscr{M}_{coh}(\mathscr{D}_\lambda, K, \eta)$ η-twisted Harish–Chandra sheaves.

Clearly, the cohomology modules of η-twisted Harish–Chandra sheaves are η-twisted Harish–Chandra modules. Moreover, the localization functor Δ_λ given by $\Delta_\lambda(V) = \mathscr{D}_\lambda \otimes_{\mathscr{U}_\theta} V$ for a $\mathscr{U}_\theta$-module V, maps η-twisted Harish–Chandra modules into η-twisted Harish–Chandra sheaves.

Assume that Σ^+ is the set of positive roots in Σ such that at any point $x \in X$ it determines the nilpotent radical of the corresponding Borel subalgebra $\mathfrak{b}_x$. Let ρ be the half-sum of roots in Σ^+. We say that $\lambda \in \mathfrak{h}^*$ is *antidominant* if $\alpha^\vee(\lambda)$ is not a positive integer for any dual root $\alpha^\vee$ of $\alpha \in \Sigma^+$. For antidominant and regular λ, the categories $\mathscr{M}_{coh}(\mathscr{D}_\lambda, K, \eta)$ and $\mathscr{M}_{fg}(\mathscr{U}_\theta, K, \eta)$ are equivalent [8, 3.9].

The next result is proved exactly as in the non-twisted case [8, 6.1].

Lemma 1.1. *Any η-twisted Harish–Chandra sheaf is holonomic.*

Proof. Let $\mathrm{F}\,\mathscr{D}_\lambda$ be the natural degree filtration of $\mathscr{D}_\lambda$.

Let $\mathscr{V}$ be an η-twisted Harish–Chandra sheaf.

First we claim that there exists a good filtration $\{\mathrm{F}_n\,\mathscr{V} \mid n \in \mathbb{Z}_+\}$ of $\mathscr{V}$ with the additional property that all $\mathrm{F}_n\,\mathscr{V}$ are K-equivariant. By twisting by a homogeneous $\mathscr{O}_X$-module $\mathscr{O}(\nu)$, for a weight ν in the weight lattice of Σ, we can assume that λ is regular and antidominant. Then $V = \Gamma(X, \mathscr{V})$ is a finitely generated $\mathscr{U}_\theta$-module with algebraic action of K. This implies that it is generated by a finite-dimensional K-invariant subspace U. Since by the equivalence of categories we have $\mathscr{V} = \mathscr{D}_\lambda \otimes_{\mathscr{U}_\theta} V$, the images $\mathrm{F}_p\,\mathscr{V}$ of the morphisms

$$\mathrm{F}_p\,\mathscr{D}_\lambda \otimes_{\mathbb{C}} U \longrightarrow \mathscr{V}$$

define an exhaustive $\mathscr{D}_\lambda$-module filtration of $\mathscr{V}$ by K-equivariant coherent $\mathscr{O}_X$-submodules. It is evident that this is a good filtration of $\mathscr{V}$.

By the K-equivariance of the filtration we see that $\xi \cdot F_p \mathscr{V} \subset F_p \mathscr{V}$ for any $\xi \in \mathfrak{k} \subset \Gamma(X, \mathscr{D}_\lambda)$. This implies that the symbols of $\xi \in \mathfrak{k}$ annihilate $\mathrm{Gr}\,\mathscr{V}$. Since they vanish on the conormal bundle to any K-orbit in X, the characteristic variety $\mathrm{Ch}(\mathscr{V})$ of $\mathscr{V}$ is contained in the union of conormal bundles of K-orbits in X. Dimension of the conormal bundle to any K-orbit in X is equal to $\dim X$. Since the number of orbits is finite, the dimension of the union of all such conormal bundles is also equal to $\dim X$. This implies that $\dim \mathrm{Ch}(\mathscr{V}) \le \dim X$.   □

In particular, this implies that twisted Harish–Chandra sheaves are of finite length. In addition we get the following consequence.

Corollary 1.2. *Any η-twisted Harish–Chandra module is of finite length.*

Proof. Assume that $\lambda \in \theta$ is antidominant. Then the localization $\Delta_\lambda(V)$ of any module V in $\mathscr{M}_{fg}(\mathscr{U}_\theta, K, \eta)$ is in $\mathscr{M}_{coh}(\mathscr{D}_\lambda, K, \eta)$. Since this Harish–Chandra sheaf is of finite length by the above remark, and $V = \Gamma(X, \Delta_\lambda(V))$ [8, 3.6], the assertion follows from the exactness of Γ and the fact that global sections of an irreducible $\mathscr{D}_\lambda$-module are irreducible or zero [8, 3.8], [7, L.4.1].   □

The first example of the twisted Harish–Chandra modules was discussed in [4, Appendix B] in relation with localization theory of Harish–Chandra modules for semisimple Lie groups with infinite center.

The second example is related to Whittaker modules [5]. In this case, $K = N$. We discuss it in more details in Sect. 4.

2   A Category of $\mathfrak{n}$-Finite Modules

Let $\mathscr{N}$ be the full subcategory of the category of $\mathfrak{g}$-modules consisting of modules which are

(i) finitely generated $\mathscr{U}(\mathfrak{g})$-modules;
(ii) $\mathscr{Z}(\mathfrak{g})$-finite;
(iii) $\mathscr{U}(\mathfrak{n})$-finite.

Let $\theta = W \cdot \lambda$ be a Weyl group orbit in $\mathfrak{h}^*$ and J_θ the corresponding maximal ideal in $\mathscr{Z}(\mathfrak{g})$.

Let $\mathscr{N}_{\hat\theta}$ be the full subcategory of $\mathscr{N}$ consisting of modules annihilated by some power of J_θ, and $\mathscr{N}_\theta$ the full subcategory of $\mathscr{N}$ consisting of modules annihilated by J_θ. Since (i) and (ii) imply that the annihilator in $\mathscr{Z}(\mathfrak{g})$ of an object in $\mathscr{N}$ is of finite codimension, we have the following result.

Lemma 2.1. $\mathcal{N} = \bigoplus_{\theta \subset \mathfrak{h}^*} \mathcal{N}_{\hat{\theta}}$.

In other words, every object in $\mathcal{N}$ is a direct sum of finitely many objects in different $\mathcal{N}_{\hat{\theta}}$.

Let V be a $\mathcal{U}(\mathfrak{n})$-finite module. For $\eta \in \mathfrak{n}^*$ we put

$$V_\eta = \{v \in V \mid (\xi - \eta(\xi))^k v = 0, \ \xi \in \mathfrak{n}, \text{ for some } k \in \mathbb{N}\}.$$

Then $V_\eta \neq 0$ implies that $\eta|[\mathfrak{n}, \mathfrak{n}] = 0$ and $V = \bigoplus_{\eta \in \mathfrak{n}^*} V_\eta$ [3, Ch. VII, §1, no. 3, Prop. 9.(i)]. If V and W are two $\mathcal{U}(\mathfrak{n})$-finite modules, it is easy to check that $V_\eta \otimes W_{\eta'} \subset (V \otimes W)_{\eta+\eta'}$ for any η and η'. Assume now that $V \in \mathcal{N}$. Since the adjoint action of $\mathfrak{n}$ on $\mathfrak{g}$ is nilpotent, we have $\mathfrak{g} = \mathfrak{g}_0$. Hence we conclude that the natural map $\mathfrak{g} \otimes V \longrightarrow V$ given by $\xi \otimes v = \xi v$ maps $\mathfrak{g} \otimes V_\eta$ into V_η, i.e., V_η is a $\mathfrak{g}$-submodule. Denote by $\mathcal{N}_\eta$ the full subcategory of modules with the property $V = V_\eta$. Then we have the following result.

Lemma 2.2. $\mathcal{N} = \bigoplus_{\eta \in \mathfrak{n}^*} \mathcal{N}_\eta$.

In other words, every object in $\mathcal{N}$ is a direct sum of finitely many objects in different $\mathcal{N}_\eta$. Put $\mathcal{N}_{\theta,\eta} = \mathcal{N}_{\hat{\theta}} \cap \mathcal{N}_\eta$. Clearly, any irreducible object in $\mathcal{N}$ is in some $\mathcal{N}_{\theta,\eta}$.

Let V be an object in $\mathcal{N}_{\theta,\eta}$. Then $V \otimes \mathbb{C}_{-\eta}$ is a $\mathcal{U}(\mathfrak{n})$-finite module and clearly $V \otimes \mathbb{C}_{-\eta} = (V \otimes \mathbb{C}_{-\eta})_0$, i.e., for any $v \in V \otimes \mathbb{C}_{-\eta}$ we have $\mathfrak{n}^k \cdot v = 0$ for sufficiently large $k \in \mathbb{N}$. Therefore, the $\mathfrak{n}$-action is the differential of an algebraic action of N on $V \otimes \mathbb{C}_{-\eta}$. Using the natural isomorphism $V \longrightarrow V \otimes \mathbb{C}_{-\eta}$ given by $v \mapsto v \otimes 1$, we get an algebraic representation of N on V with differential which differs from the original action of $\mathfrak{n}$ by η, i.e., V is in $\mathcal{M}_{fg}(\mathcal{U}_\theta, N, \eta)$. This leads us to the following result.

Lemma 2.3. $\mathcal{N}_{\theta,\eta} = \mathcal{M}_{fg}(\mathcal{U}_\theta, N, \eta)$.

In particular, the localization functor Δ_λ maps $\mathcal{N}_{\theta,\eta}$ into $\mathcal{M}_{coh}(\mathcal{D}_\lambda, N, \eta)$. Hence, from Lemma 1.1 we deduce the following result.

Theorem 2.4. Localization $\Delta_\lambda(V)$ of a module V in $\mathcal{N}_{\hat{\theta}}$ is a holonomic $\mathcal{D}_\lambda$-module.

In particular, localizations are $\mathcal{D}_\lambda$-modules of finite length. This has the following consequence originally proved in [6].

Theorem 2.5. Any module in $\mathcal{N}$ is of finite length.

Proof. By Lemma 2.1 we can assume that V is in $\mathcal{N}_{\hat{\theta}}$. Moreover, such V has a finite filtration $F V$ such that $\mathrm{Gr}\, V$ is in $\mathcal{N}_\theta$. This reduces the proof to the case of $V \in \mathcal{N}_\theta$. Let $V \in \mathcal{N}_\theta$. By Corollary 1.2 and Lemma 2.3, we see that such V is of finite length. $\qquad \square$

3 Classification of Irreducible Twisted Harish–Chandra Sheaves

Again, this is a simple variant of the results in the non-twisted case [8, §6]. Let $\mathcal{V}$ be an irreducible object in $\mathcal{M}_{fg}(\mathcal{D}_\lambda, K, \eta)$. Then its support is an irreducible K-invariant subvariety of X. Therefore, it is the closure of a K-orbit Q in X. Let i : $Q \longrightarrow X$ be the natural inclusion. Then i is an affine immersion. The twisted sheaf of differential operators $\mathcal{D}_\lambda$ on X induces the twisted sheaf of differential operators $(\mathcal{D}_\lambda)^i = \mathcal{D}_{Q,\mu}$ on Q where μ is the restriction of the specialization of $\lambda + \rho$ to $\mathfrak{k} \cap \mathfrak{b}_x$ [4, Appendix A]. By Kashiwara's equivalence of categories, the inverse image $i^!(\mathcal{V})$ is an irreducible holonomic $\mathcal{D}_{Q,\mu}$-module [8, §4]. By the compatibility condition (iii), it is also a K-homogeneous $\mathcal{O}_Q$-module such that the differential of the K-action differs from the action of $\mathfrak{k}$ through $\mathcal{D}_{Q,\mu}$ by η. Since $i^!(\mathcal{V})$ is holonomic, $i^!(\mathcal{V})$ is a connection on a dense open subset of Q and therefore a coherent $\mathcal{O}$-module there. Since it is also K-equivariant, it must be coherent everywhere on Q, hence it is a connection on Q. We put $\tau = i^!(\mathcal{V})$ and denote by $\mathcal{I}(Q, \tau)$ the direct image of τ in $\mathcal{M}(\mathcal{D}_\lambda, K, \eta)$. The module $\mathcal{I}(Q, \tau)$ is called the *standard* Harish–Chandra sheaf attached to (Q, τ). The standard Harish–Chandra sheaf $\mathcal{I}(Q, \tau)$ has a unique irreducible Harish–Chandra subsheaf $\mathcal{L}(Q, \tau)$. The module $\mathcal{L}(Q, \tau)$ is isomorphic to $\mathcal{V}$.

Moreover, the quotient $\mathcal{I}(Q, \tau)/\mathcal{L}(Q, \tau)$ is a Harish–Chandra sheaf supported in the boundary of the closure of the K-orbit Q.

It remains to describe all η-twisted irreducible $\mathcal{D}_{Q,\mu}$-connections τ on the K-orbit Q. Let $x \in Q$. Let B_x be the Borel subgroup of $\mathrm{Int}(\mathfrak{g})$ with the Lie algebra $\mathfrak{b}_x$. Any K-homogeneous $\mathcal{O}_Q$-module is completely determined by the action of the stabilizer $S_x = \phi^{-1}(\phi(K) \cap B_x)$ in the geometric fiber at x. By the compatibility, the irreducibility of τ implies also that it is irreducible as a K-homogeneous $\mathcal{O}_Q$-module. Hence, the representation of S_x in the geometric fiber of τ is irreducible. Moreover, its differential is a direct sum of a number of copies of the linear form $\mu - \eta|(\mathfrak{k} \cap \mathfrak{b}_x)$ on $\mathfrak{k} \cap \mathfrak{b}_x$.

4 Whittaker Modules

Let $K = N$. Let $\mathfrak{b}$ be the unique Borel subalgebra of $\mathfrak{g}$ containing $\mathfrak{n}$. Then N-orbits are the Bruhat cells $C(w)$, $w \in W$, with respect to $\mathfrak{b}$, i.e., each cell $C(w)$ consists of all Borel subalgebras in relative position w with respect to $\mathfrak{b}$. Let $\mathfrak{b}_w$ be one of such Borel subalgebras in $C(w)$. Fix a Cartan subalgebra $\mathfrak{c}$ of $\mathfrak{g}$ contained in $\mathfrak{b} \cap \mathfrak{b}_w$. Let R be the root system of $(\mathfrak{g}, \mathfrak{c})$ and R^+ the set of positive roots determined by $\mathfrak{n}$. Denote by $s : \mathfrak{h}^* \longrightarrow \mathfrak{c}^*$ the specialization determined by $\mathfrak{b}$ [8, §2]. Then $\mathfrak{n}_w = [\mathfrak{b}_w, \mathfrak{b}_w]$ is spanned by the root subspaces corresponding to the roots in $s^{-1}(w(\Sigma^+))$.

Now we want to discuss the compatibility condition from the end of the last section in this special case. Assume that a Bruhat cell $C(w)$ admits an irreducible

N-homogeneous connection. First, $\mathfrak{n} \cap \mathfrak{b}_w \subset \mathfrak{n}_w$, hence we have $\mu = 0$. Also, since the stabilizer of $\mathfrak{b}_w$ in N is unipotent, the only irreducible N-homogeneous $\mathcal{O}_{C(w)}$-module on $C(w)$ is $\mathcal{O}_{C(w)}$. Therefore, a connection with the properties described in Sect. 3 exists on $C(w)$ if and only if $\eta|(\mathfrak{n} \cap \mathfrak{n}_w) = 0$. Moreover, it is isomorphic to $\mathcal{O}_{C(w)}$. By abuse of notation, for $\alpha \in \Sigma$ we denote by $\mathfrak{g}_\alpha$ the root subspace in $\mathfrak{g}$ corresponding to the root $s^{-1}\alpha \in R$. Then the subalgebra $\mathfrak{n} \cap \mathfrak{n}_w$ is spanned by the root subspaces $\mathfrak{g}_\alpha$ for $\alpha \in \Sigma^+ \cap w(\Sigma^+)$. Hence $\eta|(\mathfrak{n} \cap \mathfrak{n}_w) = 0$ if and only if $\eta|\mathfrak{g}_\alpha = 0$ for $\alpha \in \Sigma^+ \cap w(\Sigma^+)$.

Let Π be the set of simple roots in Σ corresponding to Σ^+. The root subspaces $\mathfrak{g}_\alpha$, $\alpha \in \Pi$, span a complement of $[\mathfrak{n}, \mathfrak{n}]$ in $\mathfrak{n}$. Therefore, $\eta|(\mathfrak{n} \cap \mathfrak{n}_w) = 0$ if and only if $\eta|\mathfrak{g}_\alpha = 0$ for $\alpha \in \Pi \cap w(\Sigma^+)$.

Let $\ell : W \longrightarrow \mathbb{Z}_+$ be the length function on W with respect to the reflections s_α, $\alpha \in \Pi$. Then, for any $w \in W$, we have $\ell(w) = \dim C(w)$.

Let $\Theta \subset \Pi$, and let W_Θ be the subgroup of W generated by the reflections with respect to $\alpha \in \Theta$. The set of simple roots Θ determines also a standard parabolic subalgebra $\mathfrak{p}_\Theta$ containing $\mathfrak{b}$. Let P_Θ be the corresponding parabolic subgroup in $\mathrm{Int}(\mathfrak{g})$. Any P_Θ-orbit in X is a disjoint union of Bruhat cells $C(tv)$, $t \in W_\Theta$, for some $v \in W$. In this way, we obtain a bijection between the P_Θ-orbits in X and right W_Θ-cosets of W.

The following result is well known.

Lemma 4.1. *The following conditions are equivalent:*

(i) $\Theta \cap w(\Sigma^+) = \emptyset$;
(ii) $C(w)$ is the Bruhat cell open in one of the P_Θ-orbits in X;
(iii) w is the longest element in one of the right W_Θ-cosets of W.

Proof. By the above discussion, (ii) and (iii) are equivalent.

(ii)$\Rightarrow$(i) Fix a P_Θ-orbit O and let $C(w)$ be the Bruhat cell open in O. Then we have

$$\dim O = \dim C(w) = \ell(w) > \ell(s_\alpha w), \text{ for any } \alpha \in \Theta.$$

Since $\ell(v) = \mathrm{Card}(\Sigma^+ \cap (-v(\Sigma^+)))$ for any $v \in W$ [3, Ch. VI, §1, no. 6, Cor. 2 of Prop. 17], this means that

$$\mathrm{Card}(s_\alpha \Sigma^+ \cap (-w(\Sigma^+))) = \mathrm{Card}(\Sigma^+ \cap (-s_\alpha w(\Sigma^+))) < \mathrm{Card}(\Sigma^+ \cap (-w(\Sigma^+))),$$

for all $\alpha \in \Theta$. Since s_α permutes all roots in $\Sigma^+ - \{\alpha\}$, it follows that $\alpha \notin w(\Sigma^+)$ for all $\alpha \in \Theta$ and w satisfies (i).

(i)$\Rightarrow$(ii) Let Σ_Θ be the root subsystem of Σ generated by Θ. Let T be the set of roots in Σ^+ which are not in Σ_Θ. Since for any $\alpha \in \Theta$, s_α permutes the positive roots in $\Sigma^+ - \{\alpha\}$, it follows that $s_\alpha(T) \subset \Sigma^+$. On the other hand, s_α also permutes roots in the complement of Σ_Θ, i.e., $s_\alpha(T) \cap \Sigma_\Theta = \emptyset$. Therefore, $s_\alpha(T) = T$. Since W_Θ is generated by the reflections with respect to Θ, we see that T is W_Θ-invariant.

Assume that $w \in W$ satisfies (i). Then $S = \Sigma^+ \cap w(\Sigma^+)$ is disjoint from Θ. We claim that $\Sigma_\Theta \cap S = \emptyset$. Assume that $\beta \in \Sigma_\Theta \cap S$. Then $\beta \in \Sigma^+$, and it must be a sum of roots from Θ. But $\Theta \subset -w(\Sigma^+)$, hence $\beta \in -w(\Sigma^+)$ and we have a contradiction. It follows that $S \subset T$. Hence, $t(S) \subset T \subset \Sigma^+$ for any $t \in W_\Theta$. In particular, for any $t \in W_\Theta$, we have $t(S) \subset \Sigma^+ \cap tw(\Sigma^+)$. It follows that

$$\ell(tw) = \mathrm{Card}(\Sigma^+ \cap (-tw(\Sigma^+))) = \mathrm{Card}(\Sigma^+) - \mathrm{Card}(\Sigma^+ \cap tw(\Sigma^+))$$

$$\leq \mathrm{Card}(\Sigma^+) - \mathrm{Card}(S) = \ell(w)$$

for any $t \in W_\Theta$, i.e., $C(w)$ is the Bruhat cell of maximal dimension among the cells contained in O. ☐

Now let

$$\Theta = \{\alpha \in \Pi \mid \eta|_{\mathfrak{g}_\alpha} \neq 0\}.$$

As we already remarked, a compatible irreducible connection exists on $C(w)$ if and only if $\Theta \cap w(\Sigma^+) = \emptyset$. By Lemma 4.1, this is true if and only if $C(w)$ is the open Bruhat cell in one of P_Θ-orbits in X. This leads to the following result.

For a Bruhat cell $C(w)$ with the compatible irreducible connection $\mathscr{O}_{C(w)}$ denote by $\mathscr{I}(w, \lambda, \eta)$ the corresponding standard η-twisted Harish–Chandra sheaf and by $\mathscr{L}(w, \lambda, \eta)$ the corresponding irreducible η-twisted Harish–Chandra sheaf.

Theorem 4.2. *The irreducible objects in the category $\mathscr{M}_{coh}(\mathscr{D}_\lambda, N, \eta)$ are the modules $\mathscr{L}(w, \lambda, \eta)$ where $w \in W$ is such that $C(w)$ is an open Bruhat cell in a P_Θ-orbit in X.*

We can also show, that for an antidominant λ, the "costandard" Harish–Chandra sheaves $\mathscr{M}(w, \lambda, \eta)$, which are obtained from standard Harish–Chandra sheaves $\mathscr{I}(w, \lambda, \eta)$ by an appropriate holonomic dualization process, correspond under localization to the "standard" Whittaker modules studied in [6]. We are going to discuss this in a subsequent paper.

As we mentioned in the introduction, the objects in $\mathscr{M}_{coh}(\mathscr{D}_\lambda, N, \eta)$ have irregular singularities in general. This is clearly visible from the following example.

Let $\mathfrak{g} = \mathfrak{sl}(2, \mathbb{C})$ and

$$N = \left\{ \begin{pmatrix} 1 & 0 \\ x & 1 \end{pmatrix} \;\middle|\; x \in \mathbb{C} \right\}.$$

The flag variety X of $\mathfrak{g}$ is identified with $\mathbb{P}^1$. Let x be a point in X. The Borel subalgebra $\mathfrak{b}_x$ in $\mathfrak{g}$ is the stabilizer of the line in $\mathbb{C}^2$ determined by x. Therefore, $\mathfrak{n}$ is the nilpotent radical of $\mathfrak{b}_\infty$ and $\mathbb{C} \subset \mathbb{P}^1$ is the open N-orbit in X. Let $\lambda = -\rho$. Then $\mathscr{D}_{-\rho} = \mathscr{D}_X$ is the sheaf of differential operators on X. The matrix

$$\begin{pmatrix} 1 & 0 \\ x & 1 \end{pmatrix} \in N$$

moves 0 into x. Hence it defines an isomorphism of N onto $\mathbb{C}$. Also, if ∂ denotes differentiation with respect to z considered as a vector field on $\mathbb{C}$, then

$$\xi = \begin{pmatrix} 0 & 0 \\ 1 & 0 \end{pmatrix} \in \mathfrak{n}$$

corresponds to ∂ under the above isomorphism.

Let $\mathscr{I}_\eta$ be the standard η-twisted Harish–Chandra sheaf attached to the open orbit. Then the restriction of $\mathscr{I}_\eta$ to $\mathbb{C}$ is isomorphic to the quotient of $\mathscr{D}_{\mathbb{C}}$ by the left ideal generated by $\partial - \eta(\xi)$. If $\eta \neq 0$, this is a connection on $\mathbb{C}$ which has an irregular singularity at infinity.

5 The Nondegenerate Case

We say that η is *nondegenerate* if $\eta|\mathfrak{g}_\alpha \neq 0$ for $\alpha \in \Pi$. In this case $\Theta = \Pi$ and $P_\Theta = G$. Let w_0 be the longest element of the Weyl group W. By Theorem 4.2, in this case there exists only one irreducible object $\mathscr{L}_\lambda = \mathscr{L}(w_0, \lambda, \eta)$ in $\mathscr{M}_{coh}(\mathscr{D}_\lambda, N, \eta)$. Since the quotient of the corresponding standard Harish–Chandra sheaf $\mathscr{I}_\lambda$ by $\mathscr{L}_\lambda$ must be supported in the complement of the big cell, it must be zero. Hence, we conclude that $\mathscr{L}_\lambda$ is equal to the standard Harish–Chandra sheaf $\mathscr{I}_\lambda$, i.e., $\mathscr{I}_\lambda$ is irreducible.

Moreover, the space of global sections of $\mathscr{I}_\lambda$ is equal to the space $R(C(w_0))$ of regular functions on the affine variety $C(w_0)$. Therefore, from [8, 3.8], we see that $\Gamma(X, \mathscr{I}_\lambda)$ is an irreducible Whittaker module for any antidominant $\lambda \in \mathfrak{h}^*$.

This also implies that in this case there exists a unique irreducible object in the category $\mathscr{N}_{\theta, \eta}$.

Now we want to describe these modules. Clearly, the function 1 on X determines a global section of $\mathscr{I}_\lambda$ and it spans an $\mathfrak{n}$-stable subspace of $\Gamma(X, \mathscr{I}_\lambda)$ on which $\mathfrak{n}$ acts by η. Since $\mathscr{I}_\lambda$ is irreducible, this leads to an epimorphism of the coherent $\mathscr{D}_\lambda$-module $\mathscr{D}_\lambda \otimes_{\mathscr{U}(\mathfrak{n})} \mathbb{C}_\eta$ onto $\mathscr{I}_\lambda$. Clearly, with the N-action given by the tensor product of the natural action on $\mathscr{D}_\lambda$ with the trivial action on $\mathbb{C}$, $\mathscr{D}_\lambda \otimes_{\mathscr{U}(\mathfrak{n})} \mathbb{C}_\eta$ becomes an η-twisted Harish–Chandra sheaf. Moreover, since an orbit map from N into $C(w_0)$ is an isomorphism, the restriction to the big cell $C(w_0)$ is an epimorphism of the $\mathscr{O}_{C(w_0)}$-module

$$(\mathscr{D}_\lambda \otimes_{\mathscr{U}(\mathfrak{n})} \mathbb{C}_\eta) \mid C(w_0) \cong \mathscr{D}_{C(w_0)} \otimes_{\mathscr{U}(\mathfrak{n})} \mathbb{C}_\eta \cong \mathscr{O}_{C(w_0)} \otimes_{\mathbb{C}} \mathbb{C} \cong \mathscr{O}_{C(w_0)}$$

onto the irreducible connection $\mathscr{I}_\lambda | C(w_0)$. Therefore, this is an isomorphism. It follows that the kernel of the morphism of $\mathscr{D}_\lambda \otimes_{\mathscr{U}(\mathfrak{n})} \mathbb{C}_\eta$ onto $\mathscr{I}_\lambda$ is supported on the complement of the big cell. But as we remarked before, the only object which can be supported there is 0. This implies that $\mathscr{D}_\lambda \otimes_{\mathscr{U}(\mathfrak{n})} \mathbb{C}_\eta = \mathscr{I}_\lambda$. This finally proves the following result.

Theorem 5.1. *Let $\eta \in \mathfrak{n}$ be nondegenerate and $\lambda \in \mathfrak{h}^*$. Then the only irreducible object in $\mathcal{M}_{coh}(\mathcal{D}_\lambda, N, \eta)$ is $\mathcal{D}_\lambda \otimes_{\mathcal{U}(\mathfrak{n})} \mathbb{C}_\eta$.*

We also have an analogous result for $\mathcal{N}_{\theta,\eta}$. It was originally proved by Kostant in his work on Whittaker modules [5].

Theorem 5.2. *Let $\eta \in \mathfrak{n}$ be nondegenerate. Then the only irreducible module in $\mathcal{N}_{\theta,\eta}$ is $\mathcal{U}_\theta \otimes_{\mathcal{U}(\mathfrak{n})} \mathbb{C}_\eta$.*

Proof. Let $\lambda \in \theta$ be antidominant. Then

$$\Delta_\lambda(\mathcal{U}_\theta \otimes_{\mathcal{U}(\mathfrak{n})} \mathbb{C}_\eta) = \mathcal{D}_\lambda \otimes_{\mathcal{U}_\theta} (\mathcal{U}_\theta \otimes_{\mathcal{U}(\mathfrak{n})} \mathbb{C}_\eta) = \mathcal{D}_\lambda \otimes_{\mathcal{U}(\mathfrak{n})} \mathbb{C}_\eta = \mathcal{I}_\lambda.$$

Therefore, by [8, 3.6], we have

$$\Gamma(X, \mathcal{I}_\lambda) = \mathcal{U}_\theta \otimes_{\mathcal{U}(\mathfrak{n})} \mathbb{C}_\eta$$

and this is the unique irreducible object in $\mathcal{M}_{fg}(\mathcal{U}_\theta, N, \eta)$.   □

A vector in a Whittaker module which spans an $\mathfrak{n}$-stable subspace is called a Whittaker vector.

Corollary 5.3. *All Whittaker vectors in $\mathcal{U}_\theta \otimes_{\mathcal{U}(\mathfrak{n})} \mathbb{C}_\eta$ are proportional to $1 \otimes 1$.*

Proof. By the preceding argument we see that Whittaker vectors correspond exactly to N-invariant sections of $\mathcal{I}_\lambda$. These sections are exactly constant functions on the open cell $C(w_0)$.   □

Since the global sections of $\mathcal{D}_\lambda$ clearly operate faithfully on global sections of $\mathcal{I}_\lambda$ we get the following consequence [5].

Corollary 5.4. *The action of $\mathcal{U}_\theta$ on $\mathcal{U}_\theta \otimes_{\mathcal{U}(\mathfrak{n})} \mathbb{C}_\eta$ is faithful.*

Consider now an arbitrary object $\mathcal{V}$ in $\mathcal{M}_{coh}(\mathcal{D}_\lambda, N, \eta)$. Its restriction onto the big cell is an N-homogeneous connection. Since the stabilizer in N of an arbitrary point in the big cell is trivial, this connection is equal to a sum of copies of the irreducible connection on $C(w_0)$. Since the restriction is left adjoint to direct image, this implies that there exists a natural morphism φ of $\mathcal{V}$ into a sum of copies of $\mathcal{I}_\lambda$. By the preceding discussion, since the kernel and the cokernel of φ are supported in the complement of the big cell, they are equal to zero. This leads to the following results which show the extreme simplicity of the categories $\mathcal{M}_{coh}(\mathcal{D}_\lambda, N, \eta)$ and $\mathcal{N}_{\theta,\eta}$ for nondegenerate η.

Theorem 5.5. *Let $\eta \in \mathfrak{n}$ be nondegenerate. Then all objects in $\mathcal{M}_{coh}(\mathcal{D}_\lambda, N, \eta)$ are finite sums of irreducible objects $\mathcal{D}_\lambda \otimes_{\mathcal{U}(\mathfrak{n})} \mathbb{C}_\eta$.*

Theorem 5.6. *Let $\eta \in \mathfrak{n}$ be nondegenerate. Then modules in $\mathcal{N}_{\theta,\eta}$ are finite sums of irreducible modules $\mathcal{U}_\theta \otimes_{\mathcal{U}(\mathfrak{n})} \mathbb{C}_\eta$.*

Now we want to describe the structure of the category $\mathcal{N}_\eta$ for a nondegenerate $\eta \in \mathfrak{n}^*$.

We start with a simple technical result. The enveloping algebra $\mathcal{U}(\mathfrak{g})$ has a natural structure of a left $\mathscr{Z}(\mathfrak{g}) \otimes_{\mathbb{C}} \mathcal{U}(\mathfrak{n})$-module given by left multiplication. The following generalization of a classical result of Kostant must be well known.[1]

Lemma 5.7. $\mathcal{U}(\mathfrak{g})$ *is free as a* $\mathscr{Z}(\mathfrak{g}) \otimes_{\mathbb{C}} \mathcal{U}(\mathfrak{n})$-*module.*

Proof. Let $(\mathcal{U}_p(\mathfrak{g}); \ p \in \mathbb{Z}_+)$ denote the natural filtration of the enveloping algebra $\mathcal{U}(\mathfrak{g})$ of $\mathfrak{g}$.

Fix a Cartan subalgebra $\mathfrak{c}$ and a nilpotent subalgebra $\bar{\mathfrak{n}}$ opposite to $\mathfrak{n}$. Then we have $\mathfrak{g} = \mathfrak{n} \oplus \mathfrak{c} \oplus \bar{\mathfrak{n}}$, and by the Poincaré–Birkhoff–Witt theorem it follows that $\mathcal{U}(\mathfrak{g}) = \mathcal{U}(\mathfrak{n}) \otimes_{\mathbb{C}} \mathcal{U}(\mathfrak{c}) \otimes_{\mathbb{C}} \mathcal{U}(\bar{\mathfrak{n}})$ as a left $\mathcal{U}(\mathfrak{n})$-module for left multiplication. Then we define a linear space filtration $F \mathcal{U}(\mathfrak{g})$ of $\mathcal{U}(\mathfrak{g})$ via

$$F_p \mathcal{U}(\mathfrak{g}) = \mathcal{U}(\mathfrak{n}) \otimes_{\mathbb{C}} \mathcal{U}_p(\mathfrak{c}) \otimes_{\mathbb{C}} \mathcal{U}(\bar{\mathfrak{n}}).$$

Clearly, the natural filtration of $\mathcal{U}(\mathfrak{g})$ is finer than $F \mathcal{U}(\mathfrak{g})$, i.e., $\mathcal{U}_p(\mathfrak{g}) \subset F_p \mathcal{U}(\mathfrak{g})$ for all $p \in \mathbb{Z}_+$.

We define a filtration on $\mathscr{Z}(\mathfrak{g}) \otimes_{\mathbb{C}} \mathcal{U}(\mathfrak{n})$, by

$$F_p(\mathscr{Z}(\mathfrak{g}) \otimes_{\mathbb{C}} \mathcal{U}(\mathfrak{n})) = (\mathcal{U}_p(\mathfrak{g}) \cap \mathscr{Z}(\mathfrak{g})) \otimes_{\mathbb{C}} \mathcal{U}(\mathfrak{n}), \ p \in \mathbb{Z}_+.$$

In this way, $\mathscr{Z}(\mathfrak{g}) \otimes_{\mathbb{C}} \mathcal{U}(\mathfrak{n})$ becomes a filtered ring. The corresponding graded ring $\mathrm{Gr}(\mathscr{Z}(\mathfrak{g}) \otimes_{\mathbb{C}} \mathcal{U}(\mathfrak{n}))$ is equal to $\mathrm{Gr}(\mathscr{Z}(\mathfrak{g})) \otimes_{\mathbb{C}} \mathcal{U}(\mathfrak{n})$. The Harish–Chandra homomorphism $\gamma : \mathscr{Z}(\mathfrak{g}) \longrightarrow \mathcal{U}(\mathfrak{c})$ is compatible with the natural filtrations and the homomorphism $\mathrm{Gr}\,\gamma$ is an isomorphism of $\mathrm{Gr}\,\mathscr{Z}(\mathfrak{g})$ onto the subalgebra $I(\mathfrak{c})$ of all W-invariants in $S(\mathfrak{c})$ [3, Ch. VIII, §8, no. 5]. Therefore,

$$\mathrm{Gr}(\mathscr{Z}(\mathfrak{g}) \otimes_{\mathbb{C}} \mathcal{U}(\mathfrak{n})) = I(\mathfrak{c}) \otimes_{\mathbb{C}} \mathcal{U}(\mathfrak{n}).$$

Let $z \in \mathcal{U}_p(\mathfrak{g}) \cap \mathscr{Z}(\mathfrak{g})$. Then by the definition of the Harish–Chandra homomorphism, we have $z - \gamma(z) \in \mathfrak{n}\mathcal{U}_{p-1}(\mathfrak{g})$. Hence, we have

$$z\mathcal{U}_q(\mathfrak{c}) \subset \gamma(z)\mathcal{U}_q(\mathfrak{c}) + \mathfrak{n}\mathcal{U}_{p-1}(\mathfrak{g})\mathcal{U}_q(\mathfrak{c}) \subset \gamma(z)\mathcal{U}_q(\mathfrak{c}) + \mathfrak{n}\mathcal{U}_{p+q-1}(\mathfrak{g})$$

$$\subset \gamma(z)\mathcal{U}_q(\mathfrak{c}) + F_{p+q-1}\,\mathcal{U}(\mathfrak{g}) \subset F_{p+q}\,\mathcal{U}(\mathfrak{g})$$

for any $q \in \mathbb{Z}_+$. This implies first that

$$z\,F_q\,\mathcal{U}(\mathfrak{g}) \subset F_{p+q}\,\mathcal{U}(\mathfrak{g}), \ q \in \mathbb{Z}_+;$$

i.e., the filtration $F \mathcal{U}(\mathfrak{g})$ is compatible with the action of $\mathscr{Z}(\mathfrak{g}) \otimes_{\mathbb{C}} \mathcal{U}(\mathfrak{n})$. Therefore, $\mathcal{U}(\mathfrak{g})$ is a filtered $\mathscr{Z}(\mathfrak{g}) \otimes_{\mathbb{C}} \mathcal{U}(\mathfrak{n})$-module. Moreover, the corresponding graded module is

$$\mathrm{Gr}\,\mathcal{U}(\mathfrak{g}) = \mathcal{U}(\mathfrak{n}) \otimes_{\mathbb{C}} S(\mathfrak{c}) \otimes_{\mathbb{C}} \mathcal{U}(\bar{\mathfrak{n}})$$

---

[1] One of us learned this argument to prove Kostant's result from Wilfried Schmid in 1977.

with the obvious action of $I(\mathfrak{c}) \otimes_{\mathbb{C}} \mathscr{U}(\mathfrak{n})$. Since $S(\mathfrak{c})$ is a free $I(\mathfrak{c})$-module by [3, Ch. V, §5, no. 5, Thm. 4], it follows that $\mathrm{Gr}\,\mathscr{U}(\mathfrak{g})$ is a free $I(\mathfrak{c}) \otimes_{\mathbb{C}} \mathscr{U}(\mathfrak{n})$-module.

This easily implies that $\mathscr{U}(\mathfrak{g})$ is a free $\mathscr{Z}(\mathfrak{g}) \otimes_{\mathbb{C}} \mathscr{U}(\mathfrak{n})$-module [2, Ch. III, §2, no. 8, Cor. 3. of Thm. 1]. $\qquad\square$

Let U be a finite-dimensional $\mathscr{Z}(\mathfrak{g})$-module. Consider it as a $\mathscr{Z}(\mathfrak{g}) \otimes_{\mathbb{C}} \mathscr{U}(\mathfrak{n})$-module, where $\mathfrak{n}$ acts by multiplication by η. Let

$$I_\eta(U) = \mathscr{U}(\mathfrak{g}) \otimes_{\mathscr{Z}(\mathfrak{g}) \otimes_{\mathbb{C}} \mathscr{U}(\mathfrak{n})} U;$$

we consider it as a $\mathscr{U}(\mathfrak{g})$-module by left multiplication in the first factor. By the preceding lemma, the functor I_η from the category of finite-dimensional $\mathscr{Z}(\mathfrak{g})$-modules into the category of $\mathscr{U}(\mathfrak{g})$-modules is exact. It maps finite-dimensional $\mathscr{Z}(\mathfrak{g})$-modules into $\mathscr{Z}(\mathfrak{g})$-finite, finitely generated $\mathscr{U}(\mathfrak{g})$-modules. Moreover, since the action of $\mathfrak{n}$ on $I_\eta(U)$ is the quotient of action on the tensor product $\mathscr{U}(\mathfrak{g}) \otimes_{\mathbb{C}} U$ where $\mathfrak{n}$ acts on the first factor by the adjoint action, we see immediately that $I_\eta(U)$ is $\mathscr{U}(\mathfrak{n})$-finite. Hence, it is a Whittaker module. In addition, since the action of $\mathfrak{n}$ on $\mathscr{U}(\mathfrak{g})$ is nilpotent, we conclude that $I_\eta(U)$ is in $\mathscr{N}_\eta$. Therefore I_η is an exact functor from the category of finite-dimensional $\mathscr{Z}(\mathfrak{g})$-modules into the category $\mathscr{N}_\eta$.

Assume that $\dim U = 1$. Then a maximal ideal in $\mathscr{Z}(\mathfrak{g})$ annihilates U. Hence we see that $I_\eta(U) = \mathscr{U}_\theta \otimes_{\mathscr{U}(\mathfrak{n})} \mathbb{C}_\eta$ for some Weyl group orbit θ in $\mathfrak{h}^*$. Moreover, $I_\eta(U)$ is irreducible by Theorem 5.2.

By the exactness of the functor I_η, we immediately conclude that

$$\mathrm{length}\, I_\eta(U) = \dim U$$

for any finite-dimensional $\mathscr{Z}(\mathfrak{g})$-module U.

On the other hand, for a Whittaker module V in $\mathscr{N}_\eta$, let $\mathrm{Wh}(V)$ denote the space of all Whittaker vectors. Clearly, $\mathrm{Wh}(V)$ is $\mathscr{Z}(\mathfrak{g})$-invariant.

Lemma 5.8. *The functor* Wh *from the category* $\mathscr{N}_\eta$ *into the category of* $\mathscr{Z}(\mathfrak{g})$-*modules is exact.*

Proof. The subspace $\mathrm{Wh}(V)$ of Whittaker vectors in V can be identified with the module of $\mathfrak{n}$-invariants of V with respect to the ν-action. Therefore, it is enough to prove that $H^1(\mathfrak{n}, V) = 0$ for any Whittaker module V, where the cohomology is calculated with respect to the ν-action.

Consider first the case of irreducible Whittaker modules. As we remarked before an irreducible Whittaker module (with ν-action) is isomorphic to the algebra of regular functions $R(C(w_0))$ on open cell $C(w_0)$ with the natural action of N. Since an orbit map is an isomorphism of N onto the open cell, it is enough to know that groups $H^i(\mathfrak{n}, R(N)) = 0$ for $i \geq 1$. This is a well-known fact (compare [9, Lemma 1.9], for example).

Consider now an arbitrary Whittaker module V. Let V' be an irreducible submodule of V and consider the exact sequence

$$0 \longrightarrow V' \longrightarrow V \longrightarrow V'' \longrightarrow 0.$$

From the long exact sequence of Lie algebra cohomology, we see that $H^i(\mathfrak{n}, V) = H^i(\mathfrak{n}, V'')$ for $i \geq 1$. Hence, by induction on the length of V, we conclude that $H^i(\mathfrak{n}, V) = 0$ for $i \geq 1$. $\qquad\qquad\square$

By Corollary 5.3, we see that for any irreducible Whittaker module in $\mathcal{N}_\eta$, the vector space $\mathrm{Wh}(V)$ is one-dimensional. Therefore, by induction on the length of Whittaker modules and using the exactness of Wh, we immediately get

$$\dim \mathrm{Wh}(V) = \mathrm{length}(V)$$

for any V in $\mathcal{N}_\eta$. In particular, $\mathrm{Wh}(V)$ is finite-dimensional. Therefore, Wh is an exact functor from the category $\mathcal{N}_\eta$ into the category of finite-dimensional $\mathscr{L}(\mathfrak{g})$-modules. It is easy to check that

$$\mathrm{Hom}_{\mathscr{U}(\mathfrak{g})}(I_\eta(U), V) = \mathrm{Hom}_{\mathscr{L}(\mathfrak{g})}(U, \mathrm{Wh}(V)),$$

i.e., the functor Wh is the right adjoint of I_η.

Clearly, the linear map $u \longmapsto 1 \otimes u$ from U into $I_\eta(U)$ is injective, and its image is in $\mathrm{Wh}(I_\eta(U))$. Since

$$\dim \mathrm{Wh}(I_\eta(U)) = \mathrm{length}\, I_\eta(U) = \dim U,$$

it follows that the adjointness morphism $U \longrightarrow \mathrm{Wh}(I_\eta(U))$ is an isomorphism. Conversely, let V be a Whittaker module in $\mathcal{N}_\eta$. Then we have the adjointness morphism $I_\eta(\mathrm{Wh}(V)) \longrightarrow V$. Let K be its kernel and C the cokernel. Then we have the exact sequence

$$0 \longrightarrow \mathrm{Wh}(K) \longrightarrow \mathrm{Wh}(I_\eta(\mathrm{Wh}(V))) \longrightarrow \mathrm{Wh}(V) \longrightarrow \mathrm{Wh}(C) \longrightarrow 0$$

and by the preceding remark, the third arrow is an isomorphism. Hence, $\mathrm{Wh}(K) = 0$ and $\mathrm{Wh}(C) = 0$, i.e., $K = C = 0$ and $I_\eta(\mathrm{Wh}(V)) \longrightarrow V$ is an isomorphism. Therefore, we established the following result.

Theorem 5.9. *Let $\eta \in \mathfrak{n}^*$ be nondegenerate. Then the functor I_η is an equivalence of the category of finite-dimensional $\mathscr{L}(\mathfrak{g})$-modules with $\mathcal{N}_\eta$. Its quasi-inverse is the functor* Wh.

References

1. Alexander Beĭlinson and Joseph Bernstein, *Localisation de $\mathfrak{g}$-modules*, C. R. Acad. Sci. Paris Sér. I Math. **292** (1981), no. 1, 15–18.
2. Nicolas Bourbaki, *Algèbre commutative*, Mason, Paris.
3. _____ , *Groupes et algèbres de Lie*, Mason, Paris.

4. Henryk Hecht, Dragan Miličić, Wilfried Schmid, and Joseph A. Wolf, *Localization and standard modules for real semisimple Lie groups. I. The duality theorem*, Invent. Math. **90** (1987), no. 2, 297–332.
5. Bertram Kostant, *On Whittaker vectors and representation theory*, Invent. Math. **48** (1978), 101–184.
6. Edward McDowell, *On modules induced from Whittaker modules*, J. Algebra **96** (1985), no. 1, 161–177.
7. Dragan Miličić, *Localization and representation theory of reductive Lie groups*, unpublished manuscript available at http://math.utah.edu/~milicic.
8. _____, *Algebraic 𝒟-modules and representation theory of semisimple Lie groups*, The Penrose transform and analytic cohomology in representation theory (South Hadley, MA, 1992), Amer. Math. Soc., Providence, RI, 1993, pp. 133–168.
9. Dragan Miličić and Pavle Pandžić, *Equivariant derived categories, Zuckerman functors and localization*, Geometry and representation theory of real and *p*-adic groups (Córdoba, 1995), Progr. Math., Vol. 158, Birkhäuser Boston, Cambridge, MA, 1998, pp. 209–242.
10. Dragan Miličić and Wolfgang Soergel, *The composition series of modules induced from Whittaker modules*, Comment. Math. Helv. **72** (1997), no. 4, 503–520.
11. Takuro Mochizuki, *Wild harmonic bundles and wild pure twistor D-modules*, Astérisque (2011), no. 340.

Unitary Representations of Unitary Groups

Karl-Hermann Neeb

Abstract In this paper we review and streamline some results of Kirillov, Olshanski and Pickrell on unitary representations of the unitary group $U(\mathcal{H})$ of a real, complex or quaternionic separable Hilbert space and the subgroup $U_\infty(\mathcal{H})$, consisting of those unitary operators g for which $g - 1$ is compact. The Kirillov–Olshanski theorem on the continuous unitary representations of the identity component $U_\infty(\mathcal{H})_0$ asserts that they are direct sums of irreducible ones which can be realized in finite tensor products of a suitable complex Hilbert space. This is proved and generalized to inseparable spaces. These results are carried over to the full unitary group by Pickrell's theorem, asserting that the separable unitary representations of $U(\mathcal{H})$, for a separable Hilbert space $\mathcal{H}$, are uniquely determined by their restriction to $U_\infty(\mathcal{H})_0$. For the 10 classical infinite rank symmetric pairs (G, K) of non-unitary type, such as $(GL(\mathcal{H}), U(\mathcal{H}))$, we also show that all separable unitary representations are trivial.

Keywords Unitary group • Unitary representation • Restricted group • Schur modules • Bounded representation • Separable representation

Mathematics Subject Classification (2010): Primary 22E65. Secondary 22E45.

Introduction

One of the most drastic difference between the representation theory of finite-dimensional Lie groups and infinite-dimensional ones is that an infinite-dimensional Lie group G may carry many different group topologies and any such topology leads to a different class of continuous unitary representations. Another perspective on

K-H. Neeb (✉)
Department of Mathematics, Friedrich–Alexander University, Erlangen-Nuremberg, Cauerstrasse 11, 91058 Erlangen, Germany
e-mail: neeb@math.fau.de

© Springer International Publishing Switzerland 2014
G. Mason et al. (eds.), *Developments and Retrospectives in Lie Theory*,
Developments in Mathematics 37, DOI 10.1007/978-3-319-09934-7_8

the same phenomenon is that the different topologies on G lead to different completions, and the passage to a specific completion reduces the class of representations under consideration.

In the present paper we survey results and methods of A. Kirillov, G. Olshanski and D. Pickrell from the point of view of Banach–Lie groups. In the unitary representation theory of finite-dimensional Lie groups, the starting point is the representation theory of compact Lie groups and the prototypical compact Lie group is the unitary group $U(n, \mathbb{C})$ of a complex n-dimensional Hilbert space. Therefore any systematic representation theory of infinite-dimensional Banach–Lie groups should start with unitary groups of Hilbert spaces. For an infinite-dimensional Hilbert space $\mathcal{H}$, there is a large variety of unitary groups. First, there is the full unitary group $U(\mathcal{H})$, endowed with the norm topology, turning it into a simply connected Banach–Lie group with Lie algebra $\mathfrak{u}(\mathcal{H}) = \{X \in B(\mathcal{H}) : X^* = -X\}$. However, the much coarser strong operator topology also turns it into another topological group $U(\mathcal{H})_s$. The third variant of a unitary group is the subgroup $U_\infty(\mathcal{H})$ of all unitary operators g for which $g - 1$ is compact. This is a Banach–Lie group whose Lie algebra $\mathfrak{u}_\infty(\mathcal{H})$ consists of all compact operators in $\mathfrak{u}(\mathcal{H})$. If $\mathcal{H}$ is separable (which we assume in this introduction) and $(e_n)_{n \in \mathbb{N}}$ is an orthonormal basis, then we obtain natural embeddings $U(n, \mathbb{C}) \to U(\mathcal{H})$ whose union $U(\infty, \mathbb{C}) = \bigcup_{n=1}^\infty U(n, \mathbb{C})$ carries the structure of a direct limit Lie group (cf. [Gl03]). Introducing also the Banach–Lie groups $U_p(\mathcal{H})$, consisting of unitary operators g, for which $g - 1$ is of Schatten class $p \in [1, \infty]$, i.e., $\mathrm{tr}(|U - 1|^p) < \infty$, we thus obtain an infinite family of groups with continuous inclusions

$$U(\infty, \mathbb{C}) \hookrightarrow U_1(\mathcal{H}) \hookrightarrow \cdots \hookrightarrow U_p(\mathcal{H}) \hookrightarrow \cdots \hookrightarrow U_\infty(\mathcal{H}) \hookrightarrow U(\mathcal{H}) \to U(\mathcal{H})_s.$$

The representation theory of infinite-dimensional unitary groups began with I. E. Segal's paper [Se57], where he studies unitary representations of the full group $U(\mathcal{H})$, called *physical representations*. These are characterized by the condition that their differential maps finite rank hermitian projections to positive operators. Segal shows that physical representations decompose discretely into irreducible physical representations which are precisely those occurring in the decomposition of finite tensor products $\mathcal{H}^{\otimes N}$, $N \in \mathbb{N}_0$. It is not hard to see that this tensor product decomposes as in classical Schur–Weyl theory:

$$\mathcal{H}^{\otimes N} \cong \bigoplus_{\lambda \in \mathrm{Part}(N)} \mathbb{S}_\lambda(\mathcal{H}) \otimes \mathcal{M}_\lambda, \tag{1}$$

where $\mathrm{Part}(N)$ is the set of all partitions $\lambda = (\lambda_1, \ldots, \lambda_n)$ of N, $\mathbb{S}_\lambda(\mathcal{H})$ is an irreducible unitary representation of $U(\mathcal{H})$ (called a *Schur representation*), and $\mathcal{M}_\lambda$ is the corresponding irreducible representation of the symmetric group S_N, hence in particular finite-dimensional (cf. [BN12] for an extension of Schur–Weyl theory to irreducible representations of C^*-algebras). In particular, $\mathcal{H}^{\otimes N}$ is a finite sum of irreducible representations of $U(\mathcal{H})$.

The representation theory of the Banach Lie group $U_\infty(\mathcal{H})$, $\mathcal{H}$ a separable real, complex or quaternionic Hilbert space, was initiated by A. A. Kirillov in [Ki73] (which contains no proofs), and continued by G. I. Olshanski [Ol78, Thm. 1.11]. They showed that all continuous representations of $U_\infty(\mathcal{H})$ are direct sums of irreducible representations and that for $\mathbb{K} = \mathbb{C}$, all the irreducible representations are of the form $\mathbb{S}_\lambda(\mathcal{H}) \otimes \mathbb{S}_\mu(\overline{\mathcal{H}})$, where $\overline{\mathcal{H}}$ is the space $\mathcal{H}$, endowed with the opposite complex structure. They also obtained generalizations for the corresponding groups over real and quaternionic Hilbert spaces. It follows in particular that all irreducible representations $(\pi, \mathcal{H}_\pi)$ of the Banach–Lie group $U_\infty(\mathcal{H})$ are *bounded* in the sense that $\pi: U_\infty(\mathcal{H}) \to U(\mathcal{H}_\pi)$ is norm continuous, resp., a morphism of Banach–Lie groups. The classification of the bounded unitary representations of the Banach–Lie group $U_p(\mathcal{H})$ remains the same for $1 < p < \infty$, but for $p = 1$, factor representations of type II and III exist (see [Boy80] for $p = 2$, and [Ne98] for the general case). Dropping the boundedness assumptions even leads to a non-type I representation theory for $U_p(\mathcal{H})$, $p < \infty$ (cf. [Boy80, Thm. 5.5]). We also refer to [Boy93] for an approach to Kirillov's classification based on the classification of factor representations of $U(\infty, \mathbb{C})$ from [SV75].

These results clearly show that the group $U_\infty(\mathcal{H})$ is singled out among all its relatives by the fact that its unitary representation theory is well-behaved. If $\mathcal{H}$ is separable, then $U_\infty(\mathcal{H})$ is separable, so that its cyclic representations are separable as well. Hence there is no need to discuss inseparable representations for this group. This is different for the Banach–Lie group $U(\mathcal{H})$ which has many inseparable bounded irreducible unitary representations coming from irreducible representations of the Calkin algebra $B(\mathcal{H})/K(\mathcal{H})$. It was an amazing insight of D. Pickrell [Pi88] that restricting attention to representations on separable spaces tames the representation theory of $U(\mathcal{H})$ in the sense that all its separable representations are actually continuous with respect to the strong operator topology, i.e., continuous representations of $U(\mathcal{H})_s$. For analogous results on the automatic weak continuity of separable representations of W^*-algebras see [FF57, Ta60]. Since $U_\infty(\mathcal{H})_0$ is dense in $U(\mathcal{H})_s$, it follows that $U_\infty(\mathcal{H})_0$ has the same separable representation theory as $U(\mathcal{H})_s$. As we shall see below, all these results extend to unitary groups of separable real and quaternionic Hilbert spaces.

Here we won't go deeper into the still not completely developed representation theory of groups like $U_2(\mathcal{H})$ which also have a wealth of projective unitary representations corresponding to nontrivial central Lie group extensions [Boy84, Ne13]. Instead we shall discuss the regular types of unitary representation and their characterization. For the unitary groups, the natural analogs of the finite-dimensional compact groups, a regular setup is obtained by considering $U_\infty(\mathcal{H})$ or the separable representations of $U(\mathcal{H})$. For direct limit groups, such as $U(\infty, \mathbb{C})$, the same kind of regularity is introduced by Olshanski's concept of a tame representation. Here a fundamental result is that the tame unitary representation of $U(\infty, \mathbb{C})_0$ are precisely those extending to continuous representations of $U_\infty(\mathcal{H})_0$ ([Ol78]; Theorem 3.20).

The natural next step is to take a closer look at unitary representations of the Banach analogs of noncompact classical groups; we simply call them non-unitary groups. There are 10 natural families of such groups that can be realized by *-invariant groups of operators with a polar decomposition

$$G = K \exp \mathfrak{p}, \quad \text{where} \quad K = \{g \in G : g^* = g^{-1}\} \quad \text{and} \quad \mathfrak{p} = \{X \in \mathfrak{g} : X^* = X\}$$

(see the tables in Sect. 5). In particular K is the maximal unitary subgroup of G. In this context Olshanski calls a continuous unitary representation of G admissible if its restriction to K is tame. For the cases where the symmetric space G/K is of finite rank, Olshanski classifies in [Ol78, Ol84] the irreducible admissible representations and shows that they form a type I representation theory (see also [Ol89]).

The voluminous paper [Ol90] deals with the case where G/K is of infinite rank. It contains a precise conjecture about the classification of the irreducible representations and the observation that in general, there are admissible factor representations not of type I. We refer to [MN13] for recent results related to Olshanski's conjecture and to [Ne12] for the classification of the semibounded projective unitary representations of hermitian Banach–Lie groups. Both continue Olshanski's program in the context of Banach–Lie groups of operators.

In [Pi90, Prop. 7.1], Pickrell shows for the 10 classical types of symmetric pairs (G, K) of non-unitary type that for $q > 2$, all separable projective unitary representations are trivial for the restricted groups $G_{(q)} = K \exp(\mathfrak{p}_{(q)})$ with Lie algebra

$$\mathfrak{g}_{(q)} = \mathfrak{k} \oplus \mathfrak{p}_{(q)} \quad \text{and} \quad \mathfrak{p}_{(q)} := \mathfrak{p} \cap B_q(\mathcal{H}),$$

where $B_q(\mathcal{H}) \trianglelefteq B(\mathcal{H})$ is the qth Schatten ideal. This complements the observation that admissible representations often extend to the restricted groups $G_{(2)}$ [Ol90]. From these results we learn that for $q > 2$, the groups $G_{(q)}$ are too big to have nontrivial separable unitary representations and that the groups $G_{(2)}$ have just the right size for a rich nontrivial separable representation theory. An important consequence is that G itself has no non-trivial separable unitary representation. This applies in particular to the group $\mathrm{GL}_{\mathbb{K}}(\mathcal{H})$ of $\mathbb{K}$-linear isomorphisms of a $\mathbb{K}$-Hilbert space $\mathcal{H}$ and the group $\mathrm{Sp}(\mathcal{H})$ of symplectic isomorphism of the symplectic space underlying a complex Hilbert space $\mathcal{H}$.

This is naturally extended by the fact that for the 10 symmetric pairs (G, K) of unitary type and $q > 2$, all continuous unitary representations of $G_{(q)}$ extend to continuous representations of the full group G [Pi90]. This result has interesting consequences for the representation theory of mapping groups. For a compact spin manifold M of odd dimension d, there are natural homomorphisms of the group $C^\infty(M, K)$, K a compact Lie group, into $U(\mathcal{H} \oplus \mathcal{H})_{(d+1)}$, corresponding to the symmetric pair $(U(\mathcal{H} \oplus \mathcal{H}), U(\mathcal{H}))$ (cf. [PS86, Mi89, Pi89]). For $d = 1$, the rich projective representation theory of $U(\mathcal{H} \oplus \mathcal{H})_{(d+1)}$ now leads to the unitary positive energy representations of loop groups, but for $d > 1$ the (projective) unitary representations of $U(\mathcal{H} \oplus \mathcal{H})_{(d+1)}$ extend to the full unitary group $U(\mathcal{H} \oplus \mathcal{H})$, so that we do not obtain interesting unitary representations of mapping groups.

However, there are natural homomorphisms $C(M, K)$ into the motion group $\mathcal{H} \rtimes O(\mathcal{H})$ of a real Hilbert space, and this leads to the interesting class of energy representations [GGV80, AH78].

The content of the paper is as follows. In the first two sections we discuss some core ideas and methods from the work of Olshanski and Pickrell. We start in Sect. 1 with the concept of a *bounded topological group*. These are topological groups G, for which every identity neighborhood U satisfies $G \subseteq U^m$ for some $m \in \mathbb{N}$. This boundedness condition permits showing that certain subgroups of G have nonzero fixed points in unitary representations (cf. Proposition 1.6 for a typical result of this kind). We continue in Sect. 2 with Olshanski's concept of an *overgroup*. Starting with a symmetric pair (G, K) with Lie algebra $\mathfrak{g} = \mathfrak{k} \oplus \mathfrak{p}$, the overgroup $K^{\sharp}$ of K is a Lie group with the Lie algebra $\mathfrak{k} + i\mathfrak{p}$. We shall use these overgroups for the pairs $(\mathrm{GL}(\mathcal{H}), \mathrm{U}(\mathcal{H}))$, where $\mathcal{H}$ is a real, complex or quaternionic Hilbert space.

In Sect. 3 we describe Olshanski's approach to the classification of the unitary representations of $K := \mathrm{U}_\infty(\mathcal{H})_0$. Here the key idea is that any representation of this group is a direct sum of representations π generated by the fixed space V of the subgroup K_n fixing the first n basis vectors. It turns out that this space V carries a $*$-representation (ρ, V) of the involutive semigroup $C(n, \mathbb{K})$ of contractions on $\mathbb{K}^n$ which determines π uniquely by a GNS construction.

Now the main point is to understand which representations of $C(n, \mathbb{K})$ occur in this process, that they are direct sums of irreducible ones and to determine the irreducible representations. To achieve this goal, we deviate from Olshanski's approach by putting a stronger emphasis on analytic positive definite functions (cf. Appendix A).

This leads to a considerable simplification of the proof avoiding the use of zonal spherical functions and expansions with respect to orthogonal polynomials. Moreover, our technique is rather close to the setting of holomorphic induction developed in [Ne13b]. In particular, we use Theorem A.4 which is a slight generalization of [Ne12, Thm. A.7].

In Sect. 4 we provide a complete proof of Pickrell's theorem asserting that for a separable Hilbert space $\mathcal{H}$, the groups $\mathrm{U}(\mathcal{H})$ and $\mathrm{U}_\infty(\mathcal{H})$ have the same continuous separable unitary representations. Here the key result is that all continuous separable unitary representations of the quotient group $\mathrm{U}(\mathcal{H})/\mathrm{U}_\infty(\mathcal{H})$ are trivial. We show that this result carries over to the real and quaternionic case by deriving it from the complex case.

This provides a complete picture of the separable representations of $\mathrm{U}(\mathcal{H})$ and the subgroup $\mathrm{U}_\infty(\mathcal{H})$, but there are many subgroups in between. This is naturally complemented by Pickrell's result that for the 10 symmetric pairs (G, K) of unitary type for $q > 2$, all continuous unitary representations of $G_{(q)}$ extend to continuous representations of G. In Sect. 5 we show that for the pairs (G, K) of noncompact type, all separable unitary representations of G and $G_{(q)}, q > 2$, are trivial. This is also stated in [Pi90], but the proof is very sketchy. We use an argument based on Howe–Moore theory for the vanishing of matrix coefficients.

Notation and Terminology

For the nonnegative half line we write $\mathbb{R}_+ = [0, \infty[$.

In the following $\mathbb{K}$ always denotes $\mathbb{R}$, $\mathbb{C}$ or the skew field $\mathbb{H}$ of quaternions. We write $\{1, \mathcal{I}, \mathcal{J}, \mathcal{I}\mathcal{J}\}$ for the canonical basis of $\mathbb{H}$ satisfying the relations

$$\mathcal{I}^2 = \mathcal{J}^2 = -\mathbf{1} \quad \text{and} \quad \mathcal{I}\mathcal{J} = -\mathcal{J}\mathcal{I}.$$

For a real Hilbert space $\mathcal{H}$, we write $\mathcal{H}_{\mathbb{C}}$ for its complexification, and for a quaternionic Hilbert space $\mathcal{H}$, we write $\mathcal{H}^{\mathbb{C}}$ for the underlying complex Hilbert space, obtained from the complex structure $\mathcal{I} \in \mathbb{H}$. For a complex Hilbert space we likewise write $\mathcal{H}^{\mathbb{R}}$ for the underlying real Hilbert space.

For the algebra $B(\mathcal{H})$ of bounded operators on the $\mathbb{K}$-Hilbert space $\mathcal{H}$, the ideal of compact operators is denoted $K(\mathcal{H}) = B_\infty(\mathcal{H})$, and for $1 \leq p < \infty$, we write

$$B_p(\mathcal{H}) := \{A \in K(\mathcal{H}): \operatorname{tr}((A^*A)^{p/2}) = \operatorname{tr}(|A|^p) < \infty\}$$

for the *Schatten ideals*. In particular, $B_2(\mathcal{H})$ is the space of *Hilbert–Schmidt operators* and $B_1(\mathcal{H})$ the space of *trace class operators*. Endowed with the operator norm, the groups $\mathrm{GL}(\mathcal{H})$ and $\mathrm{U}(\mathcal{H})$ are Lie groups with the respective Lie algebras

$$\mathfrak{gl}(\mathcal{H}) = B(\mathcal{H}) \quad \text{and} \quad \mathfrak{u}(\mathcal{H}) := \{X \in \mathfrak{gl}(\mathcal{H}): X^* = -X\}.$$

For $1 \leq p \leq \infty$, we obtain Lie groups

$$\mathrm{GL}_p(\mathcal{H}) := \mathrm{GL}(\mathcal{H}) \cap (\mathbf{1} + B_p(\mathcal{H})) \quad \text{and} \quad \mathrm{U}_p(\mathcal{H}) := \mathrm{U}(\mathcal{H}) \cap \mathrm{GL}_p(\mathcal{H})$$

with the Lie algebras

$$\mathfrak{gl}_p(\mathcal{H}) := B_p(\mathcal{H}) \quad \text{and} \quad \mathfrak{u}_p(\mathcal{H}) := \mathfrak{u}(\mathcal{H}) \cap \mathfrak{gl}_p(\mathcal{H}).$$

To emphasize the base field $\mathbb{K}$, we sometimes write $\mathrm{U}_{\mathbb{K}}(\mathcal{H})$ for the group $\mathrm{U}(\mathcal{H})$ of $\mathbb{K}$-linear isometries of $\mathcal{H}$. We also write $\mathrm{O}(\mathcal{H}) = \mathrm{U}_{\mathbb{R}}(\mathcal{H})$.

If G is a group acting on a set X, then we write X^G for the subset of G-fixed points.

1 Bounded Groups

In this section we discuss one of Olshanski's key concepts for the approach to Kirillov's theorem on the classification of the representations of $\mathrm{U}_\infty(\mathcal{H})_0$ for a separable Hilbert space discussed in Sect. 3. As we shall see below (Lemma 4.1), this method also lies at the heart of Pickrell's theorem on the separable representations of $\mathrm{U}(\mathcal{H})$.

Definition 1.1. We call a topological group G *bounded* if, for every identity neighborhood $U \subseteq G$, there exists an $m \in \mathbb{N}$ with $G \subseteq U^m$.

Note that every locally connected bounded topological group is connected. The group $\mathbb{Q}/\mathbb{Z}$ is bounded but not connected.

Lemma 1.2. *If, for a Banach–Lie group G, there exists a $c > 0$ with*

$$G = \exp\{x \in \mathfrak{g} \colon \|x\| \le c\}, \tag{O1}$$

then G is bounded.

Proof. Let U be an identity neighborhood of G. Since the exponential function $\exp_G \colon \mathfrak{g} \to G$ is continuous, there exists an $r > 0$ with $\exp x \in U$ for $\|x\| < r$. Pick $m \in \mathbb{N}$ such that $mr > c$. For $g = \exp x$ with $\|x\| \le c$ we then have $\exp \frac{x}{m} \in U$, and therefore $g \in U^m$. $\qquad\square$

Proposition 1.3. *The following groups satisfy* (O1), *hence are bounded:*

(i) *The full unitary group $\mathrm{U}(\mathcal{H})$ of an infinite-dimensional complex or quaternionic Hilbert space.*

(ii) *The unitary group $\mathrm{U}(\mathcal{M})$ of a von Neumann algebra $\mathcal{M}$.*

(iii) *The identity component $\mathrm{U}_\infty(\mathcal{H})_0$ of $\mathrm{U}_\infty(\mathcal{H})$ for a $\mathbb{K}$-Hilbert space $\mathcal{H}$.*[1]

Proof. (i) **Case $\mathbb{K} = \mathbb{C}$:** Let $g \in \mathrm{U}(\mathcal{H})$ and let P denote the spectral measure on the unit circle $\mathbb{T} \subseteq \mathbb{C}$ with $g = \int_{\mathbb{T}} z \, dP(z)$. We consider the measurable function $L \colon \mathbb{T} \to]-\pi, \pi]i$ which is the inverse of the function $]-\pi, \pi]i \to \mathbb{T}, z \mapsto e^z$. Then

$$X := \int_{\mathbb{T}} L(z) \, dP(z) \tag{1}$$

is a skew-hermitian operator with $\|X\| \le \pi$ and $e^X = g$ (cf. [Ru73, Thm. 12.37]).

Case $\mathbb{K} = \mathbb{H}$: We consider the quaternionic Hilbert space as a complex Hilbert space $\mathcal{H}^{\mathbb{C}}$, endowed with an anticonjugation (=antilinear complex structure) $\mathcal{J}$. Then

$$\mathrm{U}_{\mathbb{H}}(\mathcal{H}) = \{g \in \mathrm{U}(\mathcal{H}^{\mathbb{C}}) \colon \mathcal{J} g \mathcal{J}^{-1} = g\}.$$

An element $g \in \mathrm{U}(\mathcal{H}^{\mathbb{C}})$ is $\mathbb{H}$-linear if and only if the relation $\mathcal{J} P(E) \mathcal{J}^{-1} = P(\overline{E})$ holds for the corresponding spectral measure P on $\mathbb{T}$.

---

[1] Actually this group is connected for $\mathbb{K} = \mathbb{C}, \mathbb{H}$ [Ne02, Cor. II.15].

Let $\mathcal{H}_0 := \ker(g+1) = P(\{-1\})\mathcal{H}$ denote the (-1)-eigenspace of g and $\mathcal{H}_1 := \mathcal{H}_0^\perp$. If X is defined by (1) and $X_1 := X|_{\mathcal{H}_1}$, then

$$\mathcal{J} X_1 \mathcal{J}^{-1} = \int_{\mathbb{T}\setminus\{-1\}} -L(z)\, dP(\bar{z}) = \int_{\mathbb{T}\setminus\{-1\}} L(z)\, dP(z) = X_1.$$

Then $X := \pi \mathcal{J}|_{\mathcal{H}_0} \oplus X_1$ on $\mathcal{H} = \mathcal{H}_0 \oplus \mathcal{H}_1$ is an element $X \in \mathfrak{u}_{\mathbb{H}}(\mathcal{H})$ with $e^X = g$ and $\|X\| \leq \pi$. Therefore (OI) is satisfied.

(ii) If $g \in U(\mathcal{M})$, then $P(E) \in \mathcal{M}$ for every measurable subset $E \subseteq \mathbb{T}$, and therefore $X \in \mathcal{M}$. Now (ii) follows as (i) for $\mathbb{K} = \mathbb{C}$.

(iii) **Case** $\mathbb{K} = \mathbb{C}, \mathbb{H}$: The operator X from (1) is compact if $g - \mathbf{1}$ is compact. Hence the group $U_{\infty,\mathbb{K}}(\mathcal{H})$ is connected and we can argue as in (i).

Case $\mathbb{K} = \mathbb{R}$: We consider $U_{\infty,\mathbb{R}}(\mathcal{H}) = O_\infty(\mathcal{H})$ as a subgroup of $U_\infty(\mathcal{H}_{\mathbb{C}})$. Let σ denote the antilinear isometry on $\mathcal{H}_{\mathbb{C}}$ whose fixed point set is $\mathcal{H}$. Then, for $g \in U(\mathcal{H}_{\mathbb{C}})$, the relation $\sigma g \sigma = g$ is equivalent to $g \in O(\mathcal{H})$. This is equivalent to the relation $\sigma P(E) \sigma = P(\overline{E})$ for the corresponding spectral measure on $\mathbb{T}$.

Next we recall from [Ne02, Cor. II.15] (see also [dlH72]) that the group $O_\infty(\mathcal{H})$ has two connected components. An element $g \in O_\infty(\mathcal{H})$ for which $g - \mathbf{1}$ is of trace class is contained in the identity component if and only if $\det(g) = 1$. From the normal form of orthogonal compact operators that follows from the spectral measure on $\mathcal{H}_{\mathbb{C}}$, it follows that $\det(g) = (-1)^{\dim \mathcal{H}^{-g}}$. Therefore the identity component of $O_\infty(\mathcal{H})$ consists of those elements g for which the (-1)-eigenspace $\mathcal{H}_0 = \mathcal{H}^{-g}$ is of even dimension. Let $J \in \mathfrak{o}(\mathcal{H}_0)$ be an orthogonal complex structure. On $\mathcal{H}_1 := \mathcal{H}_0^\perp$ the operator $X_1 := X|_{\mathcal{H}_1}$ satisfies

$$\sigma X_1 \sigma = \int_{\mathbb{T}\setminus\{-1\}} -L(z)\, dP(\bar{z}) = \int_{\mathbb{T}\setminus\{-1\}} L(z)\, dP(z) = X_1,$$

so that $X := \pi J \oplus X_1 \in \mathfrak{o}_\infty(\mathcal{H})$ satisfies $\|X\| \leq \pi$ and $e^X = g$. $\square$

Example 1.4. (a) In view of Proposition 1.3, it is remarkable that the full orthogonal group $O(\mathcal{H})$ of a real Hilbert space $\mathcal{H}$ does not satisfy (OI). Actually its exponential function is not surjective [PW52]. In fact, if $g = e^X$ for $X \in \mathfrak{o}(\mathcal{H})$, then X commutes with g, hence preserves the (-1)-eigenspace $\mathcal{H}_0 := \ker(g + 1)$. Therefore $J := e^{X/2}$ defines a complex structure on $\mathcal{H}_0$, showing that $\mathcal{H}_0$ is either infinite-dimensional or of even dimension. Therefore no element $g \in O(\mathcal{H})$ for which $\dim \mathcal{H}_0$ is odd is contained in the image of the exponential function.

(b) We shall need later that $O(\mathcal{H})$ is connected. This follows from Kuiper's theorem [Ku65], but one can give a more direct argument based on the preceding discussion. It only remains to show that elements $g \in O(\mathcal{H})$ for which the space $\mathcal{H}_0 := \ker(g + 1)$ is of finite odd dimension are contained in the identity component. We write $g = g_{-1} \oplus g_1$ with $g_{-1} = g|_{\mathcal{H}_0}$ and $g_1 := g|_{\mathcal{H}_0^\perp}$. Then g_1 lies on a one-parameter group of $O(\mathcal{H}_0^\perp)$, so that g is connected by a continuous arc to $g' := -\mathbf{1}_{\mathcal{H}_0} \oplus \mathbf{1}$. This element is connected to $g'' := -\mathbf{1}_{\mathcal{H}_0} \oplus -\mathbf{1}_{\mathcal{H}_0^\perp} = -\mathbf{1}_{\mathcal{H}}$, and this in turn to $\mathbf{1}_{\mathcal{H}}$. Therefore $O(\mathcal{H})$ is connected.

Lemma 1.5 (Olshanski Lemma). *For a group G, a subset $U \subseteq G$ and $m \in \mathbb{N}$ with $G \subseteq U^m$, we put*

$$\eta := \sqrt{1 - \frac{1}{4(m+1)^2}} \;\in\,]0, 1[.$$

If $(\pi, \mathcal{H})$ is a unitary representation with $\mathcal{H}^G = \{0\}$, then, for any non-zero $\xi \in \mathcal{H}$, there exists $u \in U$ with

$$\tfrac{1}{2}\|\xi + \pi(u)\xi\| < \eta\|\xi\|.$$

Proof ([Ol78, Lemma 1.3]). If $\|\xi - \pi(u)\xi\| \le \lambda\|\xi\|$ holds for all $u \in U$, then the triangle inequality implies

$$\|\xi - \pi(g)\xi\| \le m\lambda\|\xi\| \quad \text{for} \quad g \in U^m = G.$$

For $\lambda < \frac{1}{m}$ this implies that the closed convex hull of the orbit $\pi(G)\xi$ does not contain 0, hence contains a nonzero fixed point by the Bruhat–Tits Theorem [La99], applied to the isometric action of G on $\mathcal{H}$. This violates our assumption $\mathcal{H}^G = \{0\}$. We conclude that there exists a $u \in U$ with $\|\xi - \pi(u)\xi\| > \frac{1}{m+1}\|\xi\|$. Thus

$$2\|\xi\|^2 - 2\,\mathrm{Re}\langle\xi, \pi(u)\xi\rangle = \|\xi - \pi(u)\xi\|^2 > \frac{\|\xi\|^2}{(m+1)^2},$$

which in turn gives

$$\|\xi + \pi(u)\xi\|^2 = 2\|\xi\|^2 + 2\,\mathrm{Re}\langle\xi, \pi(u)\xi\rangle < \Big(4 - \frac{1}{(m+1)^2}\Big)\|\xi\|^2 = \eta^2\|\xi\|^2. \qquad \square$$

The following proposition is an abstraction of the proof of [Ol78, Lemma 1.4]. It will be used in two situations below, to prove Kirillov's Lemma 3.6 and in Pickrell's Lemma 4.1.

Proposition 1.6. *Let G be a bounded topological group and $(G_n)_{n\in\mathbb{N}}$ be a sequence of subgroups of G. If there exists a basis of $\mathbf{1}$-neighborhoods $U \subseteq G$ such that either*

(a) (U1) $(\exists m \in \mathbb{N})(\forall n)\, G_n \subseteq (G_n \cap U)^m$, and

    (U2) $(\forall N \in \mathbb{N})(G_N \cap U)\cdots(G_1 \cap U) \subseteq U,$

    or

(b) (V1) $(\exists m \in \mathbb{N})(\forall n)\, G_n \subseteq (G_n \cap U)^m$, and

    (V2) *there exists an increasing sequence of subgroups $(G(n))_{n\in\mathbb{N}}$ such that*

(1) $(\forall\, n \in \mathbb{N})$ *the union of* $G(m)_n := G(m) \cap G_n$, $m \in \mathbb{N}$, *is dense in* G_n.

(2) $(G(k_1)_1 \cap U)(G(k_2)_{k_1} \cap U) \cdots (G(k_N)_{k_{N-1}} \cap U) \subseteq U$ *for* $1 < k_1 < \ldots < k_N$.

Then there exists an $n \in \mathbb{N}$ *with* $\mathcal{H}^{G_n} \neq \{0\}$.

Proof. (a) We argue by contradiction and assume that $\mathcal{H}^{G_n} = \{0\}$ for every n. Let $\xi \in \mathcal{H}$ be nonzero and let $U \subseteq G$ be an identity neighborhood with $\|\pi(g)\xi - \xi\| < \frac{1}{2}\|\xi\|$ for $g \in U$ such that (U1/2) are satisfied. Let η be as in Lemma 1.5.

Since G_1 has no nonzero fixed vector, there exists an element $u_1 \in U \cap G_1$ with

$$\left\|\tfrac{1}{2}(\xi + \pi(u_1)\xi)\right\| \le \eta\|\xi\|.$$

Then $\xi_1 := \frac{1}{2}(\xi + \pi(u_1)\xi)$ satisfies $\|\xi_1 - \xi\| < \frac{1}{2}\|\xi\|$, so that $\xi_1 \neq 0$. Iterating this procedure, we obtain a sequence of vectors $(\xi_n)_{n \in \mathbb{N}}$ and elements $u_n \in U \cap G_n$ with $\xi_{n+1} := \frac{1}{2}(\xi_n + \pi(u_{n+1})\xi_n)$ and $\|\xi_{n+1}\| \le \eta\|\xi_n\|$.

We consider the probability measures $\mu_n := \frac{1}{2}(\delta_1 + \delta_{u_n})$ on G and observe that (U2) implies $\mathrm{supp}(\mu_n * \cdots * \mu_1) \subseteq U$ for every $n \in \mathbb{N}$. By construction we have $\pi(\mu_n * \cdots * \mu_1)\xi = \xi_n$, so that $\|\xi_n - \xi\| < \frac{1}{2}\|\xi\|$. On the other hand,

$$\|\pi(\mu_n * \cdots * \mu_1)\xi\| = \|\xi_n\| \le \eta^n\|\xi\| \to 0,$$

and this is a contradiction.

(b) Again, we argue by contradiction and assume that no subgroup G_n has a nonzero fixed vector. Let $\xi \in \mathcal{H}$ be a nonzero vector and U an identity neighborhood with $\|\pi(g)\xi - \xi\| < \frac{1}{2}\|\xi\|$ for $g \in U$ such that (V1) is satisfied.

Since G_1 has no nonzero fixed point in $\mathcal{H}$ and $\bigcup_{n=1}^{\infty} G(n)_1$ is dense in G_1, there exists a $k_1 \in \mathbb{N}$ and some $u_1 \in U \cap G(k_1)_1$ with $\|\frac{1}{2}(\xi + \pi(u_1)\xi)\| < \eta\|\xi\|$ (Lemma 1.5). For $\xi_1 := \frac{1}{2}(\xi + \pi(u_1)\xi)$ our construction then implies that $\|\xi_1 - \xi\| < \frac{1}{2}\|\xi\|$, so that $\xi_1 \neq 0$. Any $u \in U \cap G_{k_1}$ commutes with u_1, so that we further obtain

$$\begin{aligned}
\|\pi(u)\xi_1 - \xi_1\| &= \tfrac{1}{2}\|\pi(u)\xi - \xi + \pi(u)\pi(u_1)\xi - \pi(u_1)\xi\| \\
&< \tfrac{1}{2}\left(\tfrac{1}{2}\|\xi\| + \|\pi(u_1)\pi(u)\xi - \pi(u_1)\xi\|\right) \\
&= \tfrac{1}{2}\left(\tfrac{1}{2}\|\xi\| + \|\pi(u)\xi - \xi\|\right) < \tfrac{1}{2}\left(\tfrac{1}{2}\|\xi\| + \tfrac{1}{2}\|\xi\|\right) = \tfrac{1}{2}\|\xi\|.
\end{aligned}$$

Iterating this procedure, we obtain a strictly increasing sequence (k_n) of natural numbers, a sequence (ξ_n) in $\mathcal{H}$ and $u_n \in G(k_n)_{k_{n-1}} \cap U$ with

$$\xi_{n+1} := \tfrac{1}{2}(\xi_n + \pi(u_{n+1})\xi_n) \quad \text{and} \quad \|\xi_{n+1}\| < \eta\|\xi_n\|.$$

We consider the probability measures $\mu_n := \frac{1}{2}(\delta_1 + \delta_{u_n})$ on G. Condition (V2)(2) implies that

$$\text{supp}(\mu_n * \cdots * \mu_1) \subseteq \{\mathbf{1}, u_n\} \cdots \{\mathbf{1}, u_1\} \subseteq U \quad \text{for every} \quad n \in \mathbb{N}.$$

By construction $\pi(\mu_n * \cdots * \mu_1)\xi = \xi_n$, so that $\|\xi_n - \xi\| < \frac{1}{2}\|\xi\|$. On the other hand,

$$\|\pi(\mu_n * \cdots * \mu_1)\xi\| = \|\xi_n\| \leq \eta^n \|\xi\| \to 0,$$

and this is a contradiction.   □

2 Duality and Overgroups

Apart from the fixed point results related to bounded topological groups discussed in the preceding section, another central concept in Olshanski's approach are "overgroups". They are closely related to the duality of symmetric spaces.

Definition 2.1. A *symmetric Lie group* is a triple (G, K, τ), where τ is an involutive automorphism of the Banach–Lie group G and K is an open subgroup of the Lie subgroup G^τ of τ-fixed points in G. We write $\mathfrak{g} = \mathfrak{k} \oplus \mathfrak{p} = \mathfrak{g}^\tau \oplus \mathfrak{g}^{-\tau}$ for the eigenspace decomposition of $\mathfrak{g}$ with respect to τ and call $\mathfrak{g}^c := \mathfrak{k} \oplus i\mathfrak{p} \subseteq \mathfrak{g}_{\mathbb{C}}$ the *dual symmetric Lie algebra*.

Definition 2.2. Suppose that (G, K, τ) is a symmetric Lie group and G^c a simply connected Lie group with Lie algebra $\mathfrak{g}^c = \mathfrak{k} + i\mathfrak{p}$. Then $X + iY \mapsto X - iY$ ($X \in \mathfrak{k}$, $Y \in \mathfrak{p}$), integrates to an involution $\tilde{\tau}^c$ of G^c. Let $q_K \colon \tilde{K}_0 \to K_0$ denote the universal covering of the identity component K_0 of K and $\tilde{\iota}_K \colon \tilde{K}_0 \to G^c$ the homomorphism integrating the inclusion $\mathfrak{k} \hookrightarrow \mathfrak{g}^c$. The group $\tilde{\iota}_K(\ker q_K)$ acts trivially on $\mathfrak{g}_{\mathbb{C}}$, hence is central in G^c. If it is discrete, then we call

$$(K_0)^\sharp := G^c / \tilde{\iota}_K(\ker q_K)$$

the *overgroup of K_0*. In this case $\tilde{\iota}_K$ factors through a covering map $\iota_{K_0} \colon K_0 \to (K_0)^\sharp$ and the involution τ^c induced by $\tilde{\tau}^c$ on $(K_0)^\sharp$ leads to a symmetric Lie group $((K_0)^\sharp, \iota_{K_0}(K_0), \tau^c)$.

To extend this construction to the case where K is not connected, we first observe that $K \subseteq G$ acts naturally on the Lie algebra $\mathfrak{g}^c$, hence also on the corresponding simply connected group G^c. This action preserves $\tilde{\iota}_K(\ker q_K)$, hence induces an action on $(K_0)^\sharp$, so that we can form the semidirect product $(K_0)^\sharp \rtimes K$. In this group $N := \{(\iota_{K_0}(k), k^{-1}) \colon k \in K_0\}$ is a closed normal subgroup and we put

$$K^\sharp := ((K_0)^\sharp \rtimes K)/N, \quad \iota_K(k) := (\mathbf{1}, k)N \in K^\sharp.$$

The overgroup $K^\sharp$ has the universal property that if a morphism $\alpha \colon K \to H$ of Lie groups extends to a Lie group with Lie algebra $\mathfrak{g}^c = \mathfrak{k} + i\mathfrak{p}$, then α factors through $\tilde{\iota}_K \colon K \to K^\sharp$.

Example 2.3. (a) If $\mathcal{H}$ is a $\mathbb{K}$-Hilbert space, then the triple $(\mathrm{GL}_{\mathbb{K}}(\mathcal{H}), \mathrm{U}_{\mathbb{K}}(\mathcal{H}), \tau)$ with $\tau(g) = (g^*)^{-1}$ is a symmetric Lie group. For its Lie algebra

$$\mathfrak{g} = \mathfrak{gl}_{\mathbb{K}}(\mathcal{H}) = \mathfrak{k} \oplus \mathfrak{p} = \mathfrak{u}_{\mathbb{K}}(\mathcal{H}) \oplus \mathrm{Herm}_{\mathbb{K}}(\mathcal{H}),$$

the corresponding dual symmetric Lie algebra is

$$\mathfrak{g}^c = \mathfrak{k} + i\mathfrak{p} = \mathfrak{u}_{\mathbb{K}}(\mathcal{H}) \oplus i\,\mathrm{Herm}_{\mathbb{K}}(\mathcal{H}) \subseteq \mathfrak{u}(\mathcal{H}_{\mathbb{C}}).$$

More precisely, we have

(ℝ) $\mathfrak{gl}_{\mathbb{R}}(\mathcal{H})^c = \mathfrak{o}(\mathcal{H}) \oplus i\,\mathrm{Sym}(\mathcal{H}) \cong \mathfrak{u}(\mathcal{H}_{\mathbb{C}})$ for $\mathbb{K} = \mathbb{R}$.
(ℂ) $\mathfrak{gl}_{\mathbb{C}}(\mathcal{H})^c = \mathfrak{u}(\mathcal{H}) \oplus i\,\mathrm{Herm}(\mathcal{H}) \cong \mathfrak{u}(\mathcal{H})^2$ for $\mathbb{K} = \mathbb{C}$.
(ℍ) $\mathfrak{gl}_{\mathbb{H}}(\mathcal{H})^c = \mathfrak{u}_{\mathbb{K}}(\mathcal{H}) \oplus \mathcal{I}\,\mathrm{Herm}_{\mathbb{H}}(\mathcal{H}) \cong \mathfrak{u}(\mathcal{H}^{\mathbb{C}})$ for $\mathbb{K} = \mathbb{H}$.

Here the complex case requires additional explanation. Let I denote the given complex structure on $\mathcal{H}$. Then the maps

$$\iota_{\pm}\colon \mathcal{H} \to \mathcal{H}_{\mathbb{C}}, \quad v \mapsto \frac{1}{\sqrt{2}}(v \mp iIv)$$

are isometries to complex subspaces $\mathcal{H}_{\mathbb{C}}^{\pm}$ of $\mathcal{H}_{\mathbb{C}}$, where ι_+ is complex linear and ι_- is antilinear. We thus obtain

$$\mathcal{H}_{\mathbb{C}} = \mathcal{H}_{\mathbb{C}}^{+} \oplus \mathcal{H}_{\mathbb{C}}^{-} \cong \mathcal{H} \oplus \overline{\mathcal{H}},$$

and $\mathcal{H}_{\mathbb{C}}^{\pm}$ are the $\pm i$-eigenspaces of the complex linear extension of I to $\mathcal{H}_{\mathbb{C}}$. In particular, $\mathfrak{gl}(\mathcal{H})^c$ preserves both subspaces $\mathcal{H}_{\mathbb{C}}^{\pm}$. This leads to the isomorphism

$$\gamma\colon \mathfrak{gl}(\mathcal{H})^c \to \mathfrak{u}(\mathcal{H}_{\mathbb{C}}^{+}) \oplus \mathfrak{u}(\mathcal{H}_{\mathbb{C}}^{-}) \cong \mathfrak{u}(\mathcal{H}) \oplus \mathfrak{u}(\overline{\mathcal{H}}), \quad \gamma(X + iY) = (X + IY, X - IY).$$

Lemma 2.4. *For a $\mathbb{K}$-Hilbert space $\mathcal{H}$, let $(G, K) = (\mathrm{GL}_{\mathbb{K}}(\mathcal{H}), \mathrm{U}_{\mathbb{K}}(\mathcal{H})_0)$ and $n :=$ $\dim \mathcal{H}$. Then K is connected for $\mathbb{K} \neq \mathbb{R}$ and $n = \infty$, and*

$$(K_0)^{\sharp} \cong \begin{cases} \tilde{\mathrm{U}}(n, \mathbb{C}) & \text{for } \mathbb{K} = \mathbb{R}, n < \infty \\ \tilde{\mathrm{U}}(n, \mathbb{C})^2/\Gamma & \text{for } \mathbb{K} = \mathbb{C}, n < \infty, \ \Gamma := \{(z, z) : z \in \pi_1(\mathrm{U}(n, \mathbb{C}))\}. \\ \tilde{\mathrm{U}}(2n, \mathbb{C}) & \text{for } \mathbb{K} = \mathbb{H}, n < \infty, \\ \mathrm{U}(\mathcal{H}_{\mathbb{C}}) & \text{for } \mathbb{K} = \mathbb{R} \\ \mathrm{U}(\mathcal{H}) \oplus \mathrm{U}(\overline{\mathcal{H}}) & \text{for } \mathbb{K} = \mathbb{C} \\ \mathrm{U}(\mathcal{H}^{\mathbb{C}}) & \text{for } \mathbb{K} = \mathbb{H}. \end{cases}$$

Here $\mathcal{H}^{\mathbb{C}}$ is the complex Hilbert space underlying a quaternionic Hilbert space $\mathcal{H}$. For $\mathbb{K} = \mathbb{R}, \mathbb{H}$, the map $\iota_{\mathbb{K}}$ is the canonical inclusion, and $\iota_{\mathbb{K}}(k) = (k, k)$ for $\mathbb{K} = \mathbb{C}$.

Proof. First we consider the case where $n < \infty$. Recall that

$$\tilde{U}(n, \mathbb{C}) \cong SU(n, \mathbb{C}) \rtimes \mathbb{R}, \quad O(n, \mathbb{R})_0 = SO(n, \mathbb{R}) \subseteq SU(n, \mathbb{C}), \quad U(n, \mathbb{H}) \subseteq SU(2n, \mathbb{C}).$$

For $\mathbb{K} = \mathbb{R}$, this implies that $K = O(n, \mathbb{R})_0$ embeds into $\tilde{U}(n, \mathbb{C})$, so that $K^\sharp \cong \tilde{U}(n, \mathbb{C})$. For $\mathbb{K} = \mathbb{H}$, we see that $K = U(n, \mathbb{H})$ embeds into $\tilde{U}(2n, \mathbb{C})$, which leads to $\tilde{K} \cong \tilde{U}(2n, \mathbb{C})$.

For $\mathbb{K} = \mathbb{C}$, we have the natural inclusion

$$i_K : K = U(n, \mathbb{C}) \to U(n, \mathbb{C}) \times U(n, \mathbb{C}) \cong U(\mathbb{C}^n) \times U(\overline{\mathbb{C}^n}), \quad g \mapsto (g, g).$$

To determine $K^\sharp$, we note that the image of

$$\pi_1(i_K) : \mathbb{Z} \cong \pi_1(U(n, \mathbb{C})) \to \mathbb{Z}^2 \cong \pi_1(U(n, \mathbb{C})^2), \quad m \mapsto (m, m)$$

is Γ. Therefore $K^\sharp \cong \tilde{U}(n, \mathbb{C})^2 / \Gamma$.

If $n = \infty$, then $K = U_{\mathbb{K}}(\mathcal{H})$ is a simply connected Lie group with Lie algebra $\mathfrak{k}$ [Ku65], so that $K^\sharp$ is the simply connected Lie group with Lie algebra $\mathfrak{k}^\sharp$ and we have a natural morphism $\iota_K : K \to K^\sharp$ integrating the inclusion $\mathfrak{k} \hookrightarrow \mathfrak{k}^\sharp$. $\qquad\square$

3 The Unitary Representations of $U_\infty(\mathcal{H})_0$

In this section we completely describe the representations of the Banach–Lie groups $U_\infty(\mathcal{H})_0$ for an infinite-dimensional real, complex or quaternionic Hilbert space. In particular, we show that all continuous unitary representations are direct sums of irreducible ones and classify the irreducible ones (Theorem 3.17). Our approach is based on Olshanski's treatment in [Ol78]. We also take some short cuts that simplify the proof and put a stronger emphasis on analytic positive definite functions. This has the nice side effect that we also obtain these results for inseparable Hilbert spaces (Theorem 3.21).

3.1 Tameness as a Continuity Condition

We start with a brief discussion of Olshanski's concept of a tame representation that links representations of $U_\infty(\mathcal{H})_0$ to representations of the direct limit group $U(\infty, \mathbb{K})$.

Let K be a group and $(K_j)_{j \in J}$ a non-empty family of subgroups satisfying the following conditions

(S1) It is a *filter basis*, i.e., for $j, m \in J$, there exists an $\ell \in J$ with $K_\ell \subseteq K_j \cap K_m$.
(S2) $\bigcap_{j \in J} K_j = \{\mathbf{1}\}$.
(S3) For each $g \in K$ and $j \in J$ there exists an $m \in J$ with $g K_m g^{-1} \subseteq K_j$.

Then there exists a unique Hausdorff group topology τ on K for which $(K_j)_{j\in J}$ is a basis of $\mathbf{1}$-neighborhoods [Bou98, Ch. 4]. We call τ the *topology defined by* $(K_j)_{j\in J}$.

Definition 3.1. We call a unitary representation $(\pi, \mathcal{H})$ of K *tame* if the space

$$\mathcal{H}^T := \sum_{j\in J} \mathcal{H}^{K_j} = \bigcup_{j\in J} \mathcal{H}^{K_j},$$

is dense in $\mathcal{H}$. Note that, for $K_j \subseteq K_k \cap K_\ell$, we have $\mathcal{H}^{K_j} \supseteq \mathcal{H}^{K_k} + \mathcal{H}^{K_\ell}$, so that $\mathcal{H}^T$ is a directed union of the closed subspaces $\mathcal{H}^{K_j}$.

Lemma 3.2. *A unitary representation of K is tame if and only if it is continuous with respect to the group topology defined by the filter basis $(K_j)_{j\in J}$.*

Proof. If $(\pi, \mathcal{H})$ is a tame representation, then $\mathcal{H}^T$ obviously consists of continuous vectors for K since, for each $v \in \mathcal{H}^T$, the stabilizer is open. Hence the set of continuous vectors is dense, and therefore π is continuous.

If, conversely, $(\pi, \mathcal{H})$ is continuous and $v \in \mathcal{H}$, then the orbit map $K \to \mathcal{H}, g \mapsto gv$ is continuous. Let B_ε denote the closed ε-ball in $\mathcal{H}$. Then there exists a $j \in J$ with $\pi(K_j)v \subseteq v + B_\varepsilon$. Then $C := \overline{\mathrm{conv}(\pi(K_j)v)}$ is a closed convex invariant subset of $v + B_\varepsilon$, hence contains a K_j-fixed point by the Bruhat–Tits Theorem [La99]. This proves that $\mathcal{H}^{K_j}$ intersects $v + B_\varepsilon$, hence that $\mathcal{H}^T$ is dense in $\mathcal{H}$. $\square$

Remark 3.3. (a) For a unitary representation $(\pi, \mathcal{H})$ of a topological group, the subspace $\mathcal{H}^c$ of continuous vectors is closed and invariant. The representation π is continuous if and only if $\mathcal{H}^c = \mathcal{H}$.

(b) For a unitary representation $(\pi, \mathcal{H})$ of K, by Lemma 3.2, the space of continuous vectors coincides with $\overline{\mathcal{H}^T}$. In particular, it is K-invariant.

(c) If the representation $(\pi, \mathcal{H})$ of K is irreducible, then it is tame if and only if $\mathcal{H}^T \neq \{0\}$.

(d) If the representation $(\pi, \mathcal{H})$ of K is such that, for some n, the subspace $\mathcal{H}^{K_n}$ is cyclic, then it is tame.

Definition 3.4. Assume that the group K is the union of an increasing sequence of subgroups $(K(n))_{n\in\mathbb{N}}$. We say that the subgroups $K(n)$ are *well-complemented* by the decreasing sequence $(K_n)_{n\in\mathbb{N}}$ of subgroups of K if K_n commutes with $K(n)$ for every n and $\bigcap_{n\in\mathbb{N}} K_n = \{1\}$. For $k \in K$ and $n \in \mathbb{N}$, we then find an $m > n$ with $k \in K(m)$. Then $kK_mk^{-1} = K_m \subseteq K_n$, so that (S1-3) are satisfied and the groups $(K_n)_{n\in\mathbb{N}}$ define a group topology on K.

Example 3.5. (a) If $K = \bigoplus_{n=1}^\infty F_n$ is a direct sum of subgroups $(F_n)_{n\in\mathbb{N}}$, then the subgroups $K(n) := F_1 \times \cdots \times F_n$ are well-complemented by the subgroups $K_n := \bigoplus_{m>n} F_m$.

(b) If $K = \mathrm{U}(\infty, \mathbb{K})_0 = \bigcup_{n=1}^\infty \mathrm{U}(n, \mathbb{K})_0$ is the canonical direct limit of the compact groups $\mathrm{U}(n, \mathbb{K})_0$, then the subgroups $K(n) := \mathrm{U}(n, \mathbb{K})_0$ are well-complemented by the subgroups

$$K_n := \{g \in K : (\forall j \le n)\, g e_j = e_j\}.$$

3.2 Tame Representations of $\mathrm{U}(\infty, \mathbb{K})$

Let $\mathcal{H}$ be an infinite-dimensional separable Hilbert space over $\mathbb{K} \in \{\mathbb{R}, \mathbb{C}, \mathbb{H}\}$ and $(e_j)_{j \in \mathbb{N}}$ an orthonormal basis of $\mathcal{H}$. Accordingly, we obtain a natural dense embedding $\mathrm{U}(\infty, \mathbb{K}) \hookrightarrow \mathrm{U}_\infty(\mathcal{H})$, so that every continuous unitary representation of $\mathrm{U}_\infty(\mathcal{H})_0$ is uniquely determined by its restriction to the direct limit group $\mathrm{U}(\infty, \mathbb{K})_0$. Olshanski's approach to the classification is based on an intrinsic characterization of those representations of the direct limit group $\mathrm{U}(\infty, \mathbb{K})_0$ that extend to $\mathrm{U}_\infty(\mathcal{H})_0$. It turns out that these are precisely the tame representations (Theorem 3.20). This is complemented by the discrete decomposition and the classification of the irreducible ones (Theorem 3.17).

In the following we write $K := \mathrm{U}_\infty(\mathcal{H})_0$ for the identity component of $\mathrm{U}_\infty(\mathcal{H})$ (which is connected for $\mathbb{K} = \mathbb{C}, \mathbb{H}$, but not for $\mathbb{K} = \mathbb{R}$), and $K(n) := \mathrm{U}(n, \mathbb{K})_0 \cong \mathrm{U}(\mathcal{H}(n))_0$ for $n \in \mathbb{N}$, where $\mathcal{H}(n) = \mathrm{span}\{e_1, \dots, e_n\}$. For $n \in \mathbb{N}$, the stabilizer of $e_1, \dots, e_n$ in K is denoted K_n, and we likewise write $K(m)_n := K(m) \cap K_n$. We also write $K(\infty) := \mathrm{U}(\infty, \mathbb{K})_0 \cong \varinjlim \mathrm{U}(n, \mathbb{K})_0$ for the direct limit of the groups $\mathrm{U}(n, \mathbb{K})_0$.

We now turn to the classification of the continuous unitary representations of K. We start with an application of Proposition 1.6.

Lemma 3.6 (Kirillov's Lemma). *Let $(\pi, \mathcal{H}_\pi)$ be a continuous unitary representation of the Banach–Lie group $K = \mathrm{U}_\infty(\mathcal{H})_0$. If $\mathcal{H}_\pi \ne \{0\}$, then there exists an $n \in \mathbb{N}$, such that the stabilizer K_n of $e_1, \dots, e_n$ has a nonzero fixed point.*

Proof. We apply Proposition 1.6(b) with $G := K$, $G_n := K_n$ and $G(n) := K(n)$. Then $U_\varepsilon := \{g \in K : \|g-\mathbf{1}\| < \varepsilon\}$ provides the required basis of $\mathbf{1}$-neighborhoods in G (Proposition 1.3(iii)). Condition (V1) follows from Proposition 1.3(iii), (V2)(1) is clear, and (V2)(2) follows from the fact that, for $1 < k_1 < \dots < k_N$, elements $u_j \in K(k_j)_{k_{j-1}}$ act on pairwise orthogonal subspaces. $\square$

Proposition 3.7. *Any continuous unitary representation $(\pi, \mathcal{H})$ of $K = \mathrm{U}_\infty(\mathcal{H})_0$ restricts to a tame representation of the subgroup $K(\infty) = \mathrm{U}(\infty, \mathbb{K})_0$.*

Proof. Let $\mathcal{H}_0 \subseteq \mathcal{H}$ denote the maximal subspace on which the representation of $K(\infty)$ is tame, i.e., the space of continuous vectors for the topology defined by the subgroups $K(\infty)_n$ (Remark 3.3). Lemma 3.6 implies that $\mathcal{H}_0 \ne \{0\}$. If $\mathcal{H}_0 \ne \mathcal{H}$, then Lemma 3.6 implies the existence of nonzero continuous vectors in $\mathcal{H}_0^\perp$, which is a contradiction. $\square$

Example 3.8. The preceding proposition does not extend to the nonconnected group $\mathrm{O}_\infty(\mathcal{H})$ which has 2-connected components. The corresponding homomorphism

$$D : \mathrm{O}_\infty(\mathcal{H}) \to \{\pm 1\}$$

is nontrivial on all subgroups $O_\infty(\mathcal{H})_n$, $n \in \mathbb{N}$.

Lemma 3.9. *Let $\mathcal{H}$ be a $\mathbb{K}$-Hilbert space and $\mathcal{F} \subseteq \mathcal{H}$ be a finite-dimensional subspace with $2 \dim \mathcal{F} < \dim \mathcal{H}$.[2]*
We write $P_\mathcal{F} \colon \mathcal{H} \to \mathcal{F}$ for the orthogonal projection and

$$C(\mathcal{F}) := \{A \in B(\mathcal{F}) \colon \|A\| \leq 1\}$$

for the semigroup of contractions on $\mathcal{F}$. Then the map

$$\theta \colon K = U_\infty(\mathcal{H})_0 \to C(\mathcal{F}), \quad \theta(g) = P_\mathcal{F} g P_\mathcal{F}^*$$

is continuous, surjective and open. Its fibers are the double cosets of the pointwise stabilizer $K_\mathcal{F}$ of $\mathcal{F}$.
In particular, we obtain for $\mathcal{F} = \mathrm{span}\{e_1, \ldots, e_n\}$ a map

$$\theta \colon K \to C(n, \mathbb{K}) := \{X \in M(n, \mathbb{K}) \colon \|X\| \leq 1\}, \quad \theta(k)_{ij} := \langle k e_j, e_i \rangle,$$

which is continuous, surjective and open, and whose fibers are the double cosets $K_n k K_n$ for $k \in K$.

Proof. (i) Surjectivity: For $C \in C(\mathcal{F})$, the operator

$$U_C := \begin{pmatrix} C & \sqrt{1 - CC^*} \\ -\sqrt{1 - C^*C} & C^* \end{pmatrix} \in B(\mathcal{F} \oplus \mathcal{F})$$

is unitary. In view of $2 \dim \mathcal{F} \subseteq \mathcal{H}$, we have an isometric embedding $\mathcal{F} \oplus \mathcal{F} \hookrightarrow \mathcal{H}$, and each unitary operator on $\mathcal{F} \oplus \mathcal{F}$ extends to $\mathcal{H}$ by the identity on the orthogonal complement. To see that the resulting operator in contained in K, it remains to see that $\det U_C = 1$ if $\mathbb{K} = \mathbb{R}$. To verify this claim, we first observe that for $U_1, U_2 \in U_n(\mathbb{K})$, we have

$$U_{U_1 C U_2} = \begin{pmatrix} U_1 & 0 \\ 0 & U_2^* \end{pmatrix} U_C \begin{pmatrix} U_2 & 0 \\ 0 & U_1^* \end{pmatrix},$$

which implies in particular that $\det U_C = \det U_{U_1 C U_2}$. We may therefore assume that C is diagonal, and in this case the assertion follows from the trivial case where $\dim \mathcal{F} = 1$. This implies that θ is surjective.

(ii) θ separates the double cosets of $K_\mathcal{F}$: We may w.l.o.g. assume that $e_1, \ldots, e_n$ span $\mathcal{F}$. First we observe that for $m < 2n$, the subgroup K_m acts transitively on spheres in $\mathcal{H}(m)^\perp$.[3]

---

[2] For $\mathbb{K} = \mathbb{C}, \mathbb{H}$, the condition $2 \dim \mathcal{F} \leq \dim \mathcal{H}$ is sufficient.

[3] Our assumption implies that $\dim \mathcal{H} \geq 2$. This claim follows from the case where $\mathcal{H} = \mathbb{K}^2$. Using the diagonal inclusion $U(1, \mathbb{K})^2 \hookrightarrow U(2, \mathbb{K})$, it suffices to consider vectors with real entries,

Suppose that $\theta(k) = \theta(k')$, i.e., that the first n components of the vectors ke_j and $k'e_j$, $j = 1, \ldots, n$, coincide. Let $P: \mathcal{H} \to \mathcal{F}^{\perp}$ denote the orthogonal projection. Then $\|Pke_1\| = \|Pk'e_1\|$, so that the argument in the preceding paragraph shows that there exists a $k_1 \in K_n$ with $k_1 P k e_1 = P k' e_1$. This implies that $k_1 k e_1 = k' e_1$. Replacing k by $k_1 k$, we may now assume that $k e_1 = k' e_1$. Then $\|Pke_2\| = \|Pk'e_2\|$ and the scalar products of Pke_2 and $Pk'e_2$ with Pke_1 coincide. We therefore find an element $k_2 \in K_n$ fixing Pke_1, hence also ke_1, and satisfying $k_2 P k e_2 = P k' e_2$, i.e., $k_2 k e_2 = k' e_2$. Inductively, we thus obtain $k_1, \ldots, k_n \in K_n$ with $k_n \cdots k_1 k e_j = k' e_j$ for $j = 1, \ldots, n$, and this implies that $k' \in k_n \cdots k_1 k K_n \subseteq K_n k K_n$.

(iii) It is clear that θ is continuous. To see that it is open, let $O \subseteq K$ be an open subset. Then $\theta(O) = \theta(K_n O K_n)$, so that we may w.l.o.g. assume that $O = K_n O K_n$. From (i) and (ii) it follows that every K_n-double coset intersects $K(2n + 1)$, so that $\theta(O) = \theta(O \cap K(2n + 1))$. Therefore it is enough to observe that the restriction of θ to $K(2n + 1)$ is open, which follows from the compactness of $K(2n + 1)$ and the fact that $\theta|_{K(2n+1)}: K(2n + 1) \to C(n, \mathbb{K})$ is a quotient map. $\qquad\square$

For a continuous unitary representation $(\pi, \mathcal{H}_\pi)$ of K, let $V := \mathcal{H}_\pi^{K_n}$ denote the subspace of K_n-fixed vectors, $P: \mathcal{H}_\pi \to V$ be the orthogonal projection and $\pi_V(g) := P^* \pi(g) P$. Then π_V is a $B(V)$-valued continuous positive definite function and Lemma 3.9 implies that we obtain a well-defined continuous map

$$\rho: C(n, \mathbb{K}) \to B(V), \quad \rho(\theta(k)) := \pi_V(k) \quad \text{for} \quad k \in K.$$

The operator adjoint $*$ turns $C(n, \mathbb{K})$ into an involutive semigroup, and we obviously have $\pi_V(k)^* = \pi_V(k^*)$.

Olshanski's proof of the following lemma is based on the fact that the projection of the invariant probability measure on $\mathbb{S}^n$ to an axis for $n \to \infty$ to the Dirac measure in 0.

Lemma 3.10 ([Ol78, Lemma 1.7]). *The map ρ is a continuous $*$-representation of the involutive semigroup $C(n, \mathbb{K})$ by contractions satisfying $\rho(\mathbf{1}) = \mathbf{1}$.*

---

which reduces the problem to the transitivity of the action of $SO(2, \mathbb{R})$ on the unit circle. Since the trivial group $SO(1, \mathbb{R})$ does not act transitively on $\mathbb{S}^0 = \{\pm 1\}$, it is here where we need that $2 \dim \mathcal{F} < \dim \mathcal{H}$.

Using Zorn's Lemma, we conclude that π is a direct sum of subrepresentations for which the subspace of K_n-fixed vectors is cyclic for some $n \in \mathbb{N}$. We may therefore assume that $V = (\mathcal{H}_\pi)^{K_n}$ is cyclic in $\mathcal{H}_\pi$. Then the representation π is equivalent to the GNS-representation of K, defined by the positive definite function π_V (Remark A.3). Since the subspace $V = (\mathcal{H}_\pi)^{K_n}$ is obviously invariant under the commutant $\pi(K)'$ of $\pi(K)$, the cyclicity of V implies that we have an injective map

$$\pi(K)' \to \rho(C(n, \mathbb{K}))' \subseteq B(V)$$

which actually is an isomorphism because $\pi_V(K) = \rho(C(n, \mathbb{K}))$ is a semigroup (Proposition A.6).

Therefore the structure of π is completely encoded in the representation ρ of the semigroup $C(n, \mathbb{K})$. We therefore have to understand the $*$-representations (ρ, V) of $C(n, \mathbb{K})$ for which the $B(V)$-valued function $\rho \circ \theta \colon K \to B(V)$ is positive definite.

Definition 3.11. We call a $*$-representation (ρ, V) of $C(n, \mathbb{K})$ θ-*positive* if the corresponding function $\rho \circ \theta \colon K \to B(V)$ is positive definite.

If ρ is θ-positive, then we obtain a continuous unitary GNS-representation $(\pi_\rho, \mathcal{H}_\rho)$ of K containing a K-cyclic subspace V such that the orthogonal projection $P \colon \mathcal{H}_\rho \to V$ satisfies $P \pi_\rho(g) P^* = \rho(\theta(g))$ for $g \in K$ (cf. Remark A.3). The following lemma shows that we can recover V as the space of K_n-fixed vectors in $\mathcal{H}_\rho$.

Lemma 3.12. $(\mathcal{H}_\rho)^{K_n} = V$.

Proof. Since $\rho \circ \theta$ is K_n-biinvariant, the subspace V consists of K_n-fixed vectors because K acts in the corresponding subspace $\mathcal{H}_{\rho \circ \theta} \subseteq V^K$ by right translations (cf. Remark A.3). Let $W := (\mathcal{H}_\rho)^{K_n}$ and $Q \colon \mathcal{H}_\rho \to W$ denote the corresponding orthogonal projection. Then $\nu(\theta(g)) := Q \pi_\rho(g) Q^*$ defines a contraction representation (ν, W) of $C(n, \mathbb{K})$ (Lemma 3.10) and, for $s \in C(n, \mathbb{K})$, we have $\rho(s) = P \nu(s) P^*$.

In $\mathcal{H}$ the subspace V is K-cyclic. Therefore the subspaces $Q \pi(g) V$, $g \in K$, span W. In view of $Q \pi(g) V = \nu(\theta(g)) V$, this means that $V \subseteq W$ is cyclic for $C(n, \mathbb{K})$. Now Remark A.9 implies that $V = W$. $\square$

We subsume the results of this subsection in the following proposition.

Proposition 3.13. *Let (ρ, V) be a continuous θ-positive $*$-representation of $C(n, \mathbb{K})$ by contractions and $\varphi := \rho \circ \theta$. Then the corresponding GNS-representation $(\pi_\varphi, \mathcal{H}_\varphi)$ of K is continuous with cyclic subspace $V \cong (\mathcal{H}_\varphi)^{K_n}$ and $(\pi_\varphi)_V = \varphi = \rho \circ \theta$. This establishes a one-to-one correspondence of θ-positive continuous $*$-representation of $C(n, \mathbb{K})$ and continuous unitary representations $(\pi, \mathcal{H}_\pi)$ of K generated by the subspace $(\mathcal{H}_\pi)^{K_n}$ of K_n-fixed vectors. This correspondence preserves direct sums of representations.*

3.3 θ-Positive Representations of $C(n, \mathbb{K})$

For $\mathbb{K} = \mathbb{R}, \mathbb{H}$, let $Z :=]0, 1]\mathbf{1} \subseteq C(n, \mathbb{K})$ be the central subsemigroup of real multiples of $\mathbf{1}$. Then the continuous bounded characters of Z are of the form $\chi_s(r) := r^s$, $s \geq 0$. Any continuous $*$-representation (π, V) of Z by contractions determines a spectral measure P on $\hat{Z} := \mathbb{R}_+$ satisfying $\pi(r\mathbf{1}) = \int_0^\infty r^s \, dP(s)$ (cf. [BCR84], [Ne00, VI.2]).

For $\mathbb{K} = \mathbb{C}$, the subsemigroup $Z := \{z \in \mathbb{C}^\times \mathbf{1} : |z| \leq 1\} \cong]0, 1] \times \mathbb{T}$ is also central in $C(n, \mathbb{C})$. Its continuous bounded characters are of the form $\chi_{s,n}(re^{it}) := r^s e^{int}$, $s \geq 0$, $n \in \mathbb{Z}$. Accordingly, continuous $*$-representation of Z by contractions correspond to spectral measures on $\hat{Z} := \mathbb{R}_+ \times \mathbb{Z}$.

Let (ρ, V) be a continuous (with respect to the weak operator topology on $B(V)$) $*$-representation of $C(n, \mathbb{K})$ by contractions. Since the spectral projections for the restriction $\rho_Z := \rho|_Z$ lie in the commutant of $\rho(C(n, \mathbb{K}))$, the representation ρ is a direct sum of subrepresentations for which the support of the spectral measure of ρ_Z is a compact subset of $\hat{Z}$. We call these representations *centrally bounded*. Then the operators $\rho(r\mathbf{1})$ are invertible for $r > 0$, and this implies that

$$\hat{\rho}(rM) := \rho(r^{-1}\mathbf{1})^{-1}\rho(M) \quad \text{for} \quad r > 1, M \in C(n, \mathbb{K}),$$

yields a well-defined extension $\hat{\rho}$ of ρ to a continuous $*$-representation of the multiplicative $*$-semigroup $(M(n, \mathbb{K}), *)$ on V. For a more detailed analysis of the decomposition theory, we may therefore restrict our attention to centrally bounded representations. Decomposing further as a direct sum of cyclic representations, it even suffices to consider separable centrally bounded representations.

The following proposition contains the key new points compared with Olshanski's approach in [Ol78]. Note that it is very close to the type of reasoning used in [JN13] for the classification of the bounded unitary representations of $\mathrm{SU}_2(\mathcal{A})$.

Proposition 3.14. *For every centrally bounded contraction representation (ρ, V) of $C(n, \mathbb{K})$, the following assertions hold:*

(i) *The restriction of $\hat{\rho}$ to $\mathrm{GL}(n, \mathbb{K})$ is a norm-continuous representation whose differential $d\hat{\rho} : \mathfrak{gl}(n, \mathbb{K}) \to B(V)$ is a representation of the Lie algebra $\mathfrak{gl}(n, \mathbb{K})$ by bounded operators.*

(ii) *If ρ is θ-positive, then $\hat{\rho}$ is real analytic on $M(n, \mathbb{K})$ and extends to a holomorphic semigroup representation of the complexification*

$$M(n, \mathbb{K})_\mathbb{C} \cong \begin{cases} M(n, \mathbb{C}) & \text{for } \mathbb{K} = \mathbb{R} \\ M(n, \mathbb{C}) \oplus M(n, \mathbb{C}) \cong B(\mathbb{C}^n) \oplus B(\overline{\mathbb{C}^n}) & \text{for } \mathbb{K} = \mathbb{C} \\ M(n, M(2, \mathbb{C})) \cong M(2n, \mathbb{C}) & \text{for } \mathbb{K} = \mathbb{H}. \end{cases}$$

Proof. (i) We have already seen that $\hat{\rho} : M(n, \mathbb{K}) \to B(V)$ is locally bounded and continuous. Hence it restricts to a locally bounded continuous representation of the involutive Lie group $(\mathrm{GL}(n, \mathbb{K}), *)$. Integrating this representation to the convolution algebra $C_c^\infty(\mathrm{GL}(n, \mathbb{K}))$, we see that the subspace V^∞ of smooth vectors is dense. For the corresponding derived representation

$$\mathrm{d}\hat{\rho}: \mathfrak{gl}(n, \mathbb{K}) \to \mathrm{End}(V^\infty)$$

our construction immediately implies that the operator $\mathrm{d}\hat{\rho}(\mathbf{1})$ is bounded and $\mathrm{d}\hat{\rho}(X) \geq 0$ for $X = X^* \geq 0$ because we started with a contraction representation of $C(n, \mathbb{K})$. From $X \leq \|X\|\mathbf{1}$ we also derive $\mathrm{d}\hat{\rho}(X) \leq \|X\|\mathrm{d}\hat{\rho}(\mathbf{1})$, so that $\mathrm{d}\rho(X)$ is bounded. As $\mathfrak{u}(n, \mathbb{K}) \subseteq \mathfrak{z}(\mathfrak{gl}(n, \mathbb{K})) + [\mathrm{Herm}(n, \mathbb{K}), \mathrm{Herm}(n, \mathbb{K})]$, we conclude that $\mathrm{d}\hat{\rho}$ is a $*$-representation by bounded operators on the Hilbert space V.

For $X \in \mathfrak{gl}(n, \mathbb{K})$, we then have the relation

$$\hat{\rho}(\exp X) = e^{\mathrm{d}\hat{\rho}(X)} \quad \text{for} \quad X \in \mathfrak{gl}(n, \mathbb{K}).$$

This implies that $\hat{\rho}: \mathrm{GL}(n, \mathbb{K}) \to \mathrm{GL}(V)$ is norm-continuous.

(ii) Now we assume that $\varphi := \rho \circ \theta: K \to B(V)$ is positive definite. Since $\theta(\mathbf{1}) = \mathbf{1}$, there exists an open $\mathbf{1}$-neighborhood $U \subseteq K$ with $\theta(U) \subseteq \mathrm{GL}(n, \mathbb{K})$. For $k \in U$ we then have $\varphi(k) = \hat{\rho}(\theta(k))$, and since the representation $\hat{\rho}$ of $\mathrm{GL}(n, \mathbb{K})$ is norm continuous, hence analytic, φ is analytic on U. Now Theorem A.5 implies that φ is analytic.

Let $\Omega := \{C \in C(n, \mathbb{K}): \|C\| < 1\}$ denote the interior of $C(n, \mathbb{K})$. On this domain we have an analytic cross section of θ, given by

$$\sigma(C) := \begin{pmatrix} C & \sqrt{1 - CC^*} & 0 \\ -\sqrt{1 - C^*C} & C^* & 0 \\ 0 & 0 & 1 \end{pmatrix}.$$

Now $\varphi(\sigma(C)) = \rho(\theta(\sigma(C))) = \rho(C)$ for $C \in \Omega$ implies that $\rho|_\Omega$ is analytic. From $\hat{\rho}(rC) = \hat{\rho}(r)\rho(C)$ for $r > 0$ it now follows that $\hat{\rho}: M(n, \mathbb{K}) \to B(V)$ is analytic.

It remains to show that $\hat{\rho}$ extends to a holomorphic map on $M(n, \mathbb{K})_\mathbb{C}$. First, the analyticity of $\hat{\rho}$ implies for some $\varepsilon > 0$ the existence of a holomorphic map F on $B_\varepsilon := \{C \in M(n, \mathbb{K})_\mathbb{C}: \|C\| < \varepsilon\}$ with $F(C) = \hat{\rho}(C)$ for $C \in M(n, \mathbb{K}) \cap B_\varepsilon$. This map also satisfies $F(rC) = \hat{\rho}(r\mathbf{1})F(C)$ for $r < 1$, which implies that F extends to a holomorphic map on $r^{-1}B_\varepsilon = B_{r^{-1}\varepsilon}$ for every $r > 0$. This leads to the existence of a holomorphic extension of $\hat{\rho}$ to all of $M(n, \mathbb{K})_\mathbb{C}$. That this extension also is multiplicative follows immediately by analytic continuation. $\qquad\square$

Theorem 3.15 (Classification of Irreducible θ-Positive Representations). *Put $\mathcal{F} := \mathbb{K}^n$. Then all irreducible continuous θ-positive representations of $C(\mathcal{F}) \cong C(n, \mathbb{K})$ are of the form*

$$\begin{cases} \mathbb{S}_\lambda(\mathcal{F}_\mathbb{C}) \subseteq (\mathcal{F}_\mathbb{C})^{\otimes N}, & \text{for } \mathbb{K} = \mathbb{R}, \\ \mathbb{S}_\lambda(\mathcal{F}) \otimes \mathbb{S}_\mu(\overline{\mathcal{F}}) \subseteq \mathcal{F}^{\otimes N} \otimes \overline{\mathcal{F}}^{\otimes M}, & \text{for } \mathbb{K} = \mathbb{C}, \\ \mathbb{S}_\lambda(\mathcal{F}^\mathbb{C}) \subseteq (\mathcal{F}^\mathbb{C})^{\otimes N}, & \text{for } \mathbb{K} = \mathbb{H}, \end{cases}$$

where $\lambda \in \text{Part}(N, n), \mu \in \text{Part}(M, n)$.

Proof. Let (ρ, V) be an irreducible θ-positive representation of $C(n, \mathbb{K})$. Then $\rho(Z) \subseteq \mathbb{C}\mathbf{1}$ by Schur's Lemma, so that ρ is in particular centrally bounded and extends to a holomorphic representation $\hat{\rho} \colon M(n, \mathbb{K})_\mathbb{C} \to B(V)$ (Proposition 3.14).

For $\mathbb{K} = \mathbb{R}, \mathbb{H}$, the center of $M(n, \mathbb{K})_\mathbb{C}$ is $\mathbb{C}\mathbf{1}$. Since the only holomorphic multiplicative maps $\mathbb{C} \to \mathbb{C}$ are of the form $z \mapsto z^N$ for some $N \in \mathbb{N}_0$, it follows that $\hat{\rho}(z\mathbf{1}) = z^N\mathbf{1}$ for $z \in \mathbb{C}$. We conclude that the holomorphic map $\hat{\rho}$ is homogeneous of degree N. Hence there exists a linear map

$$\tilde{\rho} \colon S^N(M(n, \mathbb{K})_\mathbb{C}) \to B(V) \quad \text{with} \quad \tilde{\rho}(A^{\otimes N}) = \hat{\rho}(A), \quad A \in M(n, \mathbb{K})_\mathbb{C}.$$

The multiplicativity of $\hat{\rho}$ now implies that $\tilde{\rho}$ is multiplicative, hence a representation of the finite-dimensional algebra $S^N(M(n, \mathbb{K})_\mathbb{C})$.

For $\mathbb{K} = \mathbb{R}$, we have $M(n, \mathbb{R})_\mathbb{C} = M(n, \mathbb{C})$, and

$$S^N(M(n, \mathbb{C})) = (M(n, \mathbb{C})^{\otimes N})^{S_N} \cong M(nN, \mathbb{C})^{S_N} \cong B((\mathbb{C}^n)^{\otimes N})^{S_N}.$$

We conclude that $S^N(M(n, \mathbb{C}))$ is the commutant of S_N in $M(nN, \mathbb{C})$, and by Schur–Weyl theory, this algebra can be identified with the image of the group algebra $\mathbb{C}[GL(n, \mathbb{C})]$ in $B((\mathbb{C}^n)^{\otimes N})$. Therefore its irreducible representations are parametrized by the set $\text{Part}(N, n)$ of partitions of N into at most n summands. This completes the proof for $\mathbb{K} = \mathbb{R}$. For $\mathbb{K} = \mathbb{H}$, we have the same picture because $M(n, \mathbb{H})_\mathbb{C} \cong M(2n, \mathbb{C})$.

For $\mathbb{K} = \mathbb{C}$, $Z(M(n, \mathbb{C})_\mathbb{C}) \cong \mathbb{C}^2$, and the inclusion of $Z(M(n, \mathbb{C})) = \mathbb{C}\mathbf{1}$ has the form $z \mapsto (z, \bar{z})$. Hence there exist $N, M \in \mathbb{N}_0$ with $\rho_Z(z\mathbf{1}) = z^N\bar{z}^M\mathbf{1}$. Therefore the restriction of $\hat{\rho}$ to the first factor is homogeneous of degree N and the restriction to the second factor of degree M. This leads to a representation of the algebra

$$S^{N,M}(M(n, \mathbb{C})) := S^N(M(n, \mathbb{C})) \otimes S^M(M(n, \mathbb{C})),$$

so that the same arguments as in the real case apply.   □

Now that we know all irreducible θ-positive representations, we ask for the corresponding decomposition theory.

Theorem 3.16. *Every continuous θ-positive $*$-representation of $C(n, \mathbb{K})$ by contractions is a direct sum of irreducible ones, and these are finite-dimensional.*

Proof. We have already seen that ρ decomposes into a direct sum of centrally bounded representations. We may therefore assume that ρ is centrally bounded, so that ρ extends to a holomorphic representation $\hat{\rho}$ of $M(n, \mathbb{K})_\mathbb{C}$ (Proposition 3.14).

In the C^*-algebra $\mathcal{A} := M(n, \mathbb{K})_{\mathbb{C}}$, every holomorphic function is uniquely determined by its restriction to the unitary group $U(\mathcal{A})$, which is a totally real submanifold. Therefore a closed subspace $W \subseteq V$ is invariant under $\rho(C(n, \mathbb{K}))$ if and only if it is invariant under $\hat{\rho}(U(\mathcal{A}))$. Since the group $U(\mathcal{A})$ is compact, the assertion now follows from the classical fact that unitary representations of compact groups are direct sums of irreducible ones. $\square$

3.4   The Classification Theorem

We are now ready to prove the Kirillov–Olshanski Theorem [Ki73, Ol78].

Theorem 3.17 (Classification of the Representations of $U_\infty(\mathcal{H})_0$). *Let $\mathcal{H}$ be an infinite-dimensional separable $\mathbb{K}$-Hilbert space.*

(a) *The irreducible continuous unitary representations of $U_\infty(\mathcal{H})_0$ are*

$$\begin{cases} \mathbb{S}_\lambda(\mathcal{H}_{\mathbb{C}}) \subseteq (\mathcal{H}_{\mathbb{C}})^{\otimes N}, & \text{for } \mathbb{K} = \mathbb{R}, \\ \mathbb{S}_\lambda(\mathcal{H}) \otimes \mathbb{S}_\mu(\overline{\mathcal{H}}) \subseteq \mathcal{H}^{\otimes N} \otimes \overline{\mathcal{H}}^{\otimes M}, & \text{for } \mathbb{K} = \mathbb{C}, \\ \mathbb{S}_\lambda(\mathcal{H}^{\mathbb{C}}) \subseteq (\mathcal{H}^{\mathbb{C}})^{\otimes N}, & \text{for } \mathbb{K} = \mathbb{H}, \end{cases}$$

where $\lambda \in \mathrm{Part}(N), \mu \in \mathrm{Part}(M)$.

(b) *Every continuous unitary representation of $U_\infty(\mathcal{H})_0$ is a direct sum of irreducible ones.*

(c) *Every continuous unitary representation of $U_\infty(\mathcal{H})_0$ extends uniquely to a continuous unitary representations of the full unitary group $U(\mathcal{H})_s$, endowed with the strong operator topology.*

Proof. (a) In Theorem 3.15 we have classified the irreducible θ-positive representations of $C(n, \mathbb{K})$. The corresponding representations $(\pi, \mathcal{H}_\pi)$ of $K = U_\infty(\mathcal{H})_0$ can now be determined rather easily. Since the passage from ρ to π preserves direct sums, we consider the representations ρ_N of $C(n, \mathbb{K})$ on $\mathcal{F}_{\mathbb{C}}^{\otimes N}$ for $\mathbb{K} = \mathbb{R}$, on $(\mathcal{F}^{\mathbb{C}})^{\otimes N}$ for $\mathbb{K} = \mathbb{H}$, and the representation $\rho_{N,M}$ on $\mathcal{F}^{\otimes N} \otimes \overline{\mathcal{F}}^{\otimes M}$ for $\mathbb{K} = \mathbb{C}$.

We likewise have unitary representations π_N of K on $(\mathcal{H}_{\mathbb{C}})^{\otimes N}$ for $\mathbb{K} = \mathbb{R}$, on $(\mathcal{H}^{\mathbb{C}})^{\otimes N}$ for $\mathbb{K} = \mathbb{H}$, and a representation $\pi_{N,M}$ on $\mathcal{H}^{\otimes N} \otimes \overline{\mathcal{H}}^{\otimes M}$ for $\mathbb{K} = \mathbb{C}$. These are bounded continuous representations of K.

For $\mathbb{K} = \mathbb{R}, \mathbb{H}$, the space of K_n-fixed vectors in $(\mathcal{H}_{\mathbb{C}})^{\otimes N}$ obviously contains $(\mathcal{F}_{\mathbb{C}})^{\otimes N}$ and by considering the action of the subgroup of diagonal matrices, we obtain the equality $(\mathcal{F}_{\mathbb{C}})^{\otimes N} = ((\mathcal{H}_{\mathbb{C}})^{\otimes N})^{K_n}$. Therefore the representation π_N corresponds to the representation ρ_N of $C(n, \mathbb{K})$. A similar argument shows that for $\mathbb{K} = \mathbb{C}$, the K-representation $\pi_{N,M}$ corresponds to $\rho_{N,M}$.

Since the representations ρ_N and $\rho_{N,M}$ decompose into finitely many irreducible pieces, the representations π_N and $\pi_{N,M}$ decompose in precisely the same way. For $\mathbb{K} = \mathbb{R}, \mathbb{H}$, we thus obtain the Schur modules $\mathbb{S}_\lambda(\mathcal{H}_\mathbb{C})$ and $\mathbb{S}_\lambda(\mathcal{H}^\mathbb{C})$ with $\lambda \in \mathrm{Part}(N)$, respectively. For $\mathbb{K} = \mathbb{C}$, we obtain the tensor products $\mathbb{S}_\lambda(\mathcal{H}) \otimes \mathbb{S}_\mu(\overline{\mathcal{H}})$ with $\lambda \in \mathrm{Part}(N)$ and $\mu \in \mathrm{Part}(M)$.

Here the restriction to partitions consisting of at most n summands corresponds to the K-invariant subspace generated by the K_n-fixed vectors. This subspace is proper if n is small.

(b) From Theorem 3.16 we know that θ-positive contraction representations of $C(n, \mathbb{K})$ are direct sums of irreducible ones. This implies that all continuous unitary representations of K are direct sums of irreducible ones. Since the correspondence between π and ρ leads to isomorphic commutants, the irreducible subrepresentations of π and the corresponding subrepresentations of ρ have the same multiplicities.

(c) The assertion is trivial for the irreducible representations of K described under (a). Since $\mathrm{U}_\infty(\mathcal{H})_0$ is dense in $\mathrm{U}(\mathcal{H})$ with respect to the strong operator topology,[4] this extension is unique and generates the same von Neumann algebra.

$\square$

Remark 3.18 (Representations of $\mathrm{O}_\infty(\mathcal{H})$). The above classification can easily be extended to the nonconnected group $\mathrm{O}_\infty(\mathcal{H})$ (for $\mathbb{K} = \mathbb{R}$). Here the existence of a canonical extension $\overline{\pi}_\lambda$ of every irreducible representations π_λ of $\mathrm{SO}_\infty(\mathcal{H}) := \mathrm{O}_\infty(\mathcal{H})_0$ to $\mathrm{O}(\mathcal{H})$ implies that there exist precisely two extensions that differ by a twist with the canonical character $D \colon \mathrm{O}_\infty(\mathcal{H}) \to \{\pm 1\}$ corresponding to the determinant.

For a general continuous unitary representations of $\mathrm{O}_\infty(\mathcal{H})$, it follows that all $\mathrm{SO}_\infty(\mathcal{H})$-isotypic subspaces are invariant under $\mathrm{O}_\infty(\mathcal{H})$, hence of the form $\mathcal{M}_\lambda \otimes \mathcal{H}_\lambda$, where $\mathrm{O}_\infty(\mathcal{H})$ acts by $\varepsilon \otimes \overline{\pi}_\lambda$ and ε is a unitary representation of the 2-element group $\pi_0(\mathrm{O}_\infty(\mathcal{H}))$, i.e., defined by a unitary involution.

In particular, all continuous unitary representations of $\mathrm{O}_\infty(\mathcal{H})$ are direct sums of irreducible ones, which are of the form $\overline{\pi}_\lambda$ and $D \otimes \overline{\pi}_\lambda$. Here the first type extends to the full orthogonal group $\mathrm{O}(\mathcal{H})$, whereas the second type does not.

Remark 3.19 (Extension to Overgroups). (cf. [Ol84, §1.11]) Let $\mathcal{H}$ be an infinite-dimensional separable $\mathbb{K}$-Hilbert space and $K := \mathrm{U}_\mathbb{K}(\mathcal{H})$. We put

$$\mathcal{H}^\sharp := \begin{cases} \mathcal{H}_\mathbb{C} & \text{for } \mathbb{K} = \mathbb{R} \\ \mathcal{H} \oplus \overline{\mathcal{H}} & \text{for } \mathbb{K} = \mathbb{C} \\ \mathcal{H}^\mathbb{C} & \text{for } \mathbb{K} = \mathbb{H}. \end{cases}$$

---

[4]This follows from the fact that $\mathrm{U}_\infty(\mathcal{H})_0$ acts transitively on the finite orthonormal systems in $\mathcal{H}$.

For each $N \in \mathbb{N}$ we obtain a norm continuous representation

$$\pi_N : (B(\mathcal{H}^\sharp), \cdot) \to B((\mathcal{H}^\sharp)^{\otimes N}), \quad \pi_N(A) := A^{\otimes N}$$

of the multiplicative semigroup $(B(\mathcal{H}^\sharp), \cdot)$ whose restriction to $U(\mathcal{H}^\sharp)$ is unitary.
We collect some properties of this representation:

(a) Let $C(\mathcal{H}^\sharp) = \{S \in B(\mathcal{H}^\sharp) : \|S\| \leq 1\}$ denote the closed subsemigroup of contractions. Then $C(\mathcal{H}^\sharp)$ is a $*$-subsemigroup of $B(\mathcal{H}^\sharp)$ and $\pi_N|_{C(\mathcal{H}^\sharp)}$ is continuous with respect to the weak operator topology. In fact, $\pi_N(C(\mathcal{H}^\sharp))$ consists of contractions, and for the total subset of vectors of the form $v := v_1 \otimes \cdots \otimes v_N$, $w := w_1 \otimes \cdots \otimes w_N$, the matrix coefficient $S \mapsto \langle \pi_N(S)v, w \rangle = \prod_{j=1}^N \langle Sv_j, w_j \rangle$ is continuous.

(b) $K := U_{\mathbb{K}}(\mathcal{H})$ is dense in $C_{\mathbb{K}}(\mathcal{H})$ with respect to the weak operator topology. It suffices to see that for every contraction C on a finite-dimensional subspace $\mathcal{F} \subseteq \mathcal{H}$, there exists a unitary operator $U \in U_{\mathbb{K}}(\mathcal{H})$ with $P_{\mathcal{F}} U P_{\mathcal{F}}^* = C$, where $P_{\mathcal{F}} : \mathcal{H} \to \mathcal{F}$ is the orthogonal projection. Since $\mathcal{F} \oplus \mathcal{F}$ embeds isometrically into $\mathcal{H}$, this follows from the fact that the matrix

$$U := \begin{pmatrix} C & \sqrt{1 - CC^*} \\ -\sqrt{1 - C^*C} & C^* \end{pmatrix} \in M_2(B_{\mathbb{K}}(\mathcal{F})) = B_{\mathbb{K}}(\mathcal{F} \oplus \mathcal{F})$$

is unitary and satisfies $P_{\mathcal{F}} U P_{\mathcal{F}}^* = C$ (Lemma 3.9).

(c) Combining (a) and (b) implies that $\pi_N(C_{\mathbb{K}}(\mathcal{H})) \subseteq \pi_N(K)''$, and hence that $\pi_N(B_{\mathbb{K}}(\mathcal{H})) = \bigcup_{\lambda > 0} \lambda^N \pi_N(C_{\mathbb{K}}(\mathcal{H})) \subseteq \pi_N(K)''$. For the corresponding Lie algebra representation

$$d\pi_N : \mathfrak{gl}_{\mathbb{K}}(\mathcal{H}) \to B(\mathcal{H}^{\otimes N}), \quad d\pi_N(X) := \sum_{j=1}^N \mathbf{1}^{\otimes(j-1)} \otimes X \otimes \mathbf{1}^{\otimes(N-j)},$$

this implies that $d\pi_N(\mathfrak{gl}_{\mathbb{K}}(\mathcal{H})) \subseteq \pi_N(K)''$, and hence also that $d\pi_N(\mathfrak{gl}_{\mathbb{K}}(\mathcal{H}))_{\mathbb{C}} \subseteq \pi_N(K)''$. The connectedness of the group $K^\sharp$ (Lemma 2.4) now implies that $\pi_N(K^\sharp) \subseteq \pi_N(K)''$. Since the subgroup $K_\infty^\sharp$, consisting of those elements g for which $g - \mathbf{1}$ is compact, is strongly dense in $K^\sharp$ and the representation of $K^\sharp$ is continuous with respect to the strong operator topology, the representations π_N of $K^\sharp$ thus decomposes into Schur modules, as described in Theorem 3.17(a).

(d) The preceding discussion shows in particular that the representation π_N of K extends to the overgroup $K^\sharp$ without enlarging the corresponding von Neumann algebra. If ρ is the corresponding representation of $C(n, \mathbb{K}) = C(\mathcal{F})$ on $V := (\mathcal{F}^\sharp)^{\otimes N}$, where $\mathcal{F} = \mathbb{K}^n$, then ρ extends to a holomorphic representation $\hat{\rho}$ of $M(n, \mathbb{K})_{\mathbb{C}}$ and the map $\theta : K \to C(n, \mathbb{K})$ likewise extends to a holomorphic map $\hat{\theta} : B(\mathcal{H})_{\mathbb{C}} \to M(n, \mathbb{K})_{\mathbb{C}}$. Now $\hat{\rho} \circ \hat{\theta} : B(\mathcal{H})_{\mathbb{C}} \to B(V)$ is a holomorphic positive definite function corresponding to the representation of $(B(\mathcal{H})_{\mathbb{C}}, \cdot)$ on $(\mathcal{H}^{\sharp})^{\otimes N}$ whose restriction yields a unitary representation of the unitary group $\mathrm{U}(\mathcal{H})^{\sharp}$ of $B(\mathcal{H})_{\mathbb{C}}$.

We conclude this subsection with the following converse to Proposition 3.7.

Theorem 3.20. *A unitary representation of* $\mathrm{U}(\infty, \mathbb{K})_0$ *is tame if and only if it extends to a continuous unitary representation of* $\mathrm{U}_{\infty}(\mathcal{H})_0$ *for* $\mathcal{H} = \ell^2(\mathbb{N}, \mathbb{K})$.

Proof. We have already seen in Proposition 3.7 that every continuous unitary representation of $K = \mathrm{U}_{\infty}(\mathcal{H})_0$ restricts to a tame representation of $K(\infty) = \mathrm{U}(\infty, \mathbb{K})_0$.

Suppose, conversely, that $(\pi, \mathcal{H}_{\pi})$ is a tame unitary representation of $K(\infty)$. Then the same arguments as for K imply that it is a direct sum of representations generated by the subspace $V = (\mathcal{H}_{\pi})^{K(\infty)_n}$ and we obtain a representation (ρ, V) of $C(n, \mathbb{K})$ for which $\rho \circ \theta$ is positive definite on $K(\infty)$. Since it is continuous and $K(\infty)$ is dense in K, it is also positive definite on K. Now the GNS construction, applied to $\rho \circ \theta$, yields the continuous extension of π to K. $\square$

3.5 The Inseparable Case

In this subsection we show that Theorem 3.17 extends to the case where $\mathcal{H}$ is not separable.

Theorem 3.21. *Theorem 3.17 also holds if* $\mathcal{H}$ *is inseparable.*

Proof. (a) First we note that the Schur–Weyl decomposition

$$\mathcal{H}^{\otimes N} \cong \bigoplus_{\lambda \in \mathrm{Part}(N)} \mathbb{S}_{\lambda}(\mathcal{H}) \otimes \mathcal{M}_{\lambda}$$

holds for any infinite-dimensional complex Hilbert space [BN12] and that the spaces $\mathbb{S}_{\lambda}(\mathcal{H})$ carry irreducible representations of $\mathrm{U}(\mathcal{H})$ which are continuous with respect to the norm topology and the strong operator topology on $\mathrm{U}(\mathcal{H})$.

(b) To obtain the irreducible representations of $K := \mathrm{U}_{\infty}(\mathcal{H})_0$, we choose an orthonormal basis $(e_j)_{j \in J}$ of $\mathcal{H}$ and assume that $\mathbb{N} = \{1, 2, \ldots\}$ is a subset of J. Accordingly, we obtain an embedding $K(\infty) := \mathrm{U}(\infty, \mathbb{K}) \hookrightarrow \mathrm{U}_{\infty}(\mathcal{H})$ and define $K_n := \{k \in K : ke_j = e_j, j = 1, \ldots, n\}$. For a subset $M \subseteq J$, we put $K(M) := \mathrm{U}_{\infty}(\mathcal{H}_M)_0$, where $\mathcal{H}_M \subseteq \mathcal{H}$ is the closed subspace generated by $(e_j)_{j \in M}$.

(c) Kirillov's Lemma 3.6 is still valid in the inseparable case and Lemma 3.10 follows from the separable case because $\theta(K) = \theta(K(\infty))$.

(d) With the same argument as in Sect. 3.2 it follows that π is a direct sum of subrepresentations for which $(\mathcal{H}_{\pi})^{K_n}$ is cyclic. These in turn correspond to θ-positive representations (ρ, V) of $C(n, \mathbb{K})$. We claim that $\rho \circ \theta$ is positive definite on $K(\mathbb{N})$ if and only if it is positive definite on $K(M)$ for any countable subset with $\mathbb{N} \subseteq M \subseteq J$. In fact, there exists a unitary isomorphism $U_M: \mathcal{H}(\mathbb{N}) \to \mathcal{H}(M)$ fixing $e_1, \ldots, e_n$. For $k \in K(M)$ we then have $\rho(\theta(k)) = \rho(\theta(U_M^* k U_M))$, so that $\rho \circ \theta$ is positive definite on $K(M)$ if it is on $K(\mathbb{N})$.

For every finite subset $F \subseteq K$, there exists a countable subset $J_c \subseteq J$ containing $\mathbb{N}$ such that F fixes all basis elements e_j, $j \notin J_c$. Therefore $\rho \circ \theta$ is positive definite on K if and only if this is the case on $K(M)$ for every countable subset and this in turn follows from the positive definiteness on the subgroup $K(\mathbb{N})$. We conclude that the classification of the unitary representations of K is the same as for $K(\mathbb{N})$. $\qquad\square$

Remark 3.22. If $\mathcal{H}$ is inseparable, then the classification implies that all irreducible unitary representations of $U_\infty(\mathcal{H})_0$ are inseparable. In particular, all separable unitary representations of $U_\infty(\mathcal{H})_0$ are trivial because they are direct sums of irreducible ones.

Problem 3.23. It seems that the classification problem we dealt with in this section can be formulated in a more general context as follows. Let $\mathcal{A}$ be a real involutive Banach algebra and $P \in \mathcal{A}$ be a hermitian projection, so that we obtain a closed subalgebra $\mathcal{A}_P := P\mathcal{A}P$. On the unitary group.

$$U(\mathcal{A}) := \{A \in \mathcal{A}: A^* A = AA^* = \mathbf{1}\}$$

We consider the map

$$\theta: U(\mathcal{A}) \to C(\mathcal{A}_P) := \{A \in \mathcal{A}_P: \|A\| \le 1\}, \quad \theta(g) := PgP.$$

For which $*$-representations (ρ, V) of the semigroup $C(\mathcal{A}_P)$ is the function $\rho \circ \theta: U(\mathcal{A})_0 \to B(V)$ positive definite?

For $\mathcal{A} = B_\infty(\mathcal{H})$, the compact operators on the $\mathbb{K}$-Hilbert space $\mathcal{H}$ and a finite rank projection P, this problem specializes to the determination of the θ-positive representations of $C(n, \mathbb{K})$.

If P is central, then θ is a $*$-homomorphism, so that $\rho \circ \theta$ is positive definite for any representation ρ.

4 Separable Representations of $U(\mathcal{H})$

In this section we show that for the unitary group $U(\mathcal{H})$ of a separable Hilbert space $\mathcal{H}$, endowed with the norm topology, all separable representations are uniquely determined by their restrictions to the normal subgroup $U_\infty(\mathcal{H})_0$. This result of Pickrell [Pi88] extends the Kirillov–Olshanski classification to separable representations of $U(\mathcal{H})$.

4.1 Triviality of Separable Representations Modulo Compacts

Before we turn to the proof of Theorem 4.6, we need a few preparatory lemmas.

Lemma 4.1. *Let $\mathcal{H}$ be an infinite-dimensional Hilbert space, $(\pi, \mathcal{H}_\pi)$ be a continuous unitary representation of $U(\mathcal{H})$, and $\mathcal{H} = V \oplus V^\perp$ with $V \cong V^\perp$. Then $\mathcal{H}_\pi^{U(V)} \neq \{0\}$.*

Proof. Put $\mathcal{H}_1 := V$ and write $V^\perp$ as a Hilbert space direct sum $\widehat{\bigoplus}_{j=2}^\infty \mathcal{H}_j$, where each $\mathcal{H}_j$ is isomorphic to V or $\mathcal{H}$. This is possible because $|J| = |\mathbb{N} \times J|$ for every infinite set J. We claim that some $U(\mathcal{H}_j)$ has nonzero fixed points in $\mathcal{H}_\pi$. Once this claim is proved, we choose $g \in U(\mathcal{H})$ with $gV = \mathcal{H}_j$. Then $\pi(g)\mathcal{H}_\pi^{U(V)} = \mathcal{H}_\pi^{U(\mathcal{H}_j)} \neq \{0\}$ implies the assertion.

For the proof we want to use Proposition 1.6. In $G := U(\mathcal{H})$ we consider the basis of $\mathbf{1}$-neighborhoods given by $U_\varepsilon := \{g \in U(\mathcal{H}): \|g - \mathbf{1}\| < \varepsilon\}$ and the subgroups $G_j := U(\mathcal{H}_j)$. Then the proof of Proposition 1.3(i) shows that there exists an $m \in \mathbb{N}$ with $G_j \subseteq (U_\varepsilon \cap G_j)^m$ for every j, which is (U1). It is also clear that (U2) is satisfied. Therefore the assertion follows from Proposition 1.6. □

Lemma 4.2. *Let $\mathcal{F} \subseteq \mathcal{H}$ be a closed subspace of finite codimension. Then the natural morphism $U(\mathcal{F}) \to U(\mathcal{H})/U_\infty(\mathcal{H})$ is surjective, i.e., $U(\mathcal{H}) = U_\infty(\mathcal{H})U(\mathcal{F})$.*

Proof. Since the groups $U(\mathcal{H})$ and $U(\mathcal{F})$ are connected (Proposition 1.3(i)), it suffices to show that their Lie algebras satisfy

$$\mathfrak{u}(\mathcal{H}) = \mathfrak{u}(\mathcal{F}) + \mathfrak{u}_\infty(\mathcal{H}).$$

Let $P: \mathcal{H} \to \mathcal{H}$ be the orthogonal projection onto $\mathcal{F}$. Then every $X \in \mathfrak{u}(\mathcal{H})$ can be written as

$$X = PXP + (\mathbf{1} - P)XP + X(\mathbf{1} - P),$$

where $PXP \in \mathfrak{u}(\mathcal{F})$ and the other two summands are compact because $\mathbf{1} - P$ has finite range. □

Lemma 4.3. *Let $(\pi, \mathcal{H}_\pi)$ be a continuous unitary representation of $U(\mathcal{H})$ with $U_\infty(\mathcal{H}) \subseteq \ker \pi$ and $\mathcal{H} = \widehat{\bigoplus}_{j \in J} \mathcal{H}_j$ with $\mathcal{H}_j$ infinite-dimensional separable and J infinite. Then $\bigcap_{j \in J} \mathcal{H}_\pi^{U(\mathcal{H}_j)} \neq \{0\}$.*

Proof. Let $V \subseteq \mathcal{H}$ be a closed subspace of the form $V = \overline{\sum_j V_j}$, where each $V_j \subseteq \mathcal{H}_j$ is a closed subspace of codimension 1. Then $V^\perp \cong \ell^2(J, \mathbb{C}) \cong \mathcal{H} \cong V$ because $|J \times \mathbb{N}| = |J|$. According to Lemma 4.2, we then have

$$U(\mathcal{H}_j) \subseteq U(V_j)U_\infty(\mathcal{H}_j) \subseteq U(V)U_\infty(\mathcal{H}).$$

In view of Lemma 4.1, $U(V)$ has nonzero fixed points in $\mathcal{H}_\pi$, and since $U_\infty(\mathcal{H}) \subseteq \ker \pi$, any such fixed point is fixed by all the subgroups $U(\mathcal{H}_j)$. $\square$

From now on we assume that $\mathcal{H}$ is separable.

Lemma 4.4. *Let* $(\pi, \mathcal{H}_\pi)$ *be a continuous unitary representation of* $U(\mathcal{H})$ *with* $U_\infty(\mathcal{H}) \subseteq \ker \pi$ *and* $g \in U(\mathcal{H})$. *If* 1 *is contained in the essential spectrum of* g, *i.e., the image of* $g - 1$ *in the Calkin algebra* $B(\mathcal{H})/K(\mathcal{H})$ *is not invertible, then* 1 *is an eigenvalue of* $\pi(g)$.

Proof. We choose an orthogonal decomposition $\mathcal{H} = \hat{\bigoplus}_{n=1}^\infty \mathcal{H}_n$ into infinite-dimensional g-invariant subspaces of $\mathcal{H}$ as follows.

Case 1: If 1 is an eigenvalue of g of infinite multiplicity, then we put $\mathcal{H}_0 := \ker(1 - g)^\perp$. If this space is infinite-dimensional, then we put $\mathcal{H}_1 := \mathcal{H}_0$, and if this is not the case, then we pick a subspace $\mathcal{H}_0' \subseteq \mathcal{H}_0^\perp$ of infinite dimension and codimension and put $\mathcal{H}_1 := \mathcal{H}_0 \oplus \mathcal{H}_0'$. We choose all other $\mathcal{H}_n$, $n > 1$, such that $\mathcal{H}_1^\perp = \hat{\bigoplus}_{n=2}^\infty \mathcal{H}_n$ and note that $\mathcal{H}_1^\perp \subseteq \ker(1 - g)$.

Case 2: If $\ker(1 - g)$ is finite-dimensional, then let $P_\varepsilon \in B(\mathcal{H})$ be the spectral projection for g corresponding to the closed disc of radius $\varepsilon > 0$ about 1. Then

$$g_\varepsilon := P_\varepsilon \oplus (1 - P_\varepsilon)g$$

satisfies $\|g_\varepsilon - g\| \leq \varepsilon$. The noncompactness of $g - 1$ implies that if ε is small enough, then

$$g_\varepsilon - 1 = 0 \oplus (1 - P_\varepsilon)(g - 1)$$

is noncompact, and hence that $P_\varepsilon \mathcal{H}$ has infinite codimension. Further $P_\varepsilon \mathcal{H}$ is infinite-dimensional because 1 is an essential spectral value of g. Hence there exists a sequence $\varepsilon_1 > \varepsilon_2 > \ldots$ converging to 0, for which the g-invariant subspaces

$$\mathcal{H}_1 := (P_{\varepsilon_1}\mathcal{H})^\perp \quad \text{and} \quad \mathcal{H}_j := P_{\varepsilon_{j-1}}\mathcal{H} \cap (P_{\varepsilon_j}\mathcal{H})^\perp$$

are infinite-dimensional.

In both cases, we consider g_ε as an element $(g_{\varepsilon,n})$ of the product group $\prod_{n=1}^\infty U(\mathcal{H}_n) \subseteq U(\mathcal{H})$ satisfying $g_{\varepsilon,n} = 1$ for n sufficiently large. If $v \in \mathcal{H}_\pi$ is a nonzero simultaneous fixed vector for the subgroups $U(\mathcal{H}_n)$ (Lemma 4.3), we obtain $\pi(g_\varepsilon)v = v$ for every $\varepsilon > 0$, and now $v = \pi(g_\varepsilon)v \to \pi(g)v$ implies that $\pi(g)v = v$. $\square$

As an immediate consequence, we obtain:

Lemma 4.5. *Let* $(\pi, \mathcal{H}_\pi)$ *be a continuous unitary representation of* $U(\mathcal{H})$ *with* $U_\infty(\mathcal{H}) \subseteq \ker \pi$ *and* $j \in \mathbb{Z}$ *with* $\pi(\zeta\mathbf{1}) = \zeta^j \mathbf{1}$ *for* $\zeta \in \mathbb{T}$. *If* λ *is contained in the essential spectrum of* g, *then* λ^j *is an eigenvalue of* $\pi(g)$.

Proof. Lemma 4.4 implies that $\pi(\lambda^{-1}g)$ has a nonzero fixed vector v, and this means that $\pi(g)v = \lambda^j v$. $\qquad\square$

Theorem 4.6. *If $\mathcal{H}$ is a separable Hilbert space over $\mathbb{K} \in \{\mathbb{R}, \mathbb{C}, \mathbb{H}\}$, then every continuous unitary representation of $U(\mathcal{H})/U_\infty(\mathcal{H})_0$ on a separable Hilbert space is trivial.*

Proof. (a) We start with the case $\mathbb{K} = \mathbb{C}$. Let $(\pi, \mathcal{H}_\pi)$ be a separable continuous unitary representation of the Banach–Lie group $U(\mathcal{H})$ with $U_\infty(\mathcal{H}) \subseteq \ker \pi$.

Step 1: $\mathbb{T}\mathbf{1} \subseteq \ker \pi$: Decomposing the representation of the compact central subgroup $\mathbb{T}\mathbf{1}$, we may w.l.o.g. assume that $\pi(\zeta\mathbf{1}) = \zeta^j\mathbf{1}$ for some $j \in \mathbb{Z}$. Let $g \in U(\mathcal{H})$ be an element with uncountable essential spectrum. If $j \neq 0$, then Lemma 4.5 implies that $\pi(g)$ has uncountably many eigenvalues, which is impossible if $\mathcal{H}_\pi$ is separable. Therefore $j = 0$, and this means that $\mathbb{T}\mathbf{1} \subseteq \ker \pi$.

Step 2: Let $P \in B(\mathcal{H})$ be an orthogonal projection with infinite rank. Then $U(P\mathcal{H}) \cong U(\mathcal{H})$, so that Step 1 implies that $\mathbb{T}P + (\mathbf{1} - P) \subseteq \ker \pi$. If P has finite rank, then

$$\mathbb{T}P + (\mathbf{1} - P) \subseteq U_\infty(\mathcal{H}) \subseteq \ker \pi.$$

This implies that $\ker \pi$ contains all elements g with $\mathrm{Spec}(g) \subseteq \{1, \zeta\}$ for some $\zeta \in \mathbb{T}$. Since every element with finite spectrum is a finite product of such elements, it is also contained in $\ker \pi$. Finally we derive from the Spectral Theorem that the subset of elements with finite spectrum is dense in $U(\mathcal{H})$, so that π is trivial.[5]

(b) Next we consider the orthogonal group $U_{\mathbb{R}}(\mathcal{H}) = O(\mathcal{H})$ of a real Hilbert space $\mathcal{H}$. Since $\mathcal{H}$ is infinite-dimensional, there exists an orthogonal complex structure $I \in O(\mathcal{H})$. Then $\tau(g) := IgI^{-1}$ defines an involution on $O(\mathcal{H})$ whose fixed point set is the unitary group $U(\mathcal{H}, I)$ of the complex Hilbert space $(\mathcal{H}, I)$.

Let $\pi: O(\mathcal{H}) \to U(\mathcal{H}_\pi)$ be a continuous separable unitary representation with $SO_\infty(\mathcal{H}) := O_\infty(\mathcal{H})_0 \subseteq N := \ker \pi$. Applying (a) to $\pi|_{U(\mathcal{H},I)}$, it follows that $U(\mathcal{H}, I) \subseteq N$ and hence in particular that $I \in N$.

For $X^\top = -X$ and $\tau(X) = -X$ we then obtain

$$N \ni I\exp(X)I^{-1}\exp(-X) = \exp(-X)\exp(-X) = \exp(-2X).$$

---

[5]This argument simplifies Pickrell's argument that was based on the simplicity of the topological group $U(\mathcal{H})/\mathbb{T}U_\infty(\mathcal{H})$ [Ka52].

This implies that $\mathbf{L}(N) = \{X \in \mathfrak{o}(\mathcal{H}) : \exp(\mathbb{R}X) \subseteq N\} = \mathfrak{o}(\mathcal{H})$, and since $O(\mathcal{H})$ is connected by Example 1.4, it follows that $N = O(\mathcal{H})$, i.e., that π is trivial.

(c) Now let $\mathcal{H}$ be a quaternionic Hilbert space, considered as a right $\mathbb{H}$-module. Realizing $\mathcal{H}$ as $\ell^2(S, \mathbb{H})$ for some set S, we see that $\mathcal{K} := \ell^2(S, \mathbb{C})$ is a complex Hilbert space whose complex structure is given by left multiplication $\lambda_\mathcal{I}$ with the basis element $\mathcal{I} \in \mathbb{H}$ (this map is $\mathbb{H}$-linear) and we have a direct sum $\mathcal{H}^\mathbb{C} = \mathcal{K} \oplus \mathcal{K}\mathcal{J}$ of complex Hilbert spaces.

Let $\sigma: \ell^2(S, \mathbb{H}) \to \ell^2(S, \mathbb{H})$ be the real linear isometry given by $\sigma(v) = \mathcal{I}v\mathcal{I}^{-1}$ pointwise on S, so that $\mathcal{H}^\sigma = \ell^2(S, \mathbb{C}) = \mathcal{K}$ and $\mathcal{K}\mathcal{J} = \mathcal{H}^{-\sigma}$. Then $\tau(g) := \sigma g \sigma$ defines an involution on $U_\mathbb{H}(\mathcal{H})$ whose group of fixed points is isomorphic to the unitary group $U(\mathcal{K})$ of the complex Hilbert space $\mathcal{K}$, on which the complex structure is given by right multiplication with $\mathcal{I}$, which actually coincides with the left multiplication.

Let $\pi: U_\mathbb{H}(\mathcal{H}) \to U(\mathcal{H}_\pi)$ be a continuous separable unitary representation with $U_{\mathbb{H},\infty}(\mathcal{H}) \subseteq N := \ker \pi$. Applying (a) to $\pi|_{U(\mathcal{K})}$, it follows that $U(\mathcal{K}) \subseteq N$ and hence in particular that $\lambda_\mathcal{I} \in N$. On the Lie algebra level, $\mathfrak{u}(\mathcal{K})$ is complemented by

$$\{X \in \mathfrak{u}_\mathbb{H}(\mathcal{H}) : \sigma X = -X\sigma\} = \{X \in \mathfrak{u}_\mathbb{H}(\mathcal{H}) : \lambda_\mathcal{I}X = -X\lambda_\mathcal{I}\},$$

and for any element of this space we have

$$N \ni \lambda_\mathcal{I}\exp(X)\lambda_\mathcal{I}^{-1}\exp(-X) = \exp(-X)\exp(-X) = \exp(-2X).$$

This implies that $\mathbf{L}(N) = \mathfrak{u}_\mathbb{H}(\mathcal{H})$, and since $U_\mathbb{H}(\mathcal{H})$ is connected by Proposition 1.3(i), $N = U_\mathbb{H}(\mathcal{H})$, so that π is trivial. $\square$

Remark 4.7. For $\mathbb{K} = \mathbb{R}$, the group $O(\mathcal{H})/SO_\infty(\mathcal{H})$ is the 2-fold simply connected cover of the group $O(\mathcal{H})/O_\infty(\mathcal{H})$. Therefore the triviality of all separable representations of $O(\mathcal{H})/O_\infty(\mathcal{H})$ follows from the triviality of all separable representations of $O(\mathcal{H})/SO_\infty(\mathcal{H}) = O(\mathcal{H})/O_\infty(\mathcal{H})_0$.

Problem 4.8. If $\mathcal{H}$ is an inseparable Hilbert space, then we think that all separable unitary representations $(\pi, \mathcal{H})$ of $U(\mathcal{H})$ should be trivial, but we can only show that $\ker \pi$ contains all operators for which $(g - 1)\mathcal{H}$ is separable, i.e., all groups $U(\mathcal{H}_0)$, where $\mathcal{H}_0 \subseteq \mathcal{H}$ is a separable subspace.

The argument works as follows. From Remark 3.22 we know that all irreducible representations of $U_\infty(\mathcal{H})$ are inseparable. Theorem 3.21 implies that $U_\infty(\mathcal{H}) \subseteq \ker \pi$. Now Theorem 4.6 implies that $\ker \pi$ contains all subgroups $U(\mathcal{H}_0)$, where $\mathcal{H}_0$ is a separable Hilbert space, and this proves our claim.

4.2 Separable Representations of the Lie Group $U(\mathcal{H})$

Based on Pickrell's theorem on the triviality of the separable representations of the quotient Lie groups $U(\mathcal{H})/U_\infty(\mathcal{H})_0$, we can now determine all separable continuous unitary representations of the full unitary group $U(\mathcal{H})$.

Theorem 4.9. *Let $\mathcal{H}$ be a separable $\mathbb{K}$-Hilbert space. Then every separable continuous unitary representation $(\pi, \mathcal{H}_\pi)$ of the Banach–Lie group $U(\mathcal{H})$ has the following properties:*

(i) *It is continuous with respect to the strong operator topology on $U(\mathcal{H})$.*
(ii) *Its restriction to $U_\infty(\mathcal{H})_0$ has the same commutant.*
(iii) *It is a direct sum of bounded irreducible representations.*
(iv) *Every irreducible separable representation is of the form*

$$\begin{cases} \mathbb{S}_\lambda(\mathcal{H}_\mathbb{C}) \subseteq (\mathcal{H}_\mathbb{C})^{\otimes N}, \lambda \in \mathrm{Part}(N), & \text{for } \mathbb{K} = \mathbb{R}, \\ \mathbb{S}_\lambda(\mathcal{H}) \otimes \mathbb{S}_\mu(\overline{\mathcal{H}}) \subseteq \mathcal{H}^{\otimes N} \otimes \overline{\mathcal{H}}^{\otimes M}, \lambda \in \mathrm{Part}(N), \mu \in \mathrm{Part}(M), & \text{for } \mathbb{K} = \mathbb{C}, \\ \mathbb{S}_\lambda(\mathcal{H}^\mathbb{C}) \subseteq (\mathcal{H}^\mathbb{C})^{\otimes N}, \lambda \in \mathrm{Part}(N), & \text{for } \mathbb{K} = \mathbb{H}. \end{cases}$$

(v) *π extends uniquely to a strongly continuous representation of the overgroup $U(\mathcal{H})^\sharp$ with the same commutant.*

Proof. (i) From Theorem 3.17(c) we know that $\pi_\infty := \pi|_{U_\infty(\mathcal{H})}$ extends to a unique continuous unitary representation $\overline{\pi}$ of $U(\mathcal{H})_s$ on $\mathcal{H}_\pi$. In particular, the action of $U(\mathcal{H})$ on the unitary dual of the normal subgroup $U_\infty(\mathcal{H})_0$ is trivial. Hence all isotypic subspaces $\mathcal{H}_{[\lambda]}$ for π_∞ are invariant under π. We may therefore assume that π_∞ is isotypic, i.e., of the form $\mathbf{1} \otimes \rho_\lambda$, where $(\rho_\lambda, V_\lambda)$ is an irreducible representation of $U_\infty(\mathcal{H})_0$ (cf. Theorem 3.17). Then $\overline{\pi} := \mathbf{1} \otimes \overline{\rho}_\lambda$ is continuous with respect to the operator norm on $U(\mathcal{H})$ because the representations of $U(\mathcal{H})$ on the spaces $\mathcal{H}_\mathbb{C}^{\otimes N}$ are norm-continuous (Theorem 3.17(c)).
  Now

$$\beta(g) := \pi(g)\overline{\pi}(g)^{-1} \in \pi(U_\infty(\mathcal{H}))' = \overline{\pi}(U(\mathcal{H}))'$$

implies that $\beta \colon U(\mathcal{H}) \to U(\mathcal{H}_\pi)$ defines a separable norm-continuous unitary representation vanishing on $U_\infty(\mathcal{H})$. By Theorem 4.6 it is trivial, so that $\pi = \overline{\pi}$.
  (ii) follows from (i) and the density of $U_\infty(\mathcal{H})_0$ in $U(\mathcal{H})_s$.
(iii), (iv) now follow from Theorem 3.17.
  (v) In view of (iii), assertion (v) reduces to the case of irreducible representations. In this case (v) follows from the concrete classification (iv) and the description of the overgroups $U(\mathcal{H})^\sharp$ in Lemma 2.4.

$\square$

Corollary 4.10. *Let K be a quotient of a product $K_1 \times \cdots \times K_n$, where each K_j is compact, a quotient of some group $\mathrm{U}(\mathcal{H})$ or $\mathrm{U}_\infty(\mathcal{H})_0$, where $\mathcal{H}$ is a separable $\mathbb{K}$-Hilbert space. Then every separable continuous unitary representation π of K is a direct sum of irreducible representations which are bounded.*

The preceding corollary means that the separable representation theory of K is very similar to the representation theory of a compact group.

4.3 Classification of Irreducible Representations by Highest Weights

We choose an orthonormal basis $(e_j)_{j \in J}$ in the complex Hilbert space $\mathcal{H}$ and write $T \cong \mathbb{T}^J$ for the corresponding group of diagonal matrices. Characters of this group correspond to finitely supported functions $\lambda \colon J \to \mathbb{Z}$ via $\chi_\lambda(t) = \prod_{j \in J} t_j^{\lambda_j}$. For the subgroup $T(\infty)$ of those diagonal matrices t for which $t - \mathbf{1}$ has finite rank, any function $\lambda \colon J \to \mathbb{Z}$ defines a character. Accordingly, each $\lambda = (\lambda_j)_{j \in J} \in \mathbb{Z}^J$ defines a uniquely determined *unitary highest weight representation* $(\pi_\lambda, \mathcal{H}_\lambda)$ of $\mathrm{U}(\infty, \mathbb{C})$ [Ne04, Ne98]. This representation is uniquely determined by the property that its weight set with respect to the diagonal subgroup $T \cong \mathbb{T}^{(J)}$, whose character group $\widehat{T}$ is $\mathbb{Z}^J$, coincides with

$$\mathrm{conv}(\mathcal{W}\lambda) \cap (\lambda + \mathcal{Q}), \qquad \text{where} \qquad \mathcal{Q} \subseteq \widehat{T}$$

is the root group and $\mathcal{W}$ is the group of finite permutations of the set J.

Proposition 4.11. *A unitary highest weight representation $(\pi_\lambda, \mathcal{H}_\lambda)$ of $\mathrm{U}(\infty, \mathbb{C})$ is tame if and only if $\lambda \colon \mathbb{N} \to \mathbb{Z}$ is finitely supported.*

Proof. If π_λ is a tame representation, then its restriction to the diagonal subgroup is tame. Since this representation is diagonalizable, this means that each weight has finite support. It follows in particular that λ has finite support.

If, conversely, λ has finite support, then we write $\lambda = \lambda_+ - \lambda_-$, where $\lambda_\pm$ are nonnegative with finite disjoint support. Then $\mathcal{H}_\lambda$ can be embedded into $\mathbb{S}_{\lambda_+}(\mathcal{H}) \otimes \mathbb{S}_{\lambda_-}(\overline{\mathcal{H}})$ [Ne98], hence it is tame. $\square$

Example 4.12. $\mathbb{K} = \mathbb{R}$: In the infinite-dimensional real Hilbert space $\mathcal{H}$ we fix a complex structure I. Then there exists a real orthonormal basis of the form $\{e_j, Ie_j : j \in J\}$. Then the subgroup $T \subseteq \mathrm{O}(\mathcal{H})$ preserving all the planes $\mathbb{R}e_j + I\mathbb{R}e_j$ is maximal abelian. In $\mathcal{H}_\mathbb{C}$ the elements $e_j^\pm := \frac{1}{\sqrt{2}}(e_j \mp Ie_j)$ form an orthonormal basis, and we write $2J := J \times \{\pm\}$ for the corresponding index set. In $\mathrm{O}(\mathcal{H})^\sharp = \mathrm{U}(\mathcal{H}_\mathbb{C})$, the corresponding diagonal subgroup $T^\sharp \cong \mathbb{T}^{2J}$ is maximal abelian. The corresponding maximal torus $T_\mathbb{C}$ of $\mathrm{O}(\mathcal{H})_\mathbb{C} \subseteq \mathrm{GL}(\mathcal{H}_\mathbb{C})$ corresponds to diagonal matrices d acting by $de_j^\pm = d_j^{\pm 1} e_j^\pm$.

For a character χ_μ of $T^\sharp$ with $\mu: 2J \to \mathbb{Z}$, the corresponding character of T is given by the finitely supported function $\mu^\flat: J \to \mathbb{Z}$ with $\mu_j^\flat = \mu_{j,+} - \mu_{j,-}$. If $\lambda: 2J \to \mathbb{N}_0$ has finite support, then the corresponding irreducible representation of $U(\mathcal{H}_\mathbb{C})$ occurs as some $\mathbb{S}_\lambda(\mathcal{H}_\mathbb{C})$ in $\mathcal{H}_\mathbb{C}^{\otimes N}$, where $\sum_{j \in 2J} \lambda_j = N$. From the Classification Theorem 3.17 it follows that the restriction of π_λ to $O(\mathcal{H})$ is irreducible. The corresponding highest weight is $\lambda^\flat$. On the level of highest weights, it is clear that, for each finitely supported weight $\lambda: J \to \mathbb{N}_0$, we obtain by

$$\lambda_{j,+}^\sharp := \lambda_j \quad \text{and} \quad \lambda_{j,-}^\sharp := 0$$

a highest weight $\lambda^\sharp$ with $(\lambda^\sharp)^\flat = \lambda$. The irreducible representations of $O(\mathcal{H})$ are classified by orbits of the Weyl group $\mathcal{W}$ in the set of finitely supported integral weights $\lambda: J \to \mathbb{Z}$ of the root system D_{2J} (cf. [Ne98, Sect. VII]). Each orbit has a nonnegative representative, and then $\lambda^\sharp$ is the highest weight of the corresponding representation $\pi_{\lambda^\sharp}$ of $U(\mathcal{H}_\mathbb{C})$.

Example 4.13. $\mathbb{K} = \mathbb{C}$: Let $(e_j)_{j \in J}$ be an ONB of $\mathcal{H}$. In $U(\mathcal{H})^\sharp \cong U(\mathcal{H}) \times U(\overline{\mathcal{H}})$ we have the maximal abelian subgroup $T^\sharp = T \times T$, where $T \cong \mathbb{T}^J$ is the subgroup of diagonal matrices in $U(\mathcal{H})$ with respect to the ONB $(e_j)_{j \in J}$.

Let $2J := J \times \{\pm\}$, so that $T^\sharp \cong \mathbb{T}^{2J}$. For a finitely supported function $\mu: 2J \to \mathbb{Z}$, the corresponding character of T is given by $\mu^\flat: J \to \mathbb{Z}$, defined by $\mu_j^\flat = \mu_{j,+} - \mu_{j,-}$. If $\lambda: 2J \to \mathbb{N}_0$ has finite support, and $\lambda = \lambda_+ - \lambda_-$ with nonnegative summands $\lambda_\pm$ supported in $J \times \{\pm\}$, respectively, the corresponding irreducible representation π_λ lives on $\mathbb{S}_{\lambda_+}(\mathcal{H}) \otimes \mathbb{S}_{\lambda_-}(\overline{\mathcal{H}}) \subseteq \mathcal{H}^{\otimes N} \otimes \overline{\mathcal{H}}^{\otimes M}$, where $N = \sum_{\lambda_j > 0} \lambda_j$ and $M = -\sum_{\lambda_j < 0} \lambda_j$. From the Classification Theorem 3.17 it follows that the restriction of π_λ to $U(\mathcal{H})$ is irreducible. The corresponding highest weight is $\lambda^\flat = \lambda_+ - \lambda_-$. For each finitely supported weight $\lambda = \lambda_+ - \lambda_-: J \to \mathbb{N}_0$, we obtain by $\lambda_{j,\pm}^\sharp := \lambda_{\pm,j}$, $j \in J$, a highest weight $\lambda^\sharp$ with $(\lambda^\sharp)^\flat = \lambda$. The irreducible representations of $U(\mathcal{H})$ are classified by orbits of the Weyl group $\mathcal{W} \cong S_{(J)}$ in the set of finitely supported integral weights $\lambda: J \to \mathbb{Z}$ of the root system A_J (cf. [Ne98, Sect. VII]).

Example 4.14. $\mathbb{K} = \mathbb{H}$: In the quaternionic Hilbert space $\mathcal{H}$ we consider the complex structure defined by multiplication with $\mathcal{I}$, which leads to the complex Hilbert space $\mathcal{H}^\mathbb{C}$. Then there exists a complex orthonormal basis of the form $\{e_j, \mathcal{J}e_j: j \in J\}$. We write $T^\sharp \subseteq U(\mathcal{H}^\mathbb{C})$ for the corresponding diagonal subgroup. Note that $T^\sharp \cong \mathbb{T}^{2J}$ for $2J := J \times \{\pm\}$. The subgroup $T := T^\sharp \cap U(\mathcal{H}) = (T^\sharp)^{\mathcal{J}}$ acts on the basis elements $e_{j,+} := e_j$ and $e_{j,-} := \mathcal{J}e_j$ by $de_{j,\pm} = d_j^{\pm} e_{j,\pm}$.

The classification of the irreducible representations by Weyl group orbits of finitely supported functions $\lambda: J \to \mathbb{Z}$ (weights for the root system B_J) and their corresponding weights $\lambda^\sharp: 2J \to \mathbb{Z}$ is completely analogous to the situation for $\mathbb{K} = \mathbb{R}$. The irreducible representation of $U(\mathcal{H}^\mathbb{C})$ corresponding to $\lambda^\sharp$ is $\mathbb{S}_{\lambda^\sharp}(\mathcal{H}^\mathbb{C})$.

Remark 4.15 (Segal's Physical Representations). In [Se57] Segal studied unitary representations of the full group $U(\mathcal{H})$, called *physical representations*. They are characterized by the condition that their differential maps finite rank hermitian projections to positive operators. Segal shows that physical representations decompose discretely into irreducible physical representations which are precisely those occurring in the decomposition of finite tensor products $\mathcal{H}^{\otimes N}$, $N \in \mathbb{N}_0$. In view of Pickrell's theorem, this also follows from our classification of the separable representations of $U(\mathcal{H})$. Since Segal's arguments never use the separability of $\mathcal{H}$, the corresponding result remains true for inseparable spaces as well.

Problem 4.16. Theorem 3.17 implies in particular that all continuous unitary representations of $K = U_\infty(\mathcal{H})_0$ have a canonical extension to their overgroups $K^\sharp$ with the same commutant. The classification in terms of highest weights further implies that the representations of $K^\sharp$ obtained from this extension process are precisely those with nonnegative weights.

Conversely, it follows that all unitary representations of $K^\sharp$ with nonnegative weights remain irreducible when restricted to K.

One may ask a similar question for the smaller group $K^\sharp(\infty) \subseteq K^\sharp$ or its completion with respect to the trace norm. Is it true that for any unitary representation π of $K^\sharp(\infty)$ whose weights on the diagonal subgroup are nonnegative, $\pi(U(\infty, \mathbb{K}))$ has the same commutant? As we explain below, this is not true.

For the special case $\mathbb{K} = \mathbb{R}$ and $\lambda = \lambda_+ - \lambda_-$ finitely supported, the restriction of the representation $\pi_\lambda = \pi_{\lambda_+} \otimes \pi_{\lambda_-}^*$ of $K^\sharp(\infty)$ on $S_{\lambda_+}(\mathcal{H}_{\mathbb{C}}) \otimes S_{\lambda_-}(\overline{\mathcal{H}_{\mathbb{C}}})$ to the subgroup $K(\infty) = SO(\infty, \mathbb{R})$ is equivalent to the representation $\pi_{\lambda_+} \otimes \pi_{\lambda_-}$, which decomposes according to the standard Schur–Weyl theory. In particular, we obtain non-irreducible representations if λ takes positive and negative values on $K(\infty)$. That this cannot be repaired by the positivity requirement on the weights of $K^\sharp(\infty)$ follows from the fact that the determinant det: $K^\sharp(\infty) \to \mathbb{T}$ restricts to the trivial character of $K(\infty)$, but tensoring with a power of det, any bounded weight λ can be made positive.

Is it possible to characterize those irreducible highest weight representations π_λ of $K^\sharp(\infty)$ whose restriction to $K(\infty)$ is irreducible?

5 Non-existence of Separable Unitary Representations for Full Operator Groups

In this section we describe some consequences of the main results from [Pi90]. We start with the description of 10 symmetric pairs (G, K) of groups of operators, where G does not consist of unitary operators and $K \subseteq G$ is "maximal unitary". They are infinite-dimensional analogs of certain noncompact real reductive Lie groups. The dual symmetric pairs (G^c, K) have the property that G^c consists of unitary operators, hence they are analogs of certain compact matrix groups.

One of the main result of this section is that all separable unitary representations of the groups G are trivial, but there are various refinements concerning restricted groups.

5.1 The 10 Symmetric Pairs

Below we use the following notational conventions. We write $O(n) := O(n, \mathbb{R})$, $U(n) := U(n, \mathbb{C})$ and $Sp(n) := U(n, \mathbb{H})$ for $n \in \mathbb{N} \cup \{\infty\}$. For a group G, we write $\Delta_G := \{(g, g) : g \in G\}$ for the diagonal subgroup of $G \times G$.

If $\mathcal{H}$ is a complex Hilbert space, then we write $I \in B(\mathcal{H}_{\mathbb{C}})$ for the $\mathbb{C}$-linear extension of the complex structure $Iv = iv$ on $\mathcal{H}$. Then $D := -iI$ is a unitary involution that leads to the pseudo-unitary group

$$U(\mathcal{H}_{\mathbb{C}}, D) = \{g \in GL(\mathcal{H}_{\mathbb{C}}) : Dg^* D^{-1} = g^{-1}\}$$

preserving the indefinite hermitian form $\langle Dv, w \rangle$. For the isometry group of the indefinite form $h((v_1, v_2), (w_1, w_2)) := \langle v_1, w_1 \rangle - \langle v_2, w_2 \rangle$ on $\mathcal{H} \times \mathcal{H}$, we write $U(\mathcal{H}, \mathcal{H})$. Now the group

$$O^*(\mathcal{H}_{\mathbb{C}}) := U(\mathcal{H}_{\mathbb{C}}, D) \cap O(\mathcal{H})_{\mathbb{C}}$$

is a Lie group. Its Lie algebra $\mathfrak{o}^*(\mathcal{H}_{\mathbb{C}})$ satisfies $\mathfrak{o}^*(\mathcal{H}_{\mathbb{C}}) \cap \mathfrak{u}(\mathcal{H}_{\mathbb{C}}) \cong \mathfrak{u}(\mathcal{H})$ and it is a real form of $\mathfrak{o}(\mathcal{H})_{\mathbb{C}}$. It is easy to see that the symmetric pair $(O^*(\mathcal{H}_{\mathbb{C}}), U(\mathcal{H}))$ is dual to $(O(\mathcal{H}^{\mathbb{R}}), U(\mathcal{H}))$.

Non-unitary Symmetric Pairs

	Non-unit. locally finite $(G(\infty), K(\infty))$	Operator group (G, K)	K	$\mathbb{K}$
1	$(GL(\infty, \mathbb{C}), U(\infty))$	$(GL(\mathcal{H}), U(\mathcal{H}))$	$U(\mathcal{H})$	$\mathbb{C}$
2	$(SO(\infty, \mathbb{C}), SO(\infty))$	$(O(\mathcal{H})_{\mathbb{C}}, O(\mathcal{H}))$	$O(\mathcal{H})$	$\mathbb{R}$
3	$(Sp(\infty, \mathbb{C}), Sp(\infty))$	$(U_{\mathbb{H}}(\mathcal{H})_{\mathbb{C}}, U_{\mathbb{H}}(\mathcal{H}))$	$U_{\mathbb{H}}(\mathcal{H})$	$\mathbb{H}$
4	$(U(\infty, \infty), U(\infty)^2)$	$(U(\mathcal{H}, \mathcal{H}), U(\mathcal{H})^2)$	$U(\mathcal{H})^2$	$\mathbb{C}$
5	$(SO(\infty, \infty), SO(\infty)^2)$	$(O(\mathcal{H}, \mathcal{H}), O(\mathcal{H})^2)$	$O(\mathcal{H})^2$	$\mathbb{R}$
6	$(Sp(\infty, \infty), Sp(\infty)^2)$	$(U_{\mathbb{H}}(\mathcal{H}, \mathcal{H}), U_{\mathbb{H}}(\mathcal{H})^2)$	$U_{\mathbb{H}}(\mathcal{H})^2$	$\mathbb{H}$
7	$(Sp(2\infty, \mathbb{R}), U(\infty))$	$(Sp(\mathcal{H}), U(\mathcal{H}))$	$U(\mathcal{H})$	$\mathbb{C}$
8	$(SO(2\infty), U(\infty))$	$(O^*(\mathcal{H}_{\mathbb{C}}), U(\mathcal{H}))$	$U(\mathcal{H})$	$\mathbb{C}$
9	$(GL(\infty, \mathbb{R}), O(\infty))$	$(GL(\mathcal{H}), O(\mathcal{H}))$	$O(\mathcal{H})$	$\mathbb{R}$
10	$(GL(\infty, \mathbb{H}), Sp(\infty))$	$(GL_{\mathbb{H}}(\mathcal{H}), U_{\mathbb{H}}(\mathcal{H}))$	$U_{\mathbb{H}}(\mathcal{H})$	$\mathbb{H}$

Unitary Symmetric Pairs

	Unitary locally finite $(G^c(\infty), K(\infty))$	Unitary operator group (G^c, K)	K	$\mathbb{K}$
1	$(\mathrm{U}(\infty)^2, \Delta_{\mathrm{U}(\infty)})$	$(\mathrm{U}(\mathcal{H})^2, \Delta_{\mathrm{U}(\mathcal{H})})$	$\mathrm{U}(\mathcal{H})$	$\mathbb{C}$
2	$(\mathrm{SO}(\infty)^2, \Delta_{\mathrm{SO}(\infty)})$	$(\mathrm{O}(\mathcal{H})^2, \Delta_{\mathrm{O}(\mathcal{H})})$	$\mathrm{O}(\mathcal{H})$	$\mathbb{R}$
3	$(\mathrm{Sp}(\infty)^2, \Delta_{\mathrm{Sp}(\infty)})$	$(\mathrm{U}_\mathbb{H}(\mathcal{H})^2, \Delta_{\mathrm{U}_\mathbb{H}(\mathcal{H})})$	$\mathrm{U}_\mathbb{H}(\mathcal{H})$	$\mathbb{H}$
4	$(\mathrm{U}(2\infty), \mathrm{U}(\infty)^2)$	$(\mathrm{U}(\mathcal{H} \oplus \mathcal{H}), \mathrm{U}(\mathcal{H})^2)$	$\mathrm{U}(\mathcal{H})^2$	$\mathbb{C}$
5	$(\mathrm{SO}(2\infty), \mathrm{SO}(\infty)^2)$	$(\mathrm{O}(\mathcal{H} \oplus \mathcal{H}), \mathrm{O}(\mathcal{H})^2)$	$\mathrm{O}(\mathcal{H})^2$	$\mathbb{R}$
6	$(\mathrm{Sp}(2\infty), \mathrm{Sp}(\infty)^2)$	$(\mathrm{U}_\mathbb{H}(\mathcal{H} \oplus \mathcal{H}), \mathrm{U}_\mathbb{H}(\mathcal{H})^2)$	$\mathrm{U}_\mathbb{H}(\mathcal{H})^2$	$\mathbb{H}$
7	$(\mathrm{Sp}(\infty), \mathrm{U}(\infty))$	$(\mathrm{U}_\mathbb{H}(\mathcal{H} \otimes_\mathbb{C} \mathbb{H}), \mathrm{U}(\mathcal{H}))$	$\mathrm{U}(\mathcal{H})$	$\mathbb{C}$
8	$(\mathrm{SO}(2\infty), \mathrm{U}(\infty))$	$(\mathrm{O}(\mathcal{H}^\mathbb{R}), \mathrm{U}(\mathcal{H}))$	$\mathrm{U}(\mathcal{H})$	$\mathbb{C}$
9	$(\mathrm{U}(\infty), \mathrm{O}(\infty))$	$(\mathrm{U}(\mathcal{H}_\mathbb{C}), \mathrm{O}(\mathcal{H}))$	$\mathrm{O}(\mathcal{H})$	$\mathbb{R}$
10	$(\mathrm{U}(2\infty), \mathrm{Sp}(\infty))$	$(\mathrm{U}(\mathcal{H}^\mathbb{C}), \mathrm{U}_\mathbb{H}(\mathcal{H}))$	$\mathrm{U}_\mathbb{H}(\mathcal{H})$	$\mathbb{H}$

Remark 5.1. (a) The unitary symmetric pairs (1)–(3) are of group type and their non-unitary duals are complex groups.

(b) The non-unitary pairs (4)–(6) are the symmetric pairs associated to pseudo-unitary groups of indefinite hermitian forms β with the matrix $D = \begin{pmatrix} 1 & 0 \\ 0 & -1 \end{pmatrix}$ on $\mathcal{H}^2$. Accordingly, the corresponding symmetric spaces can be considered as Graßmannians of "maximal positive subspaces" for β.

(c) The symmetric spaces corresponding to (7) and (8) are spaces of complex structures on real spaces. The space $\mathrm{Sp}(\mathcal{H})/\mathrm{U}(\mathcal{H})$ is the space of positive symplectic complex structures on the real symplectic spaces $(\mathcal{H}, \omega)$, where $\omega(v, w) = \mathrm{Im}\langle v, w \rangle$. Likewise $\mathrm{O}(\mathcal{H}^\mathbb{R})/\mathrm{U}(\mathcal{H})$ is the space of orthogonal complex structures on the real Hilbert space $\mathcal{H}^\mathbb{R}$.

(d) The spaces (4), (7) and (8) are of hermitian type (cf. [Ne12]).

(e) The spaces (1), (9) and (10) are those occurring naturally for overgroups of unitary groups (cf. Example 2.3).

5.2 Restricted Symmetric Pairs

For each symmetric pair (G, K) of non-unitary type and $1 \leq q \leq \infty$, we obtain a *restricted symmetric pair* $(G_{(q)}, K)$, defined by

$$G_{(q)} := \{g \in G : \mathrm{tr}(|g^*g - \mathbf{1}|^q) < \infty\}.$$

If $\mathfrak{g} = \mathfrak{k} \oplus \mathfrak{p}$ with $\mathfrak{p} = \{X \in \mathfrak{g} : X^* = X\}$, then the Lie algebra of G^c is $\mathfrak{g}_{(q)} = \mathfrak{k} \oplus \mathfrak{p}_{(q)}$, where $\mathfrak{p}_{(q)} = \mathfrak{p} \cap B_q(\mathcal{H})$. The corresponding dual symmetric pair is $(G^c_{(q)}, K)$ with $\mathfrak{g}^c_{(q)} = \mathfrak{k} \oplus i\mathfrak{p}_{(q)}$. We also write

$$G_{\infty,(q)} := G_{(q)} \cap (1 + K(\mathcal{H})) = K_\infty \exp(\mathfrak{p}_{(q)})$$

for the closure of $G(\infty)$ in $G_{(q)}$.

Proposition 5.2. *Spherical representations of any pair* $(G(\infty), K(\infty))$ *of unitary or non-unitary type are direct integrals of irreducible ones.*

Proof. This is [Pi90, Prop. 2.4], but it also follows from the general Theorem B.3 below. $\qquad\square$

Combining the preceding proposition with two-sided estimates on the behavior of spherical functions near the identity, Pickrell proved:

Proposition 5.3 ([Pi90, Prop. 6.11]). *Irreducible real spherical functions of the direct limit pairs* $(G(\infty), K(\infty))$ *always extend to spherical functions of* $G_{(2)}$ *and, for* $q > 2$, *all spherical functions on* $G_{(q)}$ *vanish.*

If v is a C^1-spherical vector for the unitary representation $(\pi, \mathcal{H}_\pi)$ of $(G_{(q)}, K)$, then $\beta(X, Y) := \langle d\pi(X)v, d\pi(Y)v \rangle$ defines a continuous K-invariant positive semidefinite symmetric bilinear form on $\mathfrak{p}_{(q)}$. Therefore one can also show that v is fixed by the whole group $G_{(q)}$ by showing that $\beta = 0$ using the following lemma.

Lemma 5.4. *The following assertion holds for the K-action on* $\mathfrak{p}_{(q)}$:

(i) $[\mathfrak{k}, \mathfrak{p}] = \mathfrak{p}$.
(ii) $\mathfrak{p}_{(2)}$ *is an irreducible representation.*
(iii) *For* $q > 2$, *every continuous K-invariant symmetric bilinear form on* $\mathfrak{p}_{(q)}$ *vanishes.*

Proof. (i), (ii): We check these conditions for all 10 families:

(1)–(3) Then $\mathfrak{p} = i\mathfrak{k}$ with $\mathfrak{k} = \mathfrak{u}(\mathcal{H})$. Since $\mathfrak{k}$ is perfect by [Ne02, Lemma I.3], we obtain $[\mathfrak{k}, \mathfrak{p}] = i[\mathfrak{k}, \mathfrak{k}] = i\mathfrak{k} = \mathfrak{p}$.

  Here $\mathfrak{p}_{(2)} = i\mathfrak{u}_2(\mathcal{H})$, and since $\mathfrak{u}_2(\mathcal{H})$ is a simple Hilbert–Lie algebra [Sch60], (ii) follows.

(4)–(6) In these cases $\mathfrak{g}^c = \mathfrak{u}(\mathcal{H} \oplus \mathcal{H})$, $\mathfrak{k} = \mathfrak{u}(\mathcal{H}) \oplus \mathfrak{u}(\mathcal{H})$ and $\mathfrak{p} \cong \mathfrak{gl}(\mathcal{H})$ with the $\mathfrak{k}$-module structure given by $(X, Y).Z := XZ - ZY$. Since $\mathfrak{u}(\mathcal{H})$ contains invertible elements X_0, and $(X_0, 0).Z = X_0 Z$, it follows that $\mathfrak{p} = [\mathfrak{k}, \mathfrak{p}]$.

  Here $\mathfrak{p}_{(2)} \cong \mathfrak{gl}_2(\mathcal{H})$ is the space of Hilbert–Schmidt operators on $\mathcal{H}$. This immediately implies the irreducibility of the representation of $\mathbb{K} = \mathbb{R}, \mathbb{C}$. For $\mathbb{K} = \mathbb{H}$, we have $\mathfrak{p}_{(2),\mathbb{C}} \cong \mathfrak{gl}_2(\mathcal{H}^\mathbb{C})$, and since the representation of $U(\mathcal{H})$ on $\mathcal{H}^\mathbb{C}$ is irreducible (we have $U(\mathcal{H})_\mathbb{C} \cong Sp(\mathcal{H}^\mathbb{C})$), (ii) follows.

(7)–(8) In these two cases the center $\mathfrak{z} := i\mathbf{1}$ of $\mathfrak{k} \cong \mathfrak{u}(\mathcal{H})$ satisfies $\mathfrak{p} = [\mathfrak{z}, \mathfrak{p}]$, which implies (i). The Lie algebra $\mathfrak{g}_{(2)} = \mathfrak{k} \oplus \mathfrak{p}_{(2)}$ corresponds to the automorphism group of an irreducible hermitian symmetric space (cf. [Ne12, Thm. 2.6] and the subsequent discussion). This implies that the representation of K on the complex Hilbert space $\mathfrak{p}_{(2)}$ is irreducible.

(9) Here $\mathfrak{g} = \mathfrak{gl}(\mathcal{H})$, $\mathfrak{k} = \mathfrak{o}(\mathcal{H})$ and $\mathfrak{p} = \mathrm{Sym}(\mathcal{H})$. For any complex structure $I \in \mathfrak{o}(\mathcal{H})$ we then obtain

$$\mathfrak{p} = \mathrm{Herm}(\mathcal{H}, I) \oplus [I, \mathfrak{p}] = [\mathfrak{u}(\mathcal{H}, I), \mathrm{Herm}(\mathcal{H}, I)] \oplus [I, \mathfrak{p}] \subseteq [\mathfrak{k}, \mathfrak{p}].$$

This proves (i).

Next we observe that $\mathfrak{p}_{(2)} = \mathrm{Sym}_2(\mathcal{H})$ satisfies $\mathfrak{p}_{(2),\mathbb{C}} \cong \mathrm{Sym}_2(\mathcal{H}_{\mathbb{C}}) \cong S^2(\overline{\mathcal{H}_{\mathbb{C}}})$, hence is irreducible by Theorem 3.17.

(10) Here $\mathfrak{g} = \mathfrak{gl}(\mathcal{H})$, $\mathfrak{k} = \mathfrak{u}(\mathcal{H})$ and $\mathfrak{p} = \mathrm{Herm}(\mathcal{H})$ for $\mathbb{K} = \mathbb{H}$. With the aid of an orthonormal basis, we find a real Hilbert space $\mathcal{K}$ with $\mathcal{H} \cong \mathcal{K} \otimes_{\mathbb{R}} \mathbb{H}$, where $\mathbb{H}$ acts by right multiplication. This leads to an isomorphism $B_{\mathbb{H}}(\mathcal{H}) \cong B_{\mathbb{R}}(\mathcal{K}) \otimes_{\mathbb{R}} \mathbb{H}$ as real involutive algebras. In particular,

$$\mathrm{Herm}(\mathcal{H}) \cong \mathrm{Sym}(\mathcal{K}) \otimes \mathbf{1} \oplus \mathrm{Asym}(\mathcal{K}) \otimes \mathrm{Aherm}(\mathbb{H}).$$

Therefore (i) follows from $\mathrm{Aherm}(\mathbb{H}) = [\mathrm{Aherm}(\mathbb{H}), \mathrm{Aherm}(\mathbb{H})]$ and from $\mathrm{Sym}(\mathcal{K}) = [\mathfrak{o}(\mathcal{K}), \mathrm{Sym}(\mathcal{K})]$, which we derive from (9).

To verify (ii), we observe that $\mathfrak{p}_{(2)} = \mathrm{Herm}_2(\mathcal{H})$. From Kaup's classification of the real symmetric Cartan domains [Ka97] it follows that $\mathfrak{p}_{(2)}$ is a real form of the complex JH^*-triple $\mathrm{Skew}(\mathcal{H}^{\mathbb{C}})$ of skew symmetric bilinear forms on $\mathcal{H}^{\mathbb{C}}$, labelled by $II_{2n}^{\mathbb{H}}$. Since the action of $\mathrm{U}(\mathcal{H})$ on $\mathfrak{p}_{(2),\mathbb{C}} \cong \mathrm{Skew}(\mathcal{H}^{\mathbb{C}}) \cong \Lambda^2(\overline{\mathcal{H}_{\mathbb{C}}})$ extends to the overgroup $\mathrm{U}(\mathcal{H}^{\mathbb{C}})$ with the same commutant, the irreducibility of the resulting representation implies that the representation of $\mathrm{U}(\mathcal{H})$ on the real Hilbert space $\mathfrak{p}_{(2)}$ is irreducible as well.

(iii) can be derived from (i). If $\beta: \mathfrak{p}_{(q)} \times \mathfrak{p}_{(q)} \to \mathbb{R}$ is a continuous invariant symmetric bilinear form, then the same holds for its restriction to $\mathfrak{p}_{(2)}$. The simplicity of the representation on $\mathfrak{p}_{(2)}$ now implies that it is a multiple of the canonical form on $\mathfrak{p}_{(2)}$ given by the trace. But this form does not extend continuously to $\mathfrak{p}_{(q)}$ for any $q > 2$. □

Proposition 5.5. (a) *Separable unitary representations of $G_{(q)}$, $q \geq 1$, are completely determined by their restrictions to $G(\infty)$.*

(b) *Conversely, every continuous separable unitary representation of $G_{\infty,(q)}, q \geq 1$, extends to a continuous unitary representation of $G_{(q)}$.*

Proof (cf. [Pi90, Prop. 5.1]). (a) Since $\mathfrak{p}(\infty)$ is dense in $\mathfrak{p}_{(q)}$, this follows from Theorem 4.9(i), applied to K.

(b) Since K acts smoothly by conjugation on $G_{\infty,(q)}$, we can form the Lie group $G_{\infty,(q)} \rtimes K$ and note that the multiplication map to $G_{(q)}$ defines an isomorphism $(G_{\infty,(q)} \rtimes K)/K_\infty \to G_{(q)}$. Therefore the existence of the extension of $G_{(q)}$ follows from the uniqueness of the extension from K_∞ to K (Theorem 3.17(c)). □

Theorem 5.6. *If (G, K) is one of the 10 symmetric pairs of non-unitary type, then, for $q > 2$, all separable projective unitary representations of $G_{(q)}$ and all projective unitary representations of $G_{\infty,(q)}$ are trivial.*

Proof. If $\pi: G_{(q)} \to \mathrm{PU}(\mathcal{H}_\pi)$ is a continuous separable projective unitary representation, then composing with the conjugation representation of $\mathrm{PU}(\mathcal{H}_\pi)$ on the

Hilbert space $B_2(\mathcal{H}_\pi)$ leads to a separable unitary representation of $G_{(q)}$ on $B_2(\mathcal{H}_\pi)$. If we can show that this representation is trivial, then π is trivial as well. Therefore it suffices to consider unitary representations.

Since the group $G_{\infty,(q)}$ is separable, all its continuous unitary representations are direct sums of separable ones. Hence, in view of Proposition 5.5, the triviality of all continuous unitary representations of $G_{\infty,(q)}$ is equivalent to the triviality of all separable continuous unitary representations of $G_{(q)}$. We may therefore restrict our attention to separable representations of $G_{(q)}$.

Let $(\pi, \mathcal{H})$ be a continuous separable unitary representation of $G_{(q)}$. In view of Theorem 4.9, it is a direct sum of representations generated by the subspace $\mathcal{H}^{K_n}$ for some $n \in \mathbb{N}$. Any $v \in \mathcal{H}^{K_n}$ generates a spherical subrepresentation of the subgroup

$$G_{(q),n} := \{g \in G_{(q)} : ge_j = e_j, j = 1, \dots, n\}.$$

Now Theorem 5.6 implies that v is fixed by $G_{(q),n}$.

It remains to show that $G_{(q)}$ fixes v. In view of Proposition 5.5, it suffices to show that, for every $m > n$, $G(m)$ fixes v. The group $G(m)$ is reductive with maximal compact subgroup $K(m)$, and $G(m)_n$ is a non-compact subgroup.

Case 1: We first assume that the center of $G(m)_n$ is compact, which is the case for $G(m) \neq \mathrm{GL}(m, \mathbb{K})$ (this excludes 1,8 and 9). Then $G(m)$ is minimal in the sense that every continuous bijection onto a topological group is open, and this property is inherited by all its quotient groups [Ma97, Lemma 2.2]. In view of [Ma97, Prop. 3.4], all matrix coefficients of irreducible unitary representations $(\rho, \mathcal{H}_\rho)$ of quotients of $G(m)$ vanish at infinity of $G(m)/\ker \rho$. If $G(m)_n \nsubseteq \ker \rho$, then the image of $G(m)_n$ in the quotient group is noncompact, so that the only vector in $\mathcal{H}_\rho$ fixed by $G(m)_n$ is 0. Since every continuous unitary representation of $G(m)$ is a direct integral of irreducible ones, it follows that every $G(m)_n$-fixed vector in a unitary representation is fixed by $G(m)$.

Case 2: If $G(m) = \mathrm{GL}(m, \mathbb{K})$, then $Z = \mathbb{R}_+^\times \mathbf{1}$ is a noncompact subgroup of the center and the homomorphism $\chi: G \to \mathbb{R}_+^\times$, $\chi(g) := |\det_{\mathbb{R}}(g)|$ is surjective. Therefore $S(m) := \ker \chi$ has compact center and satisfies $G(m) = ZS(m)$. The preceding argument now implies that every fixed vector for $S(m)_n$ in a unitary representation of $G(m)$ is fixed by $S(m)$. Since $\chi|_{G(m)_n}$ is nontrivial, we conclude that every fixed vector for $G(m)_n$ in a unitary representation is fixed by $G(m)$.

Combining both cases, we see that in every unitary representation of $G(m)$, the subgroup $G(m)_n$ and $G(m)$ have the same fixed vectors, and this implies that every $G_{(q),n}$ fixed-vector is fixed by $G_{(q)}$.   □

Theorem 5.7 ([Pi90, Prop. 7.1]). *If (G, K) is one of the 10 symmetric pairs of unitary type, then, for $q > 2$, every separable continuous projective unitary representation of $G_{(q)}$ extends uniquely to a representation of G that is continuous with respect to the strong operator topology on G. In particular, it is a direct sum of irreducible ones which are determined by Theorem 4.9.*

Proof. With similar arguments as in the preceding proof, we see that every separable unitary representation of $G_{(q)}$ is a direct sum of representations generated by the fixed point space of some subgroup $G_{(q),n}$. Therefore its restriction to $G(\infty)$ is tame, so that Theorems 3.20 and 3.17 apply. $\qquad\square$

Theorem 5.8. *If (G, K) is one of the 10 symmetric pairs of non-unitary type, then all separable unitary representations of G are trivial.*

Proof. Let $(\pi, \mathcal{H})$ be a continuous separable unitary representation of G. We know already from Theorem 5.6 that $G_{(q)} \subseteq N := \ker \pi$ holds for $q > 2$. Now N is a closed normal subgroup containing K and its Lie algebra therefore contains $[\mathfrak{k}, \mathfrak{p}] = \mathfrak{p}$ as well (Lemma 5.4). This proves that $N = G$. $\qquad\square$

A Positive Definite Functions

In this appendix we recall some results and definitions concerning operator-valued positive definite functions.

Definition A.1. Let $\mathcal{A}$ be a C^*-algebra and X be a set. A map $Q: X \times X \to \mathcal{A}$ is called a *positive definite kernel* if, for any finite sequence $(x_1, \ldots, x_n) \in X^n$, the matrix $Q(x_i, x_j)_{i,j=1,\ldots,n} \in M(n, \mathcal{A})$ is a positive element.

For $\mathcal{A} = B(V)$, V a complex Hilbert space, this means that, for $v_1, \ldots, v_n \in V$, we always have $\sum_{i,j=1}^n \langle Q(x_i, x_j)v_j, v_i \rangle \geq 0$.

Definition A.2. Let $\mathcal{K}$ be a Hilbert space, G be a group, and $U \subseteq G$ be a subset. A function $\varphi: UU^{-1} \to B(\mathcal{K})$ is said to be *positive definite* if the kernel

$$Q_\varphi: U \times U \to B(\mathcal{K}), \quad (x, y) \mapsto \varphi(xy^{-1})$$

is positive definite. For $U = G$ we obtain the usual concept of a positive definite function on G.

Remark A.3 (Vector-Valued GNS-Construction). We briefly recall the bridge between positive definite functions and unitary representations.

(a) If $(\pi, \mathcal{H})$ is a unitary representation of G, $V \subseteq \mathcal{H}$ a closed subspace and $P_V: \mathcal{H} \to V$ the orthogonal projection on V, then $\pi_V(g) := P_V \pi(g) P_V^*$ is a $B(V)$-valued positive definite function with $\pi_V(\mathbf{1}) = \mathbf{1}$.

(b) If, conversely, $\varphi: G \to B(V)$ is positive definite with $\varphi(\mathbf{1}) = \mathbf{1}$, then there exists a unique Hilbert subspace $\mathcal{H}_\varphi$ of the space V^G of V-valued function on G for which the evaluation maps $K_g: \mathcal{H}_\varphi \to V, f \mapsto f(g)$ are continuous and satisfy $K_g K_h^* = \varphi(gh^{-1})$ for $g, h \in G$ [Ne00, Thm. I.1.4]. Then right translation by elements of G defines a unitary representation $(\pi_\varphi(g)f)(x) = f(xg)$ on this space with $K_{xg} = K_x \circ \pi(g)$. It is called the *GNS-representation associated to ρ*. Now $K_1^*: V \to \mathcal{H}_\varphi$ is an isometric embedding, so that we may identify V with a closed subspace of $\mathcal{H}_\varphi$ and K_1 with the orthogonal projection to V. This leads to $\varphi(g) = K_g K_1^* = K_1 \pi(g) K_1^*$, so that every positive definite function is of the form π_V. The construction also implies that $V \cong K_1^*(V)$ is G-cyclic in $\mathcal{H}_\varphi$.

For the following theorem, we simply note that all Banach–Lie groups are in particular Fréchet–BCH–Lie groups.

Theorem A.4. *Let G be a connected Fréchet–BCH–Lie group and $U \subseteq G$ an open connected 1-neighborhood for which the natural homomorphism $\pi_1(U, 1) \to \pi_1(G)$ is surjective. If $\mathcal{K}$ is Hilbert space and $\varphi: UU^{-1} \to B(\mathcal{K})$ an analytic positive definite function, then there exists a unique analytic positive definite function $\tilde{\varphi}: G \to B(\mathcal{K})$ extending φ.*

Proof. Let $q_G: \tilde{G} \to G$ be the universal covering morphism. The assumption that $\pi_1(U) \to \pi_1(G)$ is surjective implies that $\tilde{U} := q_G^{-1}(U)$ is connected. Now $\tilde{\varphi} := \varphi \circ q_G: \tilde{U}\tilde{U}^{-1} \to B(\mathcal{K})$ is an analytic positive definite function, hence extends by [Ne12, Thm. A.7] to an analytic positive definite function $\tilde{\varphi}$ on $\tilde{G}$. The restriction of $\tilde{\varphi}$ to $\tilde{U}$ is constant on the fibers of q_G, which are of the form $g \ker(q_G)$. Using analyticity, we conclude that $\tilde{\varphi}(gd) = \tilde{\varphi}(g)$ holds for all $g \in \tilde{G}$ and $d \in \ker(q_G)$. Therefore $\tilde{\varphi}$ factors through an analytic function $\varphi: G \to B(U)$ which is obviously positive definite. $\square$

Theorem A.5. *Let G be a connected analytic Fréchet–Lie group. Then a positive definite function $\varphi: G \to B(V)$ which is analytic in an open identity neighborhood is analytic.*

Proof. Since φ is positive definite, there exists a Hilbert space $\mathcal{H}$ and a $Q: G \to B(\mathcal{H}, V)$ with $\varphi(gh^{-1}) = Q_g Q_h^*$ for $g, h \in G$. Then the analyticity of the function φ in an open identity neighborhood of G implies that the kernel $(g, h) \mapsto Q_g Q_h^*$ is analytic on a neighborhood of the diagonal $\Delta_G \subseteq G \times G$. Therefore Q is analytic by [Ne12, Thm. A.3], and this implies that $\varphi(g) = Q_g Q_1^*$ is analytic. $\square$

The following proposition describes a natural source of operator-valued positive definite functions.

Proposition A.6. *Let $(\pi, \mathcal{H})$ be a unitary representation of the group G and $H \subseteq G$ be a subgroup. Let $V \subseteq \mathcal{H}$ be an isotypic H-subspace generating the G-module $\mathcal{H}$ and $P_V \in B(\mathcal{H})$ be the orthogonal projection onto V. Then V is invariant under the commutant $\pi(G)' = B_G(\mathcal{H})$ and the map*

$$\gamma: B_G(\mathcal{H}) \to B_H(V), \quad \gamma(A) = P_V A P_V$$

is an injective morphism of von Neumann algebras whose range is the commutant of the image of the operator-valued positive definite function

$$\pi_V: G \to B(V), \quad \pi_V(g) := P_V \pi(g) P_V.$$

In particular, if the H-representation on V is irreducible, then so is π.

Proof. That γ is injective follows from the assumption that V generates $\mathcal{H}$ under G. If the representation (ρ, V) of H is irreducible, then $\operatorname{im}(\gamma) \subseteq \mathbb{C}\mathbf{1}$ implies that $\pi(G)' = \mathbb{C}\mathbf{1}$, so that π is irreducible.

We now determine the range of γ. For any $A \in B_G(\mathcal{H})$, we have

$$P_V \pi(g) P_V P_V A P_V = P_V \pi(g) A P_V = P_V A \pi(g) P_V = P_V A P_V P_V \pi(g) P_V,$$

i.e., $\gamma(A) = P_V A P_V$ commutes with $\pi_V(G)$. Since γ is a morphism of von Neumann algebras, its range is also a von Neumann algebra of V commuting with $\pi_V(G)$. If, conversely, an orthogonal projection $Q = Q^* = Q^2 \in B_K(V)$ commutes with $\pi_V(G)$, then

$$P_V \pi(G) Q V = P_V \pi(G) P_V Q V = Q P_V \pi(G) P_V V \subseteq Q V$$

implies that the closed G-invariant subspace $\mathcal{H}_Q \subseteq \mathcal{H}$ generated by QV satisfies $P_V \mathcal{H}_Q \subseteq QV$, and therefore $\mathcal{H}_Q \cap V = QV$. For the orthogonal projection $\tilde{Q} \in B(\mathcal{H})$ onto $\mathcal{H}_Q$, which is contained in $B_G(\mathcal{H})$, this means that $\tilde{Q}|_V = Q$. This shows that $\operatorname{im}(\gamma) = \pi_V(G)'$. □

Remark A.7. The preceding proposition is particularly useful if we have specific information on the set $\pi_V(G)$. As $\pi_V(h_1 g h_2) = \rho(h_1) \pi_V(g) \rho(h_2)$, it is determined by the values of π_V on representatives of the H-double cosets in G.

(a) In the context of the lowest K-type (ρ, V) of a unitary highest weight representation (cf. [Ne00]), we can expect that $\pi_V(G) \subseteq \rho_{\mathbb{C}}(K_{\mathbb{C}})$ (by Harish–Chandra decomposition), so that $\pi_V(G)' = \rho_{\mathbb{C}}(K_{\mathbb{C}})' = \rho(K)'$ and γ is surjective.

(b) In the context of Sect. 3 and [Ol78], the representation (ρ, V) of H extends to a representation $\tilde{\rho}$ of a semigroup $S \supseteq H$ and we obtain $\pi_V(G)' = \tilde{\rho}(S)'$.

In both situations we have a certain induction procedure from representations of K and S, respectively, to G-representations which preserves the commutant but which need not be defined for every representation of K, resp., S.

Lemma A.8 ([NO13, Lemma C.3]). *Let $(S, *)$ be a unital involutive semigroup and $\varphi \colon S \to B(\mathcal{F})$ be a positive definite function with $\varphi(\mathbf{1}) = \mathbf{1}$. We write $(\pi_\varphi, \mathcal{H}_\varphi)$ for the representation on the corresponding reproducing kernel Hilbert space $\mathcal{H}_\varphi \subseteq \mathcal{F}^S$ by $(\pi_\varphi(s)f)(t) := f(ts)$. Then the inclusion*

$$\iota \colon \mathcal{F} \to \mathcal{H}_\varphi, \quad \iota(v)(s) := \varphi(s)v$$

is surjective if and only if φ is multiplicative, i.e., a representation.

Remark A.9. The preceding lemma can also be expressed without referring to positive definite functions and the corresponding reproducing kernel space. In this context it asserts the following. Let $\pi \colon S \to B(\mathcal{H})$ be a *-representation of a unital involutive semigroup $(S, *)$, $\mathcal{F} \subseteq \mathcal{H}$ a closed cyclic subspace and $P : \mathcal{H} \to \mathcal{F}$ the orthogonal projection. Then the function

$$\varphi : S \to B(\mathcal{F}), \quad \varphi(s) := P\pi(s)P^*$$

is multiplicative if and only if $\mathcal{F} = \mathcal{H}$.

B C^*-Methods for Direct Limit Groups

In this appendix we explain how to apply C^*-techniques to obtain direct integral decompositions of unitary representations of direct limit groups.

We recall that for a C^*-algebra $\mathcal{A}$, its multiplier algebra $M(\mathcal{A})$ is a C^*-algebra containing $\mathcal{A}$ as an ideal, and in every faithful representation $\mathcal{A} \hookrightarrow B(\mathcal{H})$, it is given by

$$M(\mathcal{A}) = \{M \in B(\mathcal{H}) : M\mathcal{A} + \mathcal{A}M \subseteq \mathcal{A}\}.$$

Let $G = \varinjlim G_n$ be a direct limit of locally compact groups and $\alpha_n : G_n \to G_{n+1}$ denote the connecting maps. We assume that these maps are closed embeddings. Then we have natural homomorphisms

$$\beta_n : L^1(G_n) \to M(L^1(G_{n+1}))$$

of Banach algebras, and since the action of G_n on $L^1(G_{n+1})$ is continuous, β_n is nondegenerate in the sense that $\beta(L^1(G_n)) \cdot L^1(G_{n+1})$ is dense in $L^1(G_{n+1})$. On the level of C^*-algebras we likewise obtain morphisms

$$\beta_n : C^*(G_n) \to M(C^*(G_{n+1})).$$

A state of G (=normalized continuous positive definite function) now corresponds to a sequence (φ_n) of states of the groups G_n with $\alpha_n^* \varphi_{n+1} = \varphi_n$ for every $n \in \mathbb{N}$. Passing to the C^*-algebras $C^*(G_n)$, we can view these functions also as states of the C^*-algebras. Then the compatibility condition is that the canonical extension $\tilde{\varphi}_n$ of φ_n to the multiplier algebra satisfies

$$\beta_n^* \tilde{\varphi}_{n+1} = \varphi_n.$$

Remark B.1. The ℓ^1-direct sum $\mathcal{L} := \oplus^1 L^1(G_n)$ carries the structure of a Banach-$*$-algebra (cf. [SV75]). Every unitary representation $(\pi, \mathcal{H})$ of G defines a sequence of nondegenerate representations $\pi_n : L^1(G_n) \to B(\mathcal{H}_n)$ which are compatible in the sense that

$$\pi_n = \alpha_n^* \tilde{\pi}_{n+1}.$$

Conversely, every such sequence of representations on a Hilbert space $\mathcal{H}$ leads to a sequence $\rho_n\colon G_n \to \mathrm{U}(\mathcal{H})$ of continuous unitary representations, which are uniquely determined by

$$\rho_n(g) = \tilde{\pi}_n(\eta_{G_n}(g)),$$

where $\eta_{G_n}\colon G_n \to M(L^1(G_n))$ denotes the canonical action by left multipliers. For $f \in L^1(G_n)$ and $h \in L^1(G_{n+1})$ we then have

$$\rho_{n+1}(\alpha_n(g))\pi_n(h)\pi_{n+1}(f) = \rho_{n+1}(\alpha_n(g))\pi_{n+1}(\beta_n(h)f) = \pi_{n+1}(\alpha_n(g)\beta_n(h)f)$$

$$= \pi_{n+1}(\beta_n(g*h)f) = \pi_n(g*h)\pi_{n+1}(f) = \rho_n(g)\pi_n(h)\pi_{n+1}(f),$$

which leads to

$$\rho_{n+1} \circ \alpha_n = \rho_n.$$

Therefore the sequence (ρ_n) is coherent and thus defines a unitary representation of G on $\mathcal{H}$. We conclude that the continuous unitary representations of G are in one-to-one correspondence with the coherent sequences of nondegenerate representations (π_n) of the Banach-$*$-algebras $L^1(G_n)$ (cf. [SV75, p. 60]).

Note that the nondegeneracy condition on the sequence (β_n) is much stronger than the nondegeneracy condition on the corresponding representation of the algebra $\mathcal{L}$. The group G_{n+1} does not act by multipliers on $L^1(G_n)$, so that there is no multiplier action of G on $\mathcal{L}$. However, we have a sufficiently strong structure to apply C^*-techniques to unitary representations of G.

Theorem B.2. *Let $\mathcal{A}$ be a separable C^*-algebra and $\pi\colon \mathcal{A} \to \mathcal{D}$ a homomorphism into the algebra $\mathcal{D}$ of decomposable operators on a direct integral $\mathcal{H} = \int_X^\oplus \mathcal{H}_x\, d\mu(x)$. Then there exists for each $x \in X$ a representation $(\pi_x, \mathcal{H}_x)$ of $\mathcal{A}$ such that $\pi \cong \int_X^\oplus \pi_x\, d\mu(x)$.*

If π is nondegenerate and $\mathcal{H}$ is separable, then almost all the representations π_x are nondegenerate.

Proof. The first part is [Dix64, Lemma 8.3.1] (see also [Ke78]). Suppose that π is nondegenerate and let $(E_n)_{n\in\mathbb{N}}$ be an approximate identity on $\mathcal{A}$. Then $\pi(E_n) \to \mathbf{1}$ holds strongly in $\mathcal{H}$ and [Dix69, Ch. II, no. 2.3, Prop. 4] implies the existence of a subsequence $(n_k)_{k\in\mathbb{N}}$ such that $\pi_x(E_{n_k}) \to \mathbf{1}$ holds strongly for almost every $x \in X$. For any such x, the representation π_x is non-degenerate.   □

Theorem B.3. *Let $G = \varinjlim G_n$ be a direct limit of separable locally compact groups with closed embeddings $G_n \hookrightarrow G_{n+1}$ and $(\pi, \mathcal{H})$ be a continuous separable unitary representation. For any maximal abelian subalgebra $\mathcal{A} \subseteq \pi(G)'$, we then obtain a direct integral decomposition $\pi \cong \int_X^\oplus \pi_x\, d\mu(x)$ into continuous unitary representations of G in which $\mathcal{A}$ acts by multiplication operators.*

Proof. According to the classification of commutative W^*-algebras, we have $\mathcal{A} \cong L^\infty(X, \mu)$ for a localizable measure space $(X, \mathfrak{S}, \mu)$ [Sa71, Prop. 1.18.1]. We therefore obtain a direct integral decomposition of the corresponding Hilbert space $\mathcal{H}$. To obtain a corresponding direct integral decomposition of the representation of G, we consider the C^*-algebra $\mathcal{B}$ generated by the subalgebras $\mathcal{B}_n$ which are generated by the image of the integrated representations $L^1(G_n) \to B(\mathcal{H})$. Then each $\mathcal{B}_n$ is separable and therefore $\mathcal{B}$ is also separable. Hence Theorem B.2 leads to nondegenerate representations $(\pi_x, \mathcal{H}_x)$ of $\mathcal{B}$ whose restriction to every $\mathcal{B}_n$ is nondegenerate.

In [SV75], the ℓ^1-direct sum $\mathcal{L} := \oplus^1_{n\in\mathbb{N}} L^1(G_n)$ is used as a replacement for the group algebra. From the representation $\pi: \mathcal{L} \to \mathcal{B}$ we obtain a representation ρ_x of this Banach-$*$-algebra whose restrictions to the subalgebras $L^1(G_n)$ are non-degenerate. Now the argument in [SV75, p. 60] (see also Remark B.1 above) implies that the corresponding continuous unitary representations of the subgroups G_n combine to a continuous unitary representation $(\rho_x, \mathcal{H}_x)$ of G. $\qquad\square$

Remark B.4. Let $(\pi, \mathcal{H})$ be a continuous unitary representation of the direct limit $G = \varinjlim G_n$ of locally compact groups. Let $\mathcal{A}_n := \pi_n(C^*(G_n))$ and write $\mathcal{A} := \langle \mathcal{A}_n : n \in \mathbb{N} \rangle_{C^*}$ for the C^*-algebra generated by the $\mathcal{A}_n$. Then $\mathcal{A}'' = \pi(G)''$ follows immediately from $\mathcal{A}''_n = \pi_n(G_n)''$ for each n.

From the nondegeneracy of the multiplier action of $C^*(G_n)$ on $C^*(G_{n+1})$ it follows that

$$C^*(G_n)C^*(G_{n+1}) = C^*(G_{n+1}),$$

which leads to

$$\mathcal{A}_n \mathcal{A}_{n+1} = \mathcal{A}_{n+1}.$$

We have a decreasing sequence of closed-$*$-ideals

$$\mathcal{I}_n := \overline{\sum_{k\geq n} \mathcal{A}_k} \subseteq \mathcal{A}$$

such that G_n acts continuously by multipliers on $\mathcal{I}_n$. A representation of $\mathcal{I}_n$ is non-degenerate if and only if its restriction to $\mathcal{A}_n$ is nondegenerate because $\mathcal{A}_n \mathcal{I}_n = \mathcal{I}_n$.

If a representation $(\rho, \mathcal{K})$ of $\mathcal{A}$ is nondegenerate on all these ideals, then it is non-degenerate on every $\mathcal{A}_n$, hence defines a continuous unitary representation of G.

Remark B.5. Theorem B.3 implies in particular the validity of the disintegration arguments in [Ol78, Thm. 3.6] and [Pi90, Prop. 2,4]. In [Ol84, Lemma 2.6] one also finds a very brief argument concerning the disintegration of "holomorphic" representations, namely that all the constituents are again "holomorphic". We think that this is not obvious and requires additional arguments.

Acknowledgements We thank B. Krötz for pointing out the reference [Ma97] and D. Pickrell for some notes concerning his approach to the proof of [Pi90, Prop. 7.1]. We are most grateful to D. Beltiţă, B. Janssens and C. Zellner for numerous comments on an earlier version of the manuscript.

References

[AH78] Albeverio, S., and R. J. Høegh-Krohn, *The energy representation of Sobolev–Lie groups*, Compositio Math. **36:1** (1978), 37–51.

[BN12] Beltiţă, D., and K.-H. Neeb, *Schur–Weyl Theory for C^*-algebras*, Math. Nachrichten **285:10** (2012), 1170–1198.

[BCR84] Berg, C., J.P.R. Christensen, and P. Ressel, *Harmonic analysis on semigroups*, Graduate Texts in Math., Springer Verlag, Berlin, Heidelberg, New York, 1984.

[Bou98] Bourbaki, N., *Topology: Chaps. 1–4*, Springer-Verlag, Berlin, 1998.

[Boy80] Boyer, R., *Representations of the Hilbert Lie group* $U(\mathcal{H})_2$, Duke Math. J. **47** (1980), 325–344.

[Boy84] —, *Projective representations of the Hilbert Lie group* $U(\mathcal{H})_2$ *via quasifree states on the CAR algebra*, J. Funct. Anal. **55** (1984), 277–296.

[Boy93] —, *Representation theory of infinite dimensional unitary groups*, Contemp. Math. **145** (1993), Americ. Math. Soc., 381–391.

[Dix64] Dixmier, J., *Les C^*-algèbres et leurs représentations*, Gauthier-Villars, Paris, 1964.

[Dix69] —, *Les algèbres d'opérateurs dans l'espace Hilbertien*, Gauthier-Villars, Paris, 1969.

[FF57] Feldmann, J., and J. M. G. Fell, *Separable representations of rings of operators*, Ann. of Math. **65** (1957), 241–249.

[GGV80] Gelfand, I. M., Graev, M. I., and A. M. Vershik, *Representations of the group of functions taking values in a compact Lie group*, Compositio Math. **42:2** (1980), 217–243.

[Gl03] Glöckner, H., *Direct limit Lie groups and manifolds*, J. Math. Kyoto Univ. **43** (2003), 1–26.

[dlH72] de la Harpe, P., *Classical Banach–Lie Algebras and Banach–Lie Groups of Operators in Hilbert Space*, Lecture Notes in Math. **285**, Springer-Verlag, Berlin, 1972.

[JN13] Janssens, B., and K.-H. Neeb, *Norm continuous unitary representations of Lie algebras of smooth sections*, arXiv:math.RT:1302.2535.

[Ka52] Kadison, R., *Infinite unitary groups*, Transactions of the Amer. Math. Soc. **72** (1952), 386–399.

[Ka97] Kaup, W., *On real Cartan factors*, manuscripta math. **92** (1997), 191–222.

[Ke78] Kehlet, E. T., *Disintegration theory on a constant field of nonseparable Hilbert spaces*, Math. Scand. **43:2** (1978), 353–362

[Ki73] Kirillov, A. A., *Representation of the infinite-dimensional unitary group*, Dokl. Akad. Nauk. SSSR **212** (1973), 288–290.

[Ku65] Kuiper, N. H, *The homotopy type of the unitary group of Hilbert space*, Topology **3** (1965), 19–30.

[La99] Lang, S., *Math Talks for Undergraduates*, Springer-Verlag, 1999.

[Ma97] Mayer, M., *Asymptotics of matrix coefficients and closures of Fourier Stieltjes algebras*, J. Funct. Anal. **143** (1997), 42–54.

[MN13] Merigon, S., and K.-H. Neeb, *Semibounded representations of groups of maps with values in infinite-dimensional hermitian groups*, in preparation.

[Mi89] Mickelsson, J., *Current algebras and groups*, Plenum Press, New York, 1989.

[Ne98] Neeb, K.-H., *Holomorphic highest weight representations of infinite-dimensional complex classical groups*, J. Reine Angew. Math. **497** (1998), 171–222.

[Ne00] —, *Holomorphy and Convexity in Lie Theory*, Expositions in Mathematics **28**, de Gruyter Verlag, Berlin, 2000.

[Ne02] —, Classical Hilbert–Lie groups, their extensions and their homotopy groups, in *Geometry and Analysis on Finite- and Infinite-Dimensional Lie Groups*, Eds. A. Strasburger et al., Banach Center Publications **55**, Warszawa 2002; 87–151.

[Ne04] —, Infinite dimensional Lie groups and their representations, in *Lie Theory: Lie Algebras and Representations*, Progress in Math. **228**, Ed. J. P. Anker, B. Ørsted, Birkhäuser Verlag, 2004; 213–328.

[Ne12] —, *Semibounded representations of hermitian Lie groups*, Travaux mathematiques **21** (2012), 29–109.

[Ne13] —, *Projective semibounded representations of doubly extended Hilbert–Lie groups*, in preparation.

[Ne13b] —, Holomorphic realization of unitary representations of Banach–Lie groups, *Lie Groups: Structure, Actions, and Representations– In Honor of Joseph A. Wolf on the Occasion of his 75th Birthday*, Huckleberry, A., Penkov, I., Zuckerman, G. (Eds.), Progress in Mathematics **306** (2013), 185–223.

[NO13] Neeb, K.-H., and G. Olafsson, *Reflection positive one-parameter groups and dilations*, preprint, arXiv:math.RT.1312.6161. Complex analysis and operator theory, to appear.

[Ol78] Olshanski, G. I., *Unitary representations of the infinite-dimensional classical groups* $U(p, \infty)$, $SO_0(p, \infty)$, $Sp(p, \infty)$, *and of the corresponding motion groups*, Functional Anal. Appl. **12:3** (1978), 185–195.

[Ol84] —, *Infinite-dimensional classical groups of finite $\mathbb{R}$-rank: description of representations and asymptotic properties*, Functional Anal. Appl. **18:1** (1984), 22–34.

[Ol89] —, *The method of holomorphic extension in the theory of unitary representations of infinite-dimensional classical groups*, Funct. Anal. Appl. **22** (1989), 273–285.

[Ol90] —, Unitary representations of infinite-dimensional (G, K)-pairs and the formalism of R. Howe, in *Representations of Lie Groups and Related Topics*, Eds. A. M. Vershik and D. P. Zhelobenko, Advanced Studies in Contemp. Math. **7**, Gordon and Breach Science Publ., 1990.

[Pi88] Pickrell, D., *The separable representations of $U(\mathcal{H})$*, Proc. of the Amer. Math. Soc. **102** (1988), 416–420.

[Pi89] —, *On the Mickelsson–Faddeev extension and unitary representations*, Comm. Math. Phys. **123:4** (1989), 617–625.

[Pi90] —, *Separable representations for automorphism groups of infinite symmetric spaces*, J. Funct. Anal. **90:1** (1990), 1–26.

[PS86] Pressley, A., and G. Segal, *Loop Groups*, Oxford University Press, Oxford, 1986.

[PW52] Putnam, C. R., and A. Winter, *The orthogonal group in Hilbert space*, Amer. J. Math. **74** (1952), 52–78.

[Ru73] Rudin, W., *Functional Analysis*, McGraw Hill, 1973.

[Sa71] Sakai, S., *C*-algebras and W*-algebras*, Ergebnisse der Math. und ihrer Grenzgebiete **60**, Springer-Verlag, Berlin, Heidelberg, New York, 1971.

[Sch60] Schue, J. R., *Hilbert space methods in the theory of Lie algebras*, Transactions of the Amer. Math. Soc. **95** (1960), 69–80.

[Se57] Segal, I.E., *The structure of a class of representations of the unitary group on a Hilbert space*, Proc. Amer. Math. Soc. **81** (1957), 197–203.

[SV75] Strătilă, Ş., and D. Voiculescu, *Representations of AF-algebras and of the Group $U(\infty)$*, Lecture Notes in Mathematics, Vol. 486. Springer-Verlag, Berlin - New York, 1975.

[Ta60] Takesaki, M., *On the non-separability of singular representations of operator algebra*, Kodai Math. Sem. Rep. **12** (1960), 102–108.

Weak Splittings of Quotients of Drinfeld and Heisenberg Doubles

Milen Yakimov

Abstract We investigate the fine structure of the symplectic foliations of Poisson homogeneous spaces. Two general results are proved for weak splittings of surjective Poisson submersions from Heisenberg and Drinfeld doubles. The implications of these results are that the torus orbits of symplectic leaves of the quotients can be explicitly realized as Poisson–Dirac submanifolds of the torus orbits of the doubles. The results have a wide range of applications to many families of real and complex Poisson structures on flag varieties. Their torus orbits of leaves recover important families of varieties such as the open Richardson varieties.

Key words Poisson–Lie groups • Drinfeld and Heisenberg doubles • Belavin–Drinfeld classification • Poisson–Dirac submanifolds

Mathematics Subject Classification (2010): Primary 53D17. Secondary 14M15, 22F30.

1 Introduction

The geometry of Poisson–Lie groups is well understood, both in the case of the standard Poisson structures on complex simple Lie groups [13, 20] and the general Belavin–Drinfeld Poisson structures [1, 29]. The torus orbits of symplectic leaves in the former case are the double Bruhat cells of the simple Lie group. One of the fundamental results in the theory of cluster algebras is the Berenstein–Fomin–Zelevinsky theorem [2] that their coordinate rings are upper cluster algebras.

The author was supported in part by NSF grants DMS-1001632 and DMS-1303036.

M. Yakimov (✉)

Department of Mathematics, Louisiana State University, Baton Rouge, LA 70803, USA
e-mail: yakimov@math.lsu.edu

G. Mason et al. (eds.), *Developments and Retrospectives in Lie Theory*,
Developments in Mathematics 37, DOI 10.1007/978-3-319-09934-7_9

Recently, the coordinate rings of the SL_n groups, equipped with the Cremmer–Gervais Poisson structures from [1], were also shown to be upper cluster algebras [15]. The motivation for these results is that cluster algebras give rise to Poisson structures by the work of Gekhtman et al. [14], and one attempts to go in the opposite direction using Poisson varieties from the theory of quantum groups.

On the other hand, the possible cluster algebra structures on coordinate rings of torus orbits of symplectic leaves of Poisson homogeneous spaces is much less well understood. In the special case of the standard complex Poisson structures on flag varieties, this is precisely the problem of constructing cluster algebra structures on the coordinate rings of the open Richardson varieties. These varieties have been recently studied in [4, 19, 25] in relation to Schubert calculus and total positivity. Chevalier [5] conjectured cluster algebra structures for the Richardson strata in the case when one of the two Weyl group elements is a parabolic Coxeter element. Leclerc [21] generalized this construction and showed that the coordinate ring of each open Richardson variety contains a cluster algebra whose rank is equal to the dimension of the variety. These cluster structures come from an additive categorification. Another cluster algebra structure on the Richardson–Lusztig strata of the Grassmannian was conjectured by Muller and Speyer [23] using Postnikov diagrams [24].

In this paper we prove a very general result that realizes torus orbits of symplectic leaves of a large class of Poisson homogeneous spaces as Poisson–Dirac submanifolds of torus orbits of symplectic leaves of Drinfeld and Heisenberg doubles. It applies to many important families of complex and real Poisson structures on flag varieties, double flag varieties, and their generalizations. In the context of cluster algebras, the point of this construction is that the coordinate rings of affine Poisson varieties with conjectured cluster algebra structures are realized as quotients of better understood coordinate rings of Poisson varieties, some of which are already proven to possess cluster algebra structures. The ideals defining these quotients are not Poisson but have somewhat similar properties coming from a notion of "weak splitting of surjective Poisson submersions." The construction of the latter is the main point of the paper.

To explain this in precise terms, we recall that to each point of a Poisson homogeneous space of a Poisson–Lie group Drinfeld [9] associated a Lagrangian subalgebra of the double and proved an equivariance property of this map. Motivated by this construction, Lu and Evens associated to each quadratic Lie algebra (a complex or real Lie algebra ∂ equipped with a nondegenerate invariant symmetric bilinear form $\langle.,.\rangle$) the variety of its Lagrangian subalgebras $\mathscr{L}(\partial, \langle.,.\rangle)$ and initiated its systematic study in [10]. This is a singular projective variety.

Fix a connected Lie group D with Lie algebra ∂. Given any pair of Lagrangian subalgebras $\mathfrak{g}_\pm$ such that ∂ is the vector space direct sum of $\mathfrak{g}_+$ and $\mathfrak{g}_-$ (in other words, given a Manin triple $(\partial, \mathfrak{g}_+, \mathfrak{g}_-)$ with respect to the bilinear form $\langle.,.\rangle$), one defines the r-matrix $r = \frac{1}{2} \sum_j \xi_j \wedge x_j$ where $\{x_j\}$ and $\{\xi_j\}$ is a pair of dual bases of $\mathfrak{g}_+$ and $\mathfrak{g}_-$. Using the adjoint action of D on $\mathscr{L}(\partial, \langle.,.\rangle)$, we construct the bivector field $\chi(r)$ on $\mathscr{L}(\partial, \langle.,.\rangle)$. Here and below χ refers to the infinitesimal action associated to a Lie group action. It was proved in [10] that $\chi(r)$ is Poisson. Up to minor technical details, the singular projective Poisson variety

$$(\mathcal{L}(\partial, \langle .,.\rangle), \chi(r))$$

captures the geometry of all Poisson homogeneous spaces of the Poisson–Lie groups integrating the Lie bialgebras $\mathfrak{g}_\pm$. The D-orbits on $\mathcal{L}(\partial, \langle .,.\rangle)$ are compete Poisson submanifolds, i.e., they are unions of symplectic leaves of $\chi(r)$. They have the form $D/N(\mathfrak{l})$ where $\mathfrak{l}$ is a Lagrangian subalgebra of $(\partial, \langle .,.\rangle)$ and $N(\mathfrak{l})$ is the normalizer of $\mathfrak{l}$ in D. The Belavin–Drinfeld Poisson structures [1] are coming from the case when $\partial = \mathfrak{g} \oplus \mathfrak{g}$ for a complex simple Lie algebra $\mathfrak{g}$. In this case the rank of the Poisson structure $\chi(r)$ at each point of $D/N(\mathfrak{l})$ was computed in [22, Theorem 4.10]. This describes the coarse structure of the symplectic foliations of the spaces $(D/N(\mathfrak{l}), \chi(r))$ or equivalently the variety of Lagrangian subalgebras $(\mathcal{L}(\partial, \langle .,.\rangle), \chi(r))$.

Here we address the problem of describing the fine structure of the symplectic foliations of these spaces. From the point of view of Lie theory and cluster algebras, the most important D-orbits in this picture are the orbits

$$(D/N(\mathfrak{g}_+), \chi(r)) \hookrightarrow (\mathcal{L}(\partial, \langle .,.\rangle), \chi(r)).$$

These Poisson varieties capture all examples of real and complex Poisson structures on flag varieties and double flag varieties that appeared in previous studies, see e.g., [10, 12, 16, 27]. The Poisson varieties $(D/N(\mathfrak{g}_+), \chi(r))$ also have the properties that they are quotients of *Drinfeld and Heisenberg doubles*. Recall that those are the Poisson varieties

$$(D, \pi = L(r) - R(r)) \quad \text{and} \quad (D, \pi' = L(r) + R(r)),$$

respectively. Here and below $R(.)$ and $L(.)$ refer to right and left invariant multivector fields on a Lie group. The Poisson structure π vanishes along the group

$$H := N(\mathfrak{g}_+) \cap N(\mathfrak{g}_-)$$

and as a consequence the left and right action of H on D preserves both Poisson structures π and π'. The corresponding Poisson reductions will be denoted by

$$(D/H, \pi_H) \quad \text{and} \quad (D/H, \pi'_H).$$

The canonical projections

$$\mu: (D/H, -\pi_H) \to (D/N(\mathfrak{g}_+), \chi(r)) \quad \text{and}$$

$$\mu': (D/H, \pi'_H) \to (D/N(\mathfrak{g}_+), \chi(r)) \quad (1)$$

are *surjective Poisson submersions*.

We prove that under certain general assumptions the symplectic leaves of the quotient $(D/N(\mathfrak{g}_+), \chi(r))$ can be realized as explicit symplectic submanifolds of the symplectic leaves of the (reduced) *Drinfeld double* $(D/H, -\pi_H)$ or *Heisenberg double* $(D/H, \pi'_H)$ (or even both in some cases). To be more precise, recall [6] that a Poisson manifold (X, π) is *a Poisson–Dirac submanifold admitting a Dirac projection* of a Poisson manifold (M, Π) if X is a submanifold of M, and there exists a subbundle E of $T_X M$ such that

$$E \oplus TX = T_X M \quad \text{and} \quad \Pi - \pi \in \Gamma(X, \wedge^2 E).$$

In this framework we find an explicit construction of sections of the surjective Poisson submersion $\mu \colon (D/H, -\pi_H) \to (D/N(\mathfrak{g}_+), \chi(r))$ over each $N(\mathfrak{g}_-)$-orbit whose images are Poisson–Dirac submanifolds of $(D/H, -\pi_H)$. This is the *weak splitting of the first surjective Poisson submersion* in (1) from a Drinfeld double. (We refer the reader to Sect. 2 below and [16, Sect. 2] for the definition of the notion and additional background.) As a corollary of the general construction, the symplectic leaves within each $N(\mathfrak{g}_-)$-orbit on $(D/N(\mathfrak{g}_+), \pi_{D/N(\mathfrak{g}_+)})$ are uniformly embedded as symplectic submanifolds of symplectic leaves of $(D/H, -\pi_H)$. Similarly, we construct sections of the second surjective Poisson submersion $\mu' \colon (D/H, \pi'_H) \to (D/N(\mathfrak{g}_+), \chi(r))$ in (1) over each $N(\mathfrak{g}_+)$-orbit whose images are Poisson–Dirac submanifolds of $(D/H, \pi'_H)$ admitting a Dirac projection. These constructions of weak splittings work under certain general assumptions, see Theorems 2 and 4 for details. In Sect. 5 we show that the conditions are satisfied for many important families of Poisson structures.

The above results have a wide range of applications. In the case of the standard Poisson structures on flag varieties we recover the weak splittings from [16]. Double flag varieties arise naturally as closed strata in partitions of wonderful group compactifications [7]. This gives rise to Poisson structures on them that are not products of Poisson structures on each factor [27]. The second splitting result above for Heisenberg doubles is applicable to this family of Poisson varieties. The real forms of a complex simple Lie algebra $\mathfrak{g}$ give rise to real Poisson structures on the related complex flag variety defined in [12]. Again the above second splitting is applicable for this family. Finally, the Delorme's classification result in [8] gives rise to canonical Poisson structures on products of flag varieties for complex simple Lie groups (i.e, flag varieties for a reductive group). Except for some very special cases, these Poisson structures are not products of Poisson structures on the factors. Our weak splittings are applicable for those families too.

The results in the paper are also related to the study of the spectra of the quantizations of the homogeneous coordinate rings of the above mentioned families of varieties. Currently, only the spectra of quantum flag varieties are understood [30]. We expect that a quantum version of our weak splittings of surjective Poisson submersions will be helpful in understanding the spectra of the quantizations of these families of varieties on the basis of the works on spectra of quantum groups [13, 17]. It appears that such quantum weak splittings should be also closely related to the notion of quantum folding of Berenstein and Greenstein [3].

The paper is organized as follows. In Sect. 2 we review the notion of Poisson–Dirac submanifolds of Poisson manifolds, and weak sections and weak splittings of surjective Poisson submersions. In Sect. 3 we prove two general theorems on the construction of weak sections and weak splittings for quotients of Drinfeld doubles. In Sect. 4 similar theorems are proved for quotients of Heisenberg doubles. Section 5 contains applications of these theorems.

We finish the introduction with some notation that will be used throughout the paper. Given a group G, $d \in G$ and two subgroups H_1 and H_2 of G, we will denote the H_1-orbit through $dH_2 \in G/H_2$ by

$$H_1 \cdot dH_2 \subset G/H_2$$

(to distinguish it from the double coset $H_1 d H_2 \subset G$).

For a Lie group G, we will denote by G° its identity component. For a smooth manifold X, X° will denote a connected component of X. Given a Lie group G and a subalgebra $\mathfrak{u}$ of its Lie algebra $\mathfrak{g}$, we will denote by $N(\mathfrak{u})$ the normalized of $\mathfrak{u}$ in G with respect to the adjoint action. Finally, recall that a Poisson structure π on a manifold M gives rise to the bundle map $\pi^\sharp \colon T^*M \to TM$, given by

$$\pi^\sharp(\alpha) = \alpha \otimes \mathrm{id}(\pi), \quad \alpha \in T^*_m M, m \in M.$$

2 Poisson–Dirac Submanifolds and Weak Splittings

In this section we review the notion of weak splitting of a surjective Poisson submersion from [16]. We start by recalling several facts about Poisson–Dirac submanifolds of Poisson manifolds, introduced and studied in [6, 26, 28].

Definition 1. A submanifold X of a Poisson manifold (M, Π) is called a *Poisson–Dirac submanifold* if the following conditions are satisfied:

(i) For each symplectic leaf S of (M, Π), the intersection $S \cap X$ is clean (i.e., it is smooth and $T_x(S \cap X) = T_x S \cap T_x X$ for all $x \in S \cap X$), and $S \cap X$ is a symplectic submanifold of $(S, (\Pi|_S)^{-1})$.
(ii) The family of symplectic structures $(\Pi|_S)^{-1}|_{S \cap X}$ is induced by a smooth Poisson structure π on X.

Clearly, in the setting of Definition 1, the symplectic leaves of (X, π) are the connected components of the intersections of symplectic leaves of (M, Π) with X.

An important criterion is provided by the following result.

Proposition 1 (Crainic and Fernandes [6]). *Assume that X is a submanifold of a Poisson manifold (M, Π) for which there exits a subbundle $E \subset T_X M$ such that*

(i) $E \oplus TX = T_X M$ and
(ii) $\Pi \in \Gamma(X, \wedge^2 TX \oplus \wedge^2 E)$.

Then X is Poisson–Dirac submanifold of (M, Π).

In the setting of Proposition 1 the projection of $\Pi|_X$ into $\Gamma(X, \wedge^2 TX)$ along $\wedge^2 E$ is exactly the needed Poisson structure π in Definition 1. Poisson–Dirac submanifolds with the property of Proposition 1 are called *Poisson–Dirac submanifolds admitting a Dirac projection* by Crainic–Fernandes [6] and quasi-Poisson submanifolds by Vaisman [26]. We will use the former term and call E an associated bundle of the Dirac projection.

In the presence of the condition (i) in Proposition 1, the condition (ii) is equivalent [6] to

$$\Pi_m^\sharp((T_m X)^0) \subset E_m, \quad \forall \, m \in X. \tag{2}$$

Here and below for a subspace $V \subseteq T_m M$, V^0 will denote its annihilator subspace in $T_m^* M$.

We continue with the notions of weak sections and weak splittings of surjective Poisson submersions.

Definition 2. Assume that (M, Π) and (N, π) are Poisson manifolds, X is a Poisson submanifold of (N, π), and that $p: (M, \Pi) \to (N, \pi)$ is a surjective Poisson submersion. A *weak section* of p over X is a smooth map $i: X \to M$ such that $p \circ i = \mathrm{id}_X$ and $i(X)$ is a Poisson–Dirac submanifold of (M, Π) with induced Poisson structure $i_*(\pi|_X)$.

In this situation we derive from Proposition 1 an explicit realization of all symplectic leaves of $(X, \pi|_X)$ in terms of those of (M, Π):

In the setting of Definition 2 one has that each symplectic leaf of $(X, \pi|_X)$ has the form $i^{-1}((i(X) \cap S)^\circ)$ where S is a symplectic leaf of (M, Π). In addition i realizes explicitly all leaves of $(X, \pi|_X)$ as symplectic submanifolds of symplectic leaves of (M, Π).

Remark 1. The following special case of the notion of weak section has an equivalent algebraic characterization which is of particular interest, see [16, Proposition 2.6] for details.

Let $p: (M, \Pi) \to (N, \pi)$ be a surjective Poisson submersion, let X be an open subset of N, and $i: X \to M$ a smooth map such that $p \circ i = \mathrm{id}_X$ and $i(X)$ is a smooth submanifold of M. In particular, $p^* : (C^\infty(N), \{.,.\}_\pi) \to (C^\infty(M), \{.,.\}_\Pi)$ is a homomorphism of Poisson algebras. Then i is a weak section with associated bundle equal to the tangent bundle to the fibers of p if and only if

$$i^* : (C^\infty(M), \{.,.\}_\Pi) \to (C^\infty(N), \{.,.\}_\pi)$$

is a homomorphism of Poisson $(C^\infty(N), \{.,.\}_\pi)$-modules with respect to the action on the first term coming from p^*.

Definition 3. Let (M, Π) and (N, π) be Poisson manifolds and $p: (M, \Pi) \to (N, \pi)$ a surjective Poisson submersion. A *weak splitting* of p is a partition

$$N = \bigsqcup_{a \in A} N_a$$

of (N, π) into complete Poisson submanifolds and a family of weak sections $i_a: N_a \hookrightarrow M$ of p (one for each stratum of the partition).

In the category of algebraic varieties we require M and N to be smooth algebraic varieties, X and N_a to be locally closed smooth algebraic subsets, and p, i_a to be algebraic maps.

We have

Proposition 2. *Consider a surjective Poisson submersion* $p: (M, \Pi) \to (N, \pi)$. *Let* $N = \bigsqcup_{a \in A} N_a$ *and* $i_a: N_a \to M$, $a \in A$ *define a weak splitting of* p. *Then for all* $a \in A$ *the following hold:*

(i) *Every symplectic leaf of* $(N_a, \pi|_{N_a})$ *has the form* $i_a^{-1}((i_a(N_a) \cap S)^\circ)$ *where S is a symplectic leaf of* (M, Π).

(ii) *Each symplectic leaf* S' *of* $(N_a, \pi|_{N_a})$ *is explicitly realized as a symplectic submanifold*

$$i_a: (S', \pi|_{S'}) \hookrightarrow (S, \Pi|_S)$$

of the unique symplectic leaf S of (M, Π) *that contains* $i_a(S')$.

All weak sections and weak splittings that we construct in this paper will have the property that their images are Poisson–Dirac submanifolds admitting Dirac projections, i.e., the images will satisfy the conditions in Proposition 1.

3 Weak Sections of Quotients of Drinfeld Doubles

We return to the setting of the introduction: Start with a Manin triple $(\partial, \mathfrak{g}_+, \mathfrak{g}_-)$ where ∂ is a quadratic Lie algebra with (a fixed) nondegenerate invariant symmetric bilinear form $\langle ., . \rangle$ and $\mathfrak{g}_\pm$ are two Lagrangian subalgebras such that $\partial = \mathfrak{g}_+ \oplus \mathfrak{g}_-$ as vector spaces. Let D be a connected Lie group with Lie algebra ∂ and let $G_\pm$ be the connected subgroups of D with Lie algebras $\mathfrak{g}_\pm$. Fix a pair of dual bases $\{x_j\}$ and $\{\xi_j\}$ of $\mathfrak{g}_+$ and $\mathfrak{g}_-$ with respect to $\langle ., . \rangle$. The standard r-matrix $r = \frac{1}{2} \sum \xi_j \wedge x_j$ gives rise to the Poisson structures

$$\pi = L(r) - R(r) \quad \text{and} \quad \pi' = L(r) + R(r) \tag{3}$$

on D. Then (D, π) is a Poisson–Lie group and $G_\pm$ are Poisson–Lie subgroups. Moreover, (D, π) is a Drinfeld double of $(G_\pm, \pi|_{G_\pm})$ and (D, π') is a Heisenberg double of $(G_\pm, \pi|_{G_\pm})$. Finally, $(G_-, -\pi|_{G_-})$ is a dual Poisson–Lie group of $(G_+, \pi|_{G_+})$.

Set for brevity

$$N_\pm := N(\mathfrak{g}_\pm), \quad \mathfrak{n}_\pm = \text{Lie}\,(N_\pm) = \mathfrak{n}(\mathfrak{g}_\pm). \tag{4}$$

Denote the canonical projections $p_\pm \colon \partial \to \mathfrak{g}_\pm$ along $\mathfrak{g}_\mp$. Identify ∂^* with ∂ using the form $\langle .,. \rangle$ and denote by $\alpha(y)$ the right invariant 1-form on D corresponding to $y \in \partial \cong \partial^*$.

The bundle maps $\pi^\sharp \colon T^*D \to TD$ and $(\pi')^\sharp \colon T^*D \to TD$ are given by the following formulas.

Lemma 1. *In the above setting, for all $x \in \mathfrak{g}_+$, $\xi \in \mathfrak{g}_-$ and $d \in D$,*

$$\pi^\sharp(\alpha_d(x+\xi)) = R_d(x) - L_d(p_+ \text{Ad}_d^{-1}(x+\xi))$$
$$= -R_d(\xi) + L_d(p_- \text{Ad}_d^{-1}(x+\xi)) \tag{5}$$

and

$$(\pi')^\sharp(\alpha_d(x+\xi)) = R_d(x) - L_d(p_- \text{Ad}_d^{-1}(x+\xi))$$
$$= -R_d(\xi) + L_d(p_+ \text{Ad}_d^{-1}(x+\xi)). \tag{6}$$

Equation (6) is proved in [11], Eqs. (6.2)–(6.4). Equation (5) is analogous.

The Poisson structure π vanishes on

$$H := N_+ \cap N_- = N(\mathfrak{g}_+) \cap N(\mathfrak{g}_-), \tag{7}$$

see e.g., [22, Lemma 1.12]. Denote $\mathfrak{h} = \text{Lie}\,H$. Recall that (D, π) is a Poisson Lie group and that both of its regular actions on the Heisenberg double (D, π') (given by $g \cdot g' = gg'$ and $g \cdot g' = g'g^{-1}$) are Poisson. Since $\pi|_H = 0$, the left and right actions of H on D preserve both π and π'. Denote their reductions with respect to the right action of H by

$$\pi_H \quad \text{and} \quad \pi'_H \in \Gamma(D/H, \wedge^2 T(D/H)),$$

respectively. Thus the canonical projections

$$\nu \colon (D, \pi) \to (D/H, \pi_H) \quad \text{and} \quad \nu' \colon (D, \pi') \to (D/H, \pi'_H)$$

are Poisson.

Since $\text{Lie}\,(N_+) \supseteq \mathfrak{g}_+$, it follows from the definition of the Drinfeld and Heisenberg double Poisson structures (3) that

$$\pi_{D/N_+} := \chi(r)$$

is a Poisson structure on D/N_+ and that the standard projections

$$\mu: (D, -\pi) \to (D/N_+, \pi_{D/N_+}) \quad \text{and} \quad \mu': (D, \pi') \to (D/N_+, \pi_{D/N_+})$$

are Poisson. Denote by

$$\eta: (D/H, -\pi_H) \to (D/N_+, \pi_{D/N_+}) \quad \text{and} \quad \eta': (D/H, \pi'_H) \to (D/N_+, \pi_{D/N_+})$$

the induced surjective Poisson submersions. (They are both Poisson since $\mu = \eta \nu$, $\mu' = \eta' \nu'$, ν, and ν' are surjective Poisson submersions.)

The following proposition will be used in our general construction of weak sections for η:

Proposition 3. *Assume that for a given $d \in D$ such that $dHd^{-1} \subset N_-$ there exists a subgroup Q of D with Lie algebra $\mathfrak{q}$ satisfying*

$$\mathfrak{n}_- = \mathfrak{n}_- \cap \mathrm{Ad}_d(\mathfrak{n}_+) + \mathfrak{n}_- \cap \mathrm{Ad}_d(\mathfrak{q}) \quad and \tag{8}$$

$$Q \cap N_+ = H, \ \mathfrak{n}_+ + \mathfrak{q} = \mathfrak{n}_+ + \mathfrak{q}^\perp + \mathrm{Ad}_d^{-1}(\mathfrak{n}_-) = \mathfrak{g}. \tag{9}$$

Set

$$G_d := N_- \cap dQd^{-1}. \tag{10}$$

Then

$$\tilde{E}^d \to G_d d, \ \tilde{E}^d_{gd} := R_{gd}(p_+ \mathrm{Ad}_d(\mathfrak{q}^\perp)) + L_{gd}(\mathfrak{n}_+), \ \forall \ g \in G_d \tag{11}$$

is a subbundle of $T_{G_d d} D$ such that

$$\tilde{E}^d \cap T(G_d d) = L(\mathfrak{h}), \ \tilde{E}^d + T(G_d d) = T_{G_d d} D, \tag{12}$$

and

$$\pi_{G_d d} \in \Gamma(G_d d, \wedge^2 \tilde{E}^d + \wedge^2 T(G_d d)).$$

In (12), $L(\mathfrak{h})$ denotes the subbundle of $T(G_d d)$ spanned by left invariant vector fields $L(h)$, $h \in \mathfrak{h}$. Here and below, for a subspace V of $\mathfrak{d}$, we denote $V^\perp = \{z \in \mathfrak{d} \mid \langle z, y \rangle = 0, \forall \ y \in V\}$.

The condition $dHd^{-1} \subset N_-$ ensures that $G_d d$ is stable under the right action of H. The subbundle $\tilde{E}^d$ of $T_{G_d d} D$ is equivariant with respect to this action. Indeed, if $h' \in H$, then

$$R_{h'} \tilde{E}^d_{gd} = R_{gdh'}(p_+ \mathrm{Ad}_d(\mathfrak{q}^\perp)) + L_{gd} R_{h'}(\mathfrak{n}_+)$$

$$= R_{gdh'}(p_+ \mathrm{Ad}_d(\mathfrak{q}^\perp)) + L_{gdh'}(\mathfrak{n}_+) = \tilde{E}^d_{gdh'}, \tag{13}$$

because $h' \in H \subset N_-$. Therefore, the pushforward of $\tilde{E}^d$ to $G_d \cdot dH$ is a subbundle of $T_{G_d \cdot dH}(D/H)$. As an immediate consequence of Proposition 3 we obtain the following corollary:

Corollary 1. *If, in the above setting, a subgroup Q of D and $d \in D$ satisfy (8)–(9) and $dHd^{-1} \subset N_-$, then the submanifold $G_d \cdot dH$ of the quotient $(D/H, -\pi_H)$ of the Drinfeld double $(D, -\pi)$ is a Poisson–Dirac submanifold admitting a Dirac projection with associated vector bundle equal to the pushforward $E^d = \nu_*(\tilde{E}^d)$ of $\tilde{E}^d$ to $G_d \cdot dH$.*

Proof of Proposition 3. Throughout the proof g will denote an element of G_d.

First we prove that $\tilde{E}^d$ is a subbundle of $T_{G_d d} D$ and that (12) holds. Fix $g \in G_d$. We have

$$T_{gd}(G_d d) + \tilde{E}^d_{gd} \supset L_{gd}(\mathrm{Ad}_d^{-1}\mathrm{Ad}_g^{-1}(\mathfrak{n}_- \cap \mathrm{Ad}_d(\mathfrak{q})) + L_{gd}(\mathfrak{n}_+)$$
$$= L_{gd}(\mathrm{Ad}_d^{-1}(\mathfrak{n}_-) \cap \mathfrak{q} + \mathfrak{n}_+) \supset L_{gd}(\mathrm{Ad}_d^{-1}(\mathfrak{n}_-))$$
$$= L_{gd}(\mathrm{Ad}_d^{-1}\mathrm{Ad}_g^{-1}(\mathfrak{n}_-)) = R_{gd}(\mathfrak{n}_-).$$

The second inclusion in the chain follows from (8). Thus,

$$T_{gd}(G_d d) + \tilde{E}^d_{gd} \supset R_{gd}(\mathfrak{n}_- + \mathfrak{p}_+(\mathrm{Ad}_d(\mathfrak{q}^\perp)) + L_{gd}(\mathfrak{n}_+)$$
$$\supset R_{gd}(\mathfrak{n}_- + \mathrm{Ad}_d(\mathfrak{q}^\perp)) + L_{gd}(\mathfrak{n}_+)$$
$$= L_{gd}(\mathrm{Ad}_d^{-1}\mathrm{Ad}_g^{-1}(\mathfrak{n}_-) + \mathrm{Ad}_d^{-1}\mathrm{Ad}_g^{-1}\mathrm{Ad}_d(\mathfrak{q}^\perp)) + L_{gd}(\mathfrak{n}_+)$$
$$= L_{gd}(\mathfrak{n}_+ + \mathfrak{q}^\perp + \mathrm{Ad}_d^{-1}(\mathfrak{n}_-)) = T_{gd} D,$$

where we used (8)–(9) and the fact that Q normalizes $\mathfrak{q}^\perp$. Clearly,

$$T_{gd}(G_d d) \supset L_{gd}(\mathfrak{h}) \quad \text{and} \quad \tilde{E}^d_{gd} \supset L_{gd}(\mathfrak{h}).$$

We claim that

$$\dim(\mathfrak{p}_+\mathrm{Ad}_d(\mathfrak{q}^\perp)) + \dim \mathfrak{n}_+ + \dim(\mathfrak{n}_- \cap \mathrm{Ad}_d(\mathfrak{q})) = \dim \mathfrak{g} + \dim \mathfrak{h}. \qquad (14)$$

This implies that $\tilde{E}^d$ is a subbundle of $T_{G_d d} D$ and that the first equality of (12) is satisfied. It also follows from (14) that $\tilde{E}^d$ is the direct sum of the subbundles $R(\mathfrak{p}_+\mathrm{Ad}_d(\mathfrak{q}^\perp))$ and $L(\mathfrak{n}_+)$ of $T_{G_d d} D$.

Since $\mathfrak{n}_- = \mathfrak{g}_- + \mathfrak{h}$ and $\mathrm{Ad}_d^{-1}(\mathfrak{h}) \subset \mathfrak{h} \subset \mathfrak{q}$, we have

$$\dim(\mathfrak{n}_- \cap \mathrm{Ad}_d(\mathfrak{q})) = \dim \mathfrak{n}_- - \dim \mathfrak{g}_- + \dim(\mathfrak{g}_- \cap \mathrm{Ad}_d(\mathfrak{q}))$$
$$= \dim \mathfrak{n}_- - \dim \mathfrak{g}_- + \dim \mathfrak{g} - \dim(\mathfrak{g}_-^\perp + \mathrm{Ad}_d(\mathfrak{q}^\perp))$$
$$= \dim \mathfrak{n}_- - \dim(\mathfrak{p}_+\mathrm{Ad}_d(\mathfrak{q}^\perp)),$$

taking into account that $\dim \mathfrak{n}_+ + \dim \mathfrak{n}_- = \dim \partial + \dim \mathfrak{h}$ leads to (14).

Since

$$T_{gd}G_d d = R_{gd}(\mathfrak{n}_- \cap \mathrm{Ad}_d(\mathfrak{q}))$$

and

$$(\mathfrak{n}_- \cap \mathrm{Ad}_d(\mathfrak{q}))^\perp = \mathfrak{n}_-^\perp + \mathrm{Ad}_d(\mathfrak{q}^\perp) \subset \mathfrak{g}_- + \mathfrak{p}_+(\mathrm{Ad}_d(\mathfrak{q}^\perp)),$$

we have

$$(T_{gd}G_d d)^0 \subset \{\alpha_{gd}(x + \xi) \mid x \in \mathfrak{p}_+(\mathrm{Ad}_d(\mathfrak{q}^\perp)), \xi \in \mathfrak{g}_-\}.$$

Applying (5) gives

$$\pi^\sharp((T_{gd}G_d d)^0) \subset R_{gd}(\mathfrak{p}_+(\mathrm{Ad}_d(\mathfrak{q}^\perp)) + L_{gd}(\mathfrak{g}_+) \subset \tilde{E}^d_{gd}.$$

$\square$

Observe that for $d \in D$ the conditions (8)–(9) ensure that the product $(N_- \cap dQd^{-1})(N_- \cap dN_+ d^{-1})$ is open in N_-. This implies that $(N_- \cap dQd^{-1}) \cdot dN_+$ is open in $N_- \cdot dN_+$ which is a complete Poisson submanifold of $(D/N_+, \pi_{D/N_+})$, [22, Theorem 2.3]. Thus, (8)–(9) imply that $(N_- \cap dQd^{-1}) \cdot dN_+$ is a Poisson submanifold of $(D/N_+, \pi_{D/N_+})$.

Theorem 1. *Assume that for a given $d \in D$ such that $dHd^{-1} \subset N_-$ there exists a subgroup Q of D satisfying the conditions (8)–(9). Then the smooth map $i: G_d \cdot dN_+ \to D/H$ defined by $i(gdN_+) = gdH$ for $g \in G_d := N_- \cap dQd^{-1}$ is a weak section of the surjective Poisson submersion $\eta: (D/H, -\pi_H) \to (D/N_+, \pi_{D/N_+})$ over $G_d \cdot dN_+$. Its image is a Poisson–Dirac submanifold of $(D/H, -\pi_H)$ admitting a Dirac projection with associated bundle $E^d := \nu_*(\tilde{E}^d)$ where $\tilde{E}^d$ is given by (11).*

Proof. It is straightforward to check that i is well defined: If $g_1, g_2 \in G_d$ and $g_1 dN_+ = g_2 dN_+$, then $(g_2)^{-1}g_1 \in N_- \cap d(Q \cap N_+)d^{-1} \subset dHd^{-1}$ because of (9). Thus $g_1 dH = g_2 dH$.

For $g \in G_d$, Corollary 1 implies that

$$(\pi_H)_{gdH} \in \wedge^2 T_{gdH}(G_d \cdot dH) + \wedge^2 E^d_{gdH}.$$

Observe that E^d_{gdH} contains the tangent space $\nu_*(L_{gd}(\mathfrak{n}_+))$ to the fiber of ν through gdH. Since $\eta_*(-\pi_H) = \pi_{D/N_+}$ and $\nu \circ i = \mathrm{id}_{G_d \cdot dN_+}$, we have that the projection of $(-\pi_H)|_{G_g \cdot dH}$ to $\wedge^2 T(G_d \cdot dH)$ along E^d is $i_*(\pi_{D/N_+}|_{G_d \cdot dN_+})$. This completes the proof of the theorem. $\square$

The next theorem provides a sufficient condition for the existence of a weak splitting of the surjective Poisson submersion in Theorem 1.

Theorem 2. *Assume that Q is a subgroup of D with Lie algebra $\mathfrak{q}$ such that there exists a set of representatives $\mathscr{D}$ for the (N_-, N_+)-double cosets of D satisfying*

$$N_- = (N_- \cap dQd^{-1})(N_- \cap dN_+d^{-1}) \quad and \tag{15}$$

$$dHd^{-1} \subset N_-, \; Q \cap N_+ = H, \; \mathfrak{n}_+ + \mathfrak{q} = \mathfrak{n}_+ + \mathfrak{q}^\perp + \mathrm{Ad}_d^{-1}(\mathfrak{n}_-) = \mathfrak{g} \tag{16}$$

for all $d \in \mathscr{D}$. Then the partition

$$D/N_+ = \bigsqcup_{d \in \mathscr{D}}(N_- \cap dQd^{-1}) \cdot dN_+ \tag{17}$$

and the family of smooth maps

$$i_d : (N_- \cap dQd^{-1}) \cdot dN_+ \to D/H, \quad i_d(gdN_+) := gdH, \; \forall \; g \in N_- \cap dQd^{-1}$$

is a weak splitting of the surjective Poisson submersion

$$\eta : (D/H, -\pi_H) \to (D/N_+, \pi_{D/N_+}).$$

In addition, the images of i_d are Poisson–Dirac submanifolds of $(D/H, -\pi_H)$ admitting Dirac projections with associated bundles $E^d := \nu_*(\tilde{E}^d)$ for the bundles $\tilde{E}^d$ given by (11).

Proof. The condition (15) implies that

$$(N_- \cap dQd^{-1}) \cdot dN_+ = N_- \cdot dN_+.$$

It follows from the definition of the set $\mathscr{D}$ that (17) defines a partition of G/N_+. This equality also implies that each stratum of the partition is a complete Poisson submanifold of $(D/N_+, \pi_{D/N_+})$, because this is a property of all N_+-orbits on D/N_+. The rest of the theorem follows from Theorem 1. $\qquad\square$

Proposition 2(i) implies that in the setting of Theorem 1 each symplectic leaf of $(D/N_+, \pi_{D/N_+})$ inside the stratum $(N_- \cap dQd^{-1}) \cdot dN_+ = G_d \cdot dN_+$ is of the form

$$i_d^{-1}((i_d(G_d \cdot dN_+) \cap S)^\circ)$$

for a symplectic leaf of S of $(D/H, -\pi_H)$. Furthermore, by Proposition 2(ii) each symplectic leaf of $(D/N_+, \pi_{D/N_+})$ is explicitly realized as a symplectic submanifold of a symplectic leaf of $(D/H, -\pi_H)$ via one of the maps i_d.

Remark 2. In Proposition 3, and Theorems 1 and 2 one can replace $N_\pm$ with any pair of subgroups $N'_\pm$ of D such that $N^\circ_\pm \subset N'_\pm \subset N_\pm$. The corresponding statements hold true for $H := N'_+ \cap N'_-$. Their proofs are analogous and are left to the reader.

Remark 3. If the group D and the subgroup Q in Theorems 1 and 2 are algebraic, then the constructed weak sections and splittings are algebraic.

4 Weak Sections of Quotients of Heisenberg Doubles

In this section we prove results for quotients of Heisenberg doubles that are similar to the results from the previous section for quotients of Drinfeld doubles. We use the setting and notation of the previous section. Using the second part of Lemma 1 one proves the following analog of Corollary 1 and Theorem 1 for Heisenberg doubles. We omit its proof since it is analogous to the case of Drinfeld doubles.

Theorem 3. *Let d and a subgroup Q of D with Lie algebra $\mathfrak{q}$ satisfy*

$$\mathfrak{n}_+ = \mathfrak{n}_+ \cap \mathrm{Ad}_d(\mathfrak{n}_+) + \mathfrak{n}_+ \cap \mathrm{Ad}_d(\mathfrak{q}) \quad and \tag{18}$$

$$dHd^{-1} \subset N_+, \ Q \cap N_+ = H, \ \mathfrak{n}_+ + \mathfrak{q} = \mathfrak{n}_+ + \mathfrak{q}^\perp + \mathrm{Ad}_d^{-1}(\mathfrak{n}_+) = \mathfrak{g}. \tag{19}$$

Set $G'_d = N_+ \cap dQd^{-1}$. Then the submanifold $G'_d \cdot dH$ of the quotient $(D/H, \pi'_H)$ of the Heisenberg double (D, π') is a Poisson–Dirac submanifold admitting a Dirac projection with associated vector bundle equal to the pushforward $F^d := \nu'_*(\tilde{F}^d)$ of the vector bundle

$$\tilde{F} \to G'_d d, \ \tilde{F}_{gd} := R_{gd}(p_-\mathrm{Ad}_d(\mathfrak{q}^\perp)) + L_{gd}(\mathfrak{n}_+), \ \forall g \in G'_d. \tag{20}$$

In addition, the map $i : G'_d \cdot dN_+ \to D/H$ defined by $i(gdN_+) = gdH$ for $g \in G'_d$ is a weak section of the surjective Poisson submersion $\eta' : (D/H, \pi'_H) \to (D/N_+, \pi_{D/N_+})$ over $G'_d \cdot dN_+$.

As in the previous section the theorem implies the following:

Theorem 4. *Let Q be a subgroup of D with Lie algebra $\mathfrak{q}$ for which there exists a set of representatives $\mathscr{D} \subset N(H)$ for the (N_+, N_+)-double cosets of D satisfying*

$$N_+ = (N_+ \cap dQd^{-1})(N_+ \cap dN_+d^{-1}) \tag{21}$$

and (19) for all $d \in \mathscr{D}$. Then the partition

$$D/N_+ = \sqcup_{d \in \mathscr{D}}(N_+ \cap dQd^{-1}) \cdot dN_+$$

and the family of maps

$$i'_d : (N_+ \cap dQd^{-1}) \cdot dN_+ \to D/H, \quad i'_d(gdN_+) := gdH, \ \forall g \in N_+ \cap dQd^{-1}$$

provide a weak splitting of the surjective Poisson submersion $\eta': (D/H, \pi'_H) \to$ *$(D/N_+, \pi_{D/N_+})$. In addition, the images of i'_d are Poisson–Dirac submanifolds of $(D/H, \pi'_H)$ admitting Dirac projections with associated bundles $F^d := v'_*(\tilde{F}^d)$ for the bundles $\tilde{F}^d$ given by (20) with $G'_d = N_+ \cap dQd^{-1}$.*

In light of Proposition 2, Theorem 4 provides an explicit realization of the symplectic leaves of $(D/N_+, \pi_{D/N_+})$ as symplectic submanifolds of the symplectic leaves of the Heisenberg double $(D/H, \pi'_H)$.

As in the case of Drinfeld doubles, in Theorems 3 and 4 one can replace $N_\pm$ with any pair of subgroup $N'_\pm$ of D such that $N^\circ_\pm \subset N'_\pm \subset N_\pm$ in which case one sets $H = N'_+ \cap N'_-$. The proofs of those slightly more general statements are analogous.

If the groups D and Q are algebraic, the above constructed weak sections and weak splittings are also algebraic.

5 Applications to Flag Varieties

This section contains applications of the results from the previous two sections to Poisson structures on flag varieties. Sections 5.1–5.4 deal with complex algebraic Poisson structures. There we construct weak splittings for complex surjective Poisson submersions from Drinfeld and Heisenberg doubles to flag varieties, double flag varieties and certain natural multi-flag generalizations. In Sect. 5.5 we give applications to real algebraic Poisson structures on flag varieties.

We note that all of the weak splittings that are constructed in this section provide (via Proposition 2) explicit realizations of the symplectic leaves of Poisson structures on flag varieties as symplectic submanifolds of symplectic leaves of Drinfeld and Heisenberg doubles.

Throughout the section G will denote an arbitrary connected complex simple Lie group with Lie algebra $\mathfrak{g}$. The Killing form on $\mathfrak{g}$ will be denoted by $\langle .,. \rangle$. We fix a pair of opposite Borel subgroups $B_\pm$ of G and the corresponding maximal torus $T := B_+ \cap B_-$. Let $U_\pm$ be the unipotent radicals of $B_\pm$. Set

$$\mathfrak{h} := \text{Lie } T, \quad \mathfrak{b}_\pm := \text{Lie } B_\pm, \quad \text{and} \quad \mathfrak{u}_\pm := \text{Lie } U_\pm.$$

Denote the Weyl group of G by W. For each $w \in W$, fix a representative $\dot{w}$ in the normalizer of T in G.

5.1 Standard Poisson Structures on Complex Flag Varieties

The simplest applications of Theorems 1 and 3 are to weak splittings for Poisson structures on flag varieties.

Recall the standard Manin triple

$$\partial := \mathfrak{g} \oplus \mathfrak{h}, \ \mathfrak{g}_\pm := \{(x_\pm + h, \pm h) \mid x_\pm \in \mathfrak{u}_\pm, h \in \mathfrak{h}\}$$

with respect to the invariant bilinear form on ∂ given by

$$\langle (x_1, y_1), (x_2, y_2) \rangle := \langle x_1, x_2 \rangle - \langle y_1, y_2 \rangle, \ \forall \ x_i \in \mathfrak{g}, y_i \in \mathfrak{h}.$$

The Drinfeld and Heisenberg double Poisson structures on $D := G \times T$ will be denoted by π and π', respectively. In this case $N_\pm = N(\mathfrak{g}_\pm) = B_\pm \times T$ and $H = N_+ \cap N_- = T \times T$. The reductions of $-\pi$ and π' to $G/T \cong (G \times T)/(T \times T) = D/H$ will be denoted by $-\pi_T$ and π'_T. Both structures reduce to the same Poisson structure on the flag variety $G/B_+ \cong (G \times T)/(B_+ \times T) = D/N_+$ called the standard Poisson structure. The latter will be denoted by π_{G/B_+}.

It is easy to verify that the group $Q = N_- = B_- \times T$ and the set $\mathscr{D} = \{(\dot{w}, 1) \mid w \in W\}$ satisfy the conditions in Theorem 2 for the above choice of D and $\mathfrak{g}_\pm$. This implies the following result of Goodearl and the author, proved in [16, Theorem 3.2]. For its statement we need to introduce some additional notation. Denote the vector bundle

$$\tilde{E}_w \to \big((B_- \cap wB_-w^{-1}) \times T\big)(\dot{w}, 1)$$

with fibers

$$\tilde{E}_g^w := R_g\big(p_+\mathrm{Ad}_{(\dot{w},1)}(\mathfrak{u}_- \oplus 0)\big) + L_g(\mathfrak{b}_+ \oplus \mathfrak{h})$$

for $g \in \big((B_- \cap wB_-w^{-1}) \times T\big)(\dot{w}, 1)$ where $p_+ : \partial \to \mathfrak{g}_+$ is the projection along $\mathfrak{g}_-$. Here the direct sum notation is used to denote subspaces of $\mathfrak{g} \oplus \mathfrak{h}$ identified with $\mathrm{Lie}\,(G \times T)$. Denote the canonical projection

$$v : G \times T \to (G \times T)/(T \times T) \cong G/T. \tag{22}$$

By Corollary 1 the pushforward $E^w := v_*(\tilde{E}^w)$ is a well defined vector bundle over $B_- \cdot wT \subset G/T$.

Theorem 5 ([16]). *For all connected complex simple Lie groups G, the partition of the full flag variety G/B_+ into Schubert cells*

$$G/B_+ = \sqcup_{w \in W} B_- \cdot wB_+$$

and the family of maps

$$i_w : B_- \cdot wB_+ \to G/T, \ i_w(b_- \dot{w}B_+) := b_- \dot{w}T, \ \forall \ b_- \in B_- \cap wB_-w^{-1}$$

define a weak splitting of the surjective Poisson submersion

$$(G/T, -\pi_T) \to (G/B_+, \pi_{G/B_+})$$

from a Drinfeld double to the flag variety. Furthermore, the image of each map i_w is a Poisson–Dirac submanifold of $(G/T, -\pi_T)$ admitting a Dirac projection with associated bundle E^w defined above.

Theorem 4 for weak splittings of quotients of Heisenberg doubles is also applicable to flag varieties to obtain a weak splitting for the surjective Poisson submersion

$$(G/T, -\pi'_T) \to (G/B_+, \pi_{G/B_+}).$$

It is easy to verify that the conditions of Theorem 4 are satisfied by for the same group $Q = N_- = B_- \times T$, set $\mathcal{D} = \{(\dot{w}, 1) \mid w \in W\}$ and the current choice of D and $\mathfrak{g}_\pm$. Applying the theorem leads to the following result:

Theorem 6. *For all connected complex simple Lie groups G, the partition of the full flag variety G/B_+ into Schubert cells*

$$G/B_+ = \sqcup_{w \in W} B_+ \cdot w B_+$$

and the family of maps

$$i'_w: B_+ \cdot w B_+ \to G/T, \quad i'_w(b_+ w B_+) = b_+ \dot{w} T, \quad b_+ \in B_+ \cap w B_- w^{-1}$$

is a weak splitting of the surjective Poisson submersion

$$(G/T, \pi'_T) \to (G/B_+, \pi_{G/B_+})$$

from a Heisenberg double to the flag variety. The image of each map i'_w is a Poisson–Dirac submanifold of $(G/T, \pi'_T)$ admitting a Dirac projection with associated bundle $v_*(\tilde{F}^w)$ for the pushforward bundle $v_*(\tilde{F}^w)$ with respect to (22) where

$$\tilde{F}^w \to ((B_+ \cap w B_- w^{-1}) \times T)(\dot{w}, 1)$$

is the vector bundle with fibers

$$\tilde{F}^w_g := R_g\left(p_- \mathrm{Ad}_{(\dot{w},1)}(\mathfrak{u}_- \oplus 0)\right) + L_g(\mathfrak{b}_+ \oplus \mathfrak{h})$$

for $g \in ((B_+ \cap w B_- w^{-1}) \times T)(\dot{w}, 1)$ and $p_-: \mathfrak{d} \to \mathfrak{g}_-$ is the projection along $\mathfrak{g}_+$.

Because of Proposition 2, Theorems 5 and 6 provide an explicit realization of the symplectic leaves of the flag varieties $(G/B_+, \pi_{G/B_+})$ as symplectic submanifolds of the symplectic leaves of Drinfeld and Heisenberg doubles.

We note that the partitions into Schubert cells in Theorems 5 and 6 are with respect to opposite Borel subgroups. The two results can be derived from each other. Let w_o be the longest element of W. The equivalence is shown using the facts that the translation action of $\dot{w}_o$ on $(G/B_+, \pi_{G/B_+})$ is anti-Poisson, and the left translation action of $(\dot{w}_o, 1)$ on $G \times T$ interchanges π and π'.

5.2 Double Flag Varieties

Next, we consider the standard Manin triple

$$\partial := \mathfrak{g} \oplus \mathfrak{g}, \quad \mathfrak{g}_+ := \{(x_+ + h, x_- - h) \mid x_\pm \in \mathfrak{u}_\pm, h \in \mathfrak{h}\},$$

$$\mathfrak{g}_- := \{(x, x) \mid x \in \mathfrak{g}\} \quad (23)$$

with respect to the invariant bilinear form on ∂

$$\langle (x_1, y_1), (x_2, y_2) \rangle := \langle x_1, x_2 \rangle - \langle y_1, y_2 \rangle, \quad x_i, y_i \in \mathfrak{g}.$$

Let $D := G \times G$. For this setting we have

$$N_+ = N(\mathfrak{g}_+) = B_+ \times B_-, \quad N_-^\circ = N(\mathfrak{g}_-)^\circ = G_\Delta, \quad \text{and} \quad N_+ \cap N_-^\circ = T_\Delta$$

where G_Δ and T_Δ are the diagonal subgroups of $G \times G$ and $T \times T$, respectively.

The Drinfeld and Heisenberg double Poisson structures on $G \times G$ will be again denoted by π and π'. The group T_Δ acts on the left and right on $(G \times G, \pi)$ and $(G \times G, \pi')$ by Poisson automorphisms. The reductions of π and π' to $(G \times G)/T_\Delta$ will be denoted by π_{T_Δ} and π'_{T_Δ}. The pushforwards of $-\pi_{T_\Delta}$ and π'_{T_Δ} to $(G \times G)/N_+ \cong G/B_+ \times G/B_-$ are well defined and are equal to each other. Denote the resulting Poisson structure by π_{df}. The Poisson manifold

$$(G/B_+ \times G/B_-, \pi_{df})$$

is the double flag variety studied in [27]. Theorem 2 cannot be applied to the Poisson submersion $((G \times G)/T_\Delta, -\pi_{T_\Delta}) \to (G/B_+ \times G/B_-, \pi_{df})$, but Theorem 4 can be applied for the following choice of a group Q and a set $\mathscr{D}$:

$$Q = (U_- \times U_+)T_\Delta, \quad \mathscr{D} = \{(\dot{w}, \dot{v}) \mid w, v \in W\},$$

and the above D and $\mathfrak{g}_\pm$. This gives a weak splitting of the surjective Poison submersion $((G \times G)/T_\Delta, \pi'_{T_\Delta}) \to (G/B_+ \times G/B_-, \pi_{df})$. We will need the following notation to state the result. Denote the projection

$$v' : G \times G \to (G \times G)/T_\Delta. \quad (24)$$

Consider the vector bundle

$$\tilde{F}^{w,v} \to ((U_+ \cap wU_-w^{-1}) \times (U_- \cap vU_+v^{-1}))(\dot{w}, \dot{v})T_\Delta$$

with fibers

$$\tilde{F}^{w,v}_g := R_g(p_-\mathrm{Ad}_{(\dot{w},\dot{v})}(\mathfrak{u}_- \oplus \mathfrak{u}_+ + \mathfrak{t}_{a\Delta})) + L_g(\mathfrak{b}_+ \oplus \mathfrak{b}_-) \qquad (25)$$

for $g \in ((U_+\cap wU_-w^{-1})\times(U_-\cap vU_+v^{-1}))(\dot{w}, \dot{v})T_\Delta$ where the direct sum notation is used for subspaces of $\partial = \mathfrak{g}\oplus\mathfrak{g}$ identified with $\mathrm{Lie}\,(G \times G)$, $p_-: \partial \to \mathfrak{g}_-$ denotes the projection along $\mathfrak{g}_+$, and $\mathfrak{t}_{a\Delta}$ is the antidiagonal of $\mathfrak{t} \oplus \mathfrak{t}$.

Theorem 7. *For all connected complex simple Lie groups G, the partition of the double flag variety into $B_+ \times B_-$-Schubert cells*

$$G/B_+ \times G/B_- = \sqcup_{w,v\in W} B_+ \cdot wB_+ \times B_- \cdot vB_-$$

and the family of maps

$$i'_{w,v}: B_+ \cdot wB_+ \times B_- \cdot vB_- \to (G \times G)/T_\Delta$$

given by

$$i'_{w,v}(u_+wB_+, u_-vB_-) := (u_+\dot{w}, u_-\dot{v})T_\Delta$$

for all $u_+ \in U_+ \cap wU_-w^{-1}$ and $u_- \in U_- \cap vU_+v^{-1}$ is a weak splitting of the surjective Poisson submersion

$$((G \times G)/T_\Delta, \pi'_\Delta) \to (G/B_+ \times G/B_-, \pi_{df})$$

from a Heisenberg double to the double flag variety. The image of each map $i'_{w,v}$ is a Poisson–Dirac submanifold of $((G \times G)/T_\Delta, \pi'_\Delta)$ admitting a Dirac projection with associated vector bundle $v'_*(\tilde{F}^{w,v})$, cf. (24) and (25).

5.3 Partial Flag Varieties

All results in Sect. 5 on weak splittings for full flag varieties have analogs to partial flag varieties. We will provide full details in the case of double partial flag varieties. The generalizations of the results in Sects. 5.1, 5.4, and 5.5 are analogous.

For a subset I of simple roots of $\mathfrak{g}$, denote by $P^I_\pm \supseteq B_\pm$ the corresponding parabolic subgroups of G and by W_I the subgroup of the Weyl group. Let W^I be the sets of minimal length representatives for the cosets in W/W_I.

Fix two subsets I_1, I_2 of simple roots of $\mathfrak{g}$. The pushforward of π_{df} under the canonical projection

$$G/B_+ \times G/B_- \to G/P_+^{I_1} \times G/P_-^{I_2}$$

is a well-defined Poisson structure since $P_+^{I_1} \times P_-^{I_2}$ is a Poisson–Lie subgroup of $(G \times G, \pi)$. Denote this pushforward by $\pi_{df}^{I_1,I_2}$. The map

$$(G/B_+ \times G/B_-, \pi_{df}) \to (G/P_+^{I_1} \times G/P_-^{I_2}, \pi_{df}^{I_1,I_2})$$

is a surjective Poisson submersion and its restrictions

$$(B_+ \cdot wB_+ \times B_- \cdot vB_-, \pi_{df}) \to (B_+ \cdot wP_+^{I_1} \times B_- \cdot vP_-^{I_2}, \pi_{df}^{I_1,I_2})$$

are Poisson isomorphisms for all $w \in W^{I_1}$, $v \in W^{I_2}$. (Similar Poisson isomorphisms are constructed in the settings of Sects. 5.1, 5.4, and 5.5. This produces the generalizations of those results to the cases of partial flag varieties.) Taking inverses of the above Poisson isomorphisms and composing them with the maps $i'_{w,v}$ in Theorem 7 leads to the following result:

Corollary 2. *For all connected complex simple Lie groups G and subsets of simple roots I_1 and I_2, the partition of the corresponding double partial flag variety into $B_+ \times B_-$-Schubert cells*

$$G/P_+^{I_1} \times G/P_-^{I_2} = \bigsqcup_{w \in W^{I_1}, v \in W^{I_2}} B_+ \cdot wP_+^{I_1} \times B_- \cdot vP_-^{I_2}$$

and the family of maps

$$j'_{w,v} : B_+ \cdot wP_+^{I_1} \times B_- \cdot vP_-^{I_2} \to (G \times G)/T_\Delta$$

given by

$$j'_{w,v}(u_+wP_+^{I_1}, u_-vP_-^{I_2}) := (u_+\dot{w}, u_-\dot{v})T_\Delta$$

for all $u_+ \in U_+ \cap wU_-w^{-1}$ and $u_- \in U_- \cap vU_+v^{-1}$ is a weak splitting of the surjective Poisson submersion

$$((G \times G)/T_\Delta, \pi'_\Delta) \to (G/P_+^{I_1} \times G/P_-^{I_2}, \pi_{df}^{I_1,I_2})$$

from a Heisenberg double to the double partial flag variety. The image of each map $j'_{w,v}$ (which is the same as the image of the map $i'_{w,v}$) is a Poisson–Dirac submanifold of $((G \times G)/T_\Delta, \pi'_\Delta)$ admitting a Dirac projection with associated vector bundle $v'_*(\tilde{F}^{w,v})$, see (24) and (25).

5.4 Multiple Flag Varieties

The results in Sects. 5.1–5.2 can be generalized to a very large class of Poisson structures on multiple flag varieties. Those are Cartesian products of flag varieties for complex simple Lie groups (i.e., flag varieties for reductive Lie groups) with Poisson structures which in general are not products of Poisson structures on the factors. Since the arguments are similar, we only state the results leaving the details to the reader.

We start with any reductive Lie algebra ∂ and an invariant bilinear form $\langle .,. \rangle$ on it. All Lagrangian subalgebras and Manin triples in this situation were classified by Delorme [8] up to the action of the adjoint group of ∂. Let $\overline{b}_\pm$ be a pair of opposite Borel subalgebras of ∂ and let $\overline{t} := \overline{b}_+ \cap \overline{b}_-$ be the corresponding Cartan subalgebra of ∂. Denote by $\overline{u}_\pm$ the nilradicals of $\overline{b}_\pm$. Let D be a connected reductive Lie group with Lie algebra ∂, and let $\overline{B}_\pm$ and $\overline{T}$ be its Borel subgroups and maximal torus corresponding to $\overline{b}_\pm$ and $\overline{t}$.

Consider any Manin triple

$$(\partial, g_+, g_-) \quad \text{such that} \quad g_+ \subset \overline{b}_+. \tag{26}$$

The results of Delorme imply that after a conjugation by an element of $\overline{B}_+$, one has

$$g_+ = \overline{u}_+ + g_+ \cap \overline{t}$$

(in particular, $N(g_+) = \overline{B}_+$) and the group $H := N(g_+) \cap N(g_-)^\circ$ satisfies

$$H = \overline{B}_+ \cap N(g_-)^\circ = \overline{T} \cap N(g_-)^\circ. \tag{27}$$

One can write an explicit formula for the subgroup H of the maximal torus $\overline{T}$ in terms of generalized Belavin–Drinfeld triples in the setting of [8]. We leave the details to the reader since this requires extra notation. For the rest we will assume that the conjugation by an element of $\overline{B}_+$ is performed so that the above conditions are satisfied.

Denote the Drinfeld and Heisenberg double Poisson structures on D corresponding to a Manin triple of the type (26) by π and π'. By the general facts in Sect. 3, the left and right regular actions of H on (G, π') are Poisson. Denote the reduction $(D/H, \pi'_H)$ for the right action and the surjective Poisson submersion

$$\nu' : (D, \pi') \to (D/H, \pi'_H). \tag{28}$$

The Drinfeld and Heisenberg Poisson structures π and π' descend to the same Poisson structure on the multiple flag variety $D/N(g_+) = D/\overline{B}_+$ which will be denoted by $\pi_{D/\overline{B}_+}$. The Poisson structures in Sects. 5.1–5.2 are special cases of this construction when

$$D = G \times T \quad \text{or} \quad D = G \times G$$

for a complex simple Lie group G and a maximal torus T of G.

The canonical projection

$$(D/H, \pi'_H) \to (D/\overline{B}_+, \pi_{D/\overline{B}_+})$$

is Poisson, because $\overline{B}_+ = N(\mathfrak{g}_+)$ is a Poisson–Lie subgroup of (D, π). Denote the connected subgroups of D with Lie algebras $\overline{\mathfrak{u}}_\pm$ by $\overline{U}_\pm$. Let $\overline{W}$ be the Weyl group of D. For each of $w \in \overline{W}$, fix a representative $\dot{w}$ in the normalizer of the maximal torus $\overline{T}$ of D.

A simple computation shows that the conditions of Theorem 4 are satisfied for the group $Q = H\overline{U}_-$ and the set $\mathcal{D} = \{\dot{w} \mid w \in \overline{W}\}$. We have

Theorem 8. *For all Manin triples for a connected reductive algebraic group D of the form* (26), *the partition of the multiple flag variety $D/\overline{B}_+$ into Schubert cells*

$$D/\overline{B}_+ = \bigsqcup_{w \in \overline{W}} \overline{B}_+ \cdot w\overline{B}_+$$

and the family of maps

$$i'_w \colon \overline{B}_+ \cdot w\overline{B}_+ \to D/H, \quad i'_w(u_+ w\overline{B}_+) := u_+ \dot{w} H, \quad \forall\, u_+ \in \overline{U}_+ \cap w\overline{U}_- w^{-1}$$

(recall (27)*) is a weak splitting of the surjective Poisson submersion*

$$(D/H, \pi'_H) \to (D/\overline{B}_+, \pi_{D/\overline{B}_+})$$

from a Heisenberg double to the multiple flag variety. The image of each map i'_w is a Poisson–Dirac submanifold of $(D/H, \pi'_H)$ admitting a Dirac projection with associated bundle $v'_*(\tilde{F}^w)$ for the pushforward with respect to (28) *of the vector bundle*

$$\tilde{F}^w \to (\overline{U}_+ \cap w\overline{U}_- w^{-1})\dot{w} H$$

with fibers

$$\tilde{F}^w_g := R_g\left(p_- \mathrm{Ad}_{\dot{w}}(\overline{\mathfrak{u}}_- + \mathfrak{h}^\perp)\right) + L_g(\overline{\mathfrak{b}}_+)$$

for $g \in (\overline{U}_+ \cap w\overline{U}_- w^{-1})\dot{w} H$. Here $\mathfrak{h}^\perp$ denotes the orthogonal complement to $\mathfrak{h} := \mathrm{Lie}\, H$ in $\overline{\mathfrak{t}}$ with respect to $\langle ., .\rangle$, and $p_- \colon \mathfrak{d} \to \mathfrak{g}_-$ is the projection along $\mathfrak{g}_+$.

5.5 Real Poisson Structures on Flag Varieties

All results in Sects. 5.1–5.4 remain valid when all complex groups are replaced with their real split forms. These provide many examples of weak splittings of surjective Poisson submersions to real flag varieties.

We continue with certain nonsplit analogs of the results in Sects. 5.1 and 5.4 which concern the real Poisson structures on complex flag varieties introduced by Foth and Lu in [12]. Let $\mathfrak{g}$ be a complex simple Lie algebra and G a connected, simply connected Lie group with Lie algebra $\mathfrak{g}$. Consider $\mathfrak{g}$ as a real quadratic Lie algebra with the nondegenerate bilinear form

$$x, y \in \mathfrak{g} \mapsto \mathrm{Im}\langle x, y \rangle \in \mathbb{R}. \tag{29}$$

Each Vogan diagram v for $\mathfrak{g}$ gives rise to a complex conjugate linear involution τ_v of $\mathfrak{g}$ and to the real form $\mathfrak{g}_v := \mathfrak{g}^{\tau_v}$ of $\mathfrak{g}$, see [12, 18] for details. The map τ_v is defined using a choice of root vectors of $\mathfrak{g}$. We will denote by $\mathfrak{t}$ the corresponding Cartan subalgebra of $\mathfrak{g}$ (which is stable under τ_v) and by $\mathfrak{u}_\pm$ the nilradicals of the corresponding Borel subalgebras of $\mathfrak{g}$. The following is a (real) Manin triple:

$$(\partial := \mathfrak{g}, \mathfrak{g}_+ := \mathfrak{t}^{-\tau_v} + \mathfrak{u}_+, \mathfrak{g}_- := \mathfrak{g}_v),$$

see [12, Sect. 2]. Let $D := G$ be considered as a real Lie group. We have

$$N_+ = N(\mathfrak{g}_+) = B_+ \quad \text{and} \quad N_-^\circ = N(\mathfrak{g}_-) = G_v$$

where G_v is the connected subgroup of G with Lie algebra $\mathfrak{g}_v$ and B_+ is the Borel subgroup of G with Lie algebra $\mathfrak{t} + \mathfrak{u}_+$. The group $H := N_+ \cap N_-^\circ$ is the connected subgroup of T with Lie algebra

$$\mathrm{Lie}\, H = \mathfrak{t}^{\tau_v}. \tag{30}$$

Denote once again the associated Drinfeld and Heisenberg double Poisson structures on G by π_v and π_v'. We have $\pi|_H = 0$, and thus H acts by Poisson automorphisms on (G, π') on the left and right. Denote the reduced Poisson structure on G/H by $\pi_{H,v}'$ and the Poisson projection

$$v' : (G, \pi_v') \to (G/H, \pi_{H,v}'). \tag{31}$$

The pushforwards of $-\pi_v$ and π_v' under the canonical projection $G \to G/B_+$ are well defined and are equal to each other, because $B_+ = N(\mathfrak{g}_+)$. Denote the corresponding real Poisson structure on the complex flag variety G/B_+ by $\pi_{G/B_+,v}$. It has very interesting properties, for example the intersections of the orbits of the Borel subgroup B_+ and the real form G_v are regular complete Poisson submanifolds. Theorem 4 is applicable to construct a weak splitting of the real surjective Poisson submersion

$$(G/H, \pi'_{H,v}) \to (G/B_+, \pi_{G/B_+,v}).$$

Once again it is easy to verify that the group $Q = HU_-$ and the set $\mathscr{D} = \{\dot{w} \mid w \in W\}$ satisfy the conditions of Theorem 4. This leads to the following result:

Theorem 9. *For all connected, simply connected complex simple groups G and Vogan diagrams v, the partition of the flag variety G/B_+ into Schubert cells*

$$G/B_+ = \sqcup_{w \in W} B_+ \cdot wB_+$$

and the family of maps

$$i'_w : B_+ \cdot wB_+ \to G/H, \quad i'_w(u_+ wB_+) = u_+ \dot{w} H, \quad u_+ \in U_+ \cap wU_- w^{-1}$$

is a weak splitting of the real surjective Poisson submersion

$$(G/H, \pi'_{H,v}) \to (G/B_+, \pi_{G/B_+,v})$$

from a Heisenberg double to the flag variety, where the group H is given by (30). The image of each map i'_w is a Poisson–Dirac submanifold of $(G/H, \pi'_H)$ admitting a Dirac projection with associated bundle $v'_*(\tilde{F}^w)$ for the pushforward with respect to (31) of the vector bundle

$$\tilde{F}^w \to (U_+ \cap wU_- w^{-1})\dot{w}H$$

with fibers

$$\tilde{F}^w_g := R_g\left(p_- \mathrm{Ad}_{\dot{w}}(\mathfrak{t}^{\tau_v} + \mathfrak{u}_-)\right) + L_g(\mathfrak{b}_+)$$

for $g \in (U_+ \cap wU_- w^{-1})\dot{w}H$. Here $p_- : \mathfrak{g} \to \mathfrak{g}_-$ denotes the projection along $\mathfrak{g}_+$.

Acknowledgements I would like to thank Bernard Leclerc and Jiang-Hua Lu for their valuable comments and suggestions on the first version of the preprint. I am also grateful to I. Dimitrov, G. Mason, S. Montgomery, I. Penkov, V.S. Varadarajan, and J. Wolf for the opportunity to present these and related results at the Lie theory meetings at UCB, UCLA, UCSC, and USC.

References

1. A. A. Belavin and V. G. Drinfeld, *Triangular equations and simple Lie algebras*, Math. Phys. Rev. **4** (1984), 93–165.
2. A. Berenstein, S. Fomin, and A. Zelevinsky, *Cluster algebras III, Upper bounds and double Bruhat cells*, Duke Math. J. **126** (2005), 1–52.
3. A. Berenstein and J. Greenstein, *Quantum folding*, Int. Math. Res. Not. IMRN **2011**, no. 21, 4821–4883.

4. S. Billey and I. Coskun, *Singularities of generalized Richardson varieties*, Comm. Algebra **40** (2012), 1466–1495.

5. N. Chevalier, *Algèbres amassées et positivité totale*, Ph.D. thesis, Univ. of Caen, 2012.

6. M. Crainic and R. L. Fernandes, *Integrability of Poisson brackets*, J. Diff. Geom. **66** (2004), 71–137.

7. C. De Concini and C. Procesi, *Complete symmetric varieties*, in: Invariant theory (Montecatini, 1982), Lect. Notes Math. **996** (1983), 1–44.

8. P. Delorme, *Classification des triples de Manin pour les algèbres de Lie réductives complex*, with an appendix by G. Macey, J. Algebra **246** (2001), 97–174.

9. V. G. Drinfeld, *On Poisson homogeneous spaces of Poisson–Lie groups*, Theor. and Math. Phys. **95** (1993), 524–525.

10. S. Evens and J.-H. Lu, *On the variety of Lagrangian subalgebras. I*, Ann. Sci. École Norm. Sup. (4), **34** (2001), 631–668.

11. S. Evens and J.-H. Lu, *Poisson geometry of the Grothendieck resolution of a complex semisimple Lie group*, Mosc. Math. J. **7** (2007), 613–642.

12. Ph. Foth and J.-H. Lu, *Poisson structures on complex flag manifolds associated with real forms*, Trans. Amer. Math. Soc. **358** (2006), 1705–1714.

13. T. J. Hodges and T. Levasseur, *Primitive ideals of $\mathbb{C}_q[SL(3)]$*, Comm. Math. Phys. **156** (1993), 581–605.

14. M. Gekhtman, M. Shapiro, and A. Vainshtein, *Cluster algebras and Poisson geometry*, Mosc. Math. J. **3** (2003), 899–934

15. M. Gekhtman, M. Shapiro, and A. Vainshtein, *Exotic cluster structures on SL_n: the Cremmer–Gervais case*, preprint arXiv:1307.1020.

16. K. R. Goodearl and M. Yakimov, *Poisson structures on affine spaces and flag varieties. II*, Trans. Amer. Math. Soc. **361** (2009), 5753–5780.

17. A. Joseph, *On the prime and primitive spectra of the algebra of functions on a quantum group*, J. Algebra **169** (1994), 441–511.

18. A. W. Knapp, *Lie Groups Beyond an Introduction*, 2nd ed., Progress in Math. **140**, Birkhäuser, Cambridge, 2002.

19. A. Knutson, T. Lam, and D. E. Speyer, *Projections of Richardson varieties*, J. Reine Angew. Math. **687** (2014), 133–157.

20. M. Kogan and A. Zelevinsky, *On symplectic leaves and integrable systems in standard complex semisimple Poisson-Lie groups*, Int. Math. Res. Not. **2002**, no. 32, 1685–1702.

21. B. Leclerc, *Cluster structures on strata of flag varieties*, preprint arXiv:1402.4435.

22. J.-H. Lu and M. Yakimov, *Group orbits and regular partitions of Poisson manifolds*, Comm. Math. Phys. **283** (2008), 729–748.

23. G. Muller and D. E. Speyer, *Cluster algebras of Grassmannians are locally acyclic*, preprint arXiv:1401.5137.

24. A. Postnikov, *Total positivity, Grassmannians, and networks*, preprint arXiv:math/0609764.

25. K. Rietsch and L. Williams, *The totally nonnegative part of G/P is a CW complex*, Transform. Groups **13** (2008), 839–853.

26. I. Vaisman, *Dirac submanifolds of Jacobi manifolds*, in: The breadth of symplectic and Poisson geometry, 603–622, Progr. Math. 232, Birkhäuser Boston, Cambridge, MA, 2005.

27. B. Webster and M. Yakimov, *A Deodhar-type stratification on the double flag variety*, Transform. Groups **12** (2007), 769–785.

28. P. Xu, *Dirac submanifolds and Poisson involutions*, Ann. Sci. École Norm. Sup. (4) **36** (2003), 403–430.

29. M. Yakimov, *Symplectic leaves of complex reductive Poisson–Lie groups*, Duke Math. J. **112** (2002), 453–509.

30. M. Yakimov, *A classification of H-primes of quantum partial flag varieties*, Proc. Amer. Math. Soc. **138** (2010), 1249–1261.

Printed in the United States
By Bookmasters